Water Quality Concepts, Sampling, and Analyses

Water Quality Concepts, Sampling, and Analyses

Edited by
Yuncong Li and Kati Migliaccio

CRC Press is an imprint of the
Taylor & Francis Group, an **informa** business

CRC Press
Taylor & Francis Group
6000 Broken Sound Parkway NW, Suite 300
Boca Raton, FL 33487-2742

Printed in the United States of America on acid-free paper
10 9 8 7 6 5 4 3 2 1

International Standard Book Number: 978-1-4200-9266-0 (Hardback)

Library of Congress Cataloging-in-Publication Data

Water quality concepts, sampling, and analyses / editors: Yuncong Li and Kati Migliaccio. -- 1st ed.
p. cm.
Includes bibliographical references and index.
ISBN 978-1-4200-9266-0 (alk. paper)
1. Water quality--Measurement. I. Li, Yuncong. II. Migliaccio, Kati W.

TD367.W3765 2010
628.1'61--dc22 2010017447

Visit the Taylor & Francis Web site at
http://www.taylorandfrancis.com

and the CRC Press Web site at
http://www.crcpress.com

Contents

Preface

The intent of this book, *Water Quality Concepts, Sampling, and Analyses,* is to provide practical information for those who are planning, conducting, or evaluating water quality monitoring programs. During our professional activities in this field, we have noticed that many do not have the training or tools to successfully sample water quality, and we are often contacted regarding basic water sampling, laboratory, and data analysis questions. This book is meant to provide applied information that can be used to improve water quality monitoring programs leading to better management of our water resources.

The book provides a wide range of water quality topics, from water quality regulations and criteria (Chapters 2 and 3) to project planning (Chapter 4) and sampling activities (Chapters 5 to 7). Water quality analyses are also detailed including field measurement techniques (Chapter 8) and laboratory techniques (Chapters 9 and 10). More contemporary issues in water quality analysis, such as emerging pollutants (Chapter 11) and uncertainty in measured data (Chapter 12) are also presented. Unlike many other water quality books, our book includes statistical analysis of data (Chapter 13), examples of water quality monitoring programs (Chapter 14), and discussions of a video of sampling activities (Chapter 15) are also provided to complement the material presented in earlier chapters.

As water quality becomes a leading concern for people and ecosystems worldwide, it is critical to properly assess water quality in order to protect water resources for current and future generations. Water quality assessment depends on effective sampling and analytical procedures, as well as appropriate data analysis and presentation. All components of this process should be conducted with the highest of quality and reproducibility. While this book does not contain every detail, it does provide the framework and information needed to further explore different components in the water quality assessment process.

We would like to sincerely acknowledge all chapter authors and those that contributed to this book in many different ways. In particular, we would like to thank Sam Allen and Malcolm Sumner for reviewing most of the chapters; Waldy Klassen, Guodong Liu, Guanliang Liu, and Yin Chen for reviewing one or more chapters; Xiaohui Fan, Yanli Nie, and Huiqin Ren for formatting chapters and checking references; and Randy Brehm, Pat Roberson, and Gail Renard from Taylor & Francis, for their support and friendship. We would also specifically like to thank contributors for Chapter 15's video who were photographer Ian Maguire and DVD editor Shaun Wright, as well as the organizers of the water quality workshop (Ed Hanlon, Pamela Fletcher, Teresa Olczyk, and Qingren Wang), presenters at the workshop (Nick Aumen, Joffre Castro, Ramon Garza, Jim Hendee, Ed Hanlon, Robert Johnson, Lee Massey, Miguel McKinney, Teresa Olczyk, Larry Parson, John Proni, Forrest Shaw, Dave Struve, Van Waddill, Qingren Wang, Dave Wanless, and Rod Zika), instructors for laboratory analyses during the workshop (Rick Armstrong, Tina Dispenza,

Ed Hanlon, Guodong Liu, Kelly Morgan, Yun Qian, Janzhong Qiao, Laura Rosado, Grurpal Toor, and Guiqin Yu), and all workshop participants.

We especially thank our spouses and children for their understanding and patience.

Yuncong Li
Kati W. Migliaccio

Contributors

David A. Alvarez
Columbia Environmental Research Center
U.S. Geological Survey
Columbia, Missouri

Mary Jane Angelo
Levin College of Law
University of Florida
Gainesville, Florida

Joffre Castro
Everglades National Park, South Florida Ecosystem Office
Homestead, Florida

Pamela J. Fletcher
Florida Sea Grant, National Oceanic and Atmospheric Administration
Miami, Florida

Adam Foster
U.S. Geological Survey
Fort Lauderdale, Florida

Brian E. Haggard
Agricultural and Biological Engineering
University of Arkansas
Fayetteville, Arkansas

Daren Harmel
U.S. Department of Agriculture
Agricultural Research Service
Grassland Soil and Water Research Laboratory
Temple, Texas

Delia Ivanoff
Vegetation and Land Management Department
South Florida Water Management District, MS 5233
West Palm Beach, Florida

Tammy L. Jones-Lepp
Environmental Sciences Division
National Exposure Research Laboratory
Office of Research and Development
U.S. Environmental Protection Agency
Las Vegas, Nevada

Nicholas Kiggundu
Department of Agricultural and Biological Engineering
Tropical Research and Education Center
University of Florida
Homestead, Florida

Yuncong Li
Department of Soil and Water Sciences
Tropical Research and Education Center, IFAS
University of Florida
Homestead, Florida

Lena Q. Ma
Department of Soil and Water Science, IFAS
University of Florida
Gainesville, Florida

Renuka R. Mathur
Cornell Nutrient Analysis Laboratory
Department of Crop and Soil Sciences
Cornell University
Ithaca, New York

Kati W. Migliaccio
Department of Agricultural and Biological Engineering
Tropical Research and Education Center, IFAS
University of Florida
Homestead, Florida

Sapna Mulki
Tropical Research and Education Center, University of Florida
Homestead, Florida

Rafael Muñoz-Carpena
Agricultural and Biological Engineering Department, IFAS
University of Florida
Gainesville, Florida

J. Thad Scott
Crop, Soil and Environmental Sciences Department
University of Arkansas
Fayetteville, Arkansas

Peter C. Smiley, Jr.
U.S. Department of Agriculture
Agricultural Research Service
Soil Drainage Research Unit
Columbus, Ohio

Patricia Smith
Biological and Agricultural Engineering
Texas A&M University
College Station, Texas

David Struve
Analytical Services Division
Restoration Sciences Department
South Florida Water Management District (SFWMD)
West Palm Beach, Florida

Qingren Wang
Department of Soil and Water Science
Tropical Research and Education Center, IFAS
University of Florida
Homestead, Florida

Jianqiang Zhao
Bureau of Agricultural Environmental Laboratories
Division of Agricultural Environmental Services
Florida Department of Agriculture and Consumer Services
Tallahassee, Florida

Meifang Zhou
Analytical Services Division
Restoration Sciences Department
South Florida Water Management District (SFWMD)
West Palm Beach, Florida

1 Introduction

Yuncong Li and Kati W. Migliaccio

CONTENTS

1.1 WHAT IS WATER QUALITY?

Water quality is a term used to describe the chemical, physical, and biological characteristics of water. These attributes affect water suitability for human consumption (drinking, irrigation, industrial use) and ecosystem health. The chemical constituents of water are substances that dissolve in water, including gases (e.g., oxygen and carbon dioxide), metals (e.g., iron and lead), nutrients (e.g., nitrogen and phosphorus), pesticides (e.g., atrazine and endosulfan), and other organic compounds (e.g., polychlorinated biphenyls). The most common physical characteristics of water are color, odor, temperature, taste, and turbidity, while biological constituents of water are living organisms including bacteria (e.g., *Escherichia coli),* viruses, protozoans (e.g., *Cryptosporidiosis*), phytoplankton (i.e., microscopic algae), zooplankton (i.e., tiny animals), insects, plants, and fish.

The origins of the term *water quality* are not certain, and its first usage to describe suitability of water is unknown. The concept of water quality likely started at the beginning of civilization. Our ancestors settled near water resources such as rivers and lakes and probably evaluated water quality solely based on physical or aesthetic properties of water such as color, smell, and taste. Historical evidence shows that early humans developed methods to improve water quality as early as 4000 BC (USEPA, 2000). Ancient Sanskrit and Greek writings documented water treatment methods indicating that "impure water should be purified by being boiled over a fire, or being heated in the sun, or by dipping a heated iron into it, or it may be purified by filtration through sand and coarse gravel and then allowed to cool" (Jesperson, 2009). Egyptians used a chemical (alum) to remove suspended particles by flocculation as early as 1500 BC.

Modern concepts of water quality and water treatment began in the 1700s when new knowledge of microbiology, physics, and chemistry lead to better understanding of drinking water contamination. In order to achieve better water quality, sand filtration was established as an effective means of removing particles and was being used regularly in Europe in the 1800s. By the early 1900s, water treatment not only focused on aesthetic problems but also on pathogens. Ozone was used to treat water in Europe, and chlorine was first used in Jersey City, New Jersey. Since then, treatment of water to improve water quality has become more commonplace as numerous water treatment technologies and water quality-related policies, regulations, and standards have been developed. Chapters 2 and 3 provide a detailed discussion on the development of water quality regulations and standards.

The interpretation of water quality as either "good" or "poor" differs depending on the use of the water. For example, high concentration of nitrate in drinking water has been shown to cause a potentially fatal blood disorder in infants called *methemoglobinemia* or *blue-baby syndrome*; thus, water with a high nitrate concentration is not suitable for drinking water and would be designated as "poor" water quality for this use. However, high concentrations of nitrate in irrigation water would contribute to crop growth and would be considered "good" water quality for this use. Similarly, water with high phosphorus is beneficial as irrigation but is detrimental in phosphorus-limited ecosystems such as the Florida Everglades. Therefore, water quality should be linked with specific water uses as most water quality standards vary, depending on the use of the water.

It is estimated that 8% of worldwide water use is for household purposes (drinking water, bathing, cooking, sanitation, and gardening), 22% for industrial uses (mainly hydropower or nuclear power), and 70% for crop irrigation (Sterling and Vintinner, 2008). Irrigation land provides almost half of the world's food and accounts for the majority of water use. However, few people have concerns regarding the quality of irrigation water except farmers. Unless reclaimed water is used for irrigation, there are no national standards or regulations for irrigation water quality. If crops are irrigated with reclaimed water, some environmental guidelines have to be followed. Reclaimed water must meet disinfection standards to reduce the concentrations of constituents that may affect public health and/or limit human contact with reclaimed water. Reclaimed water intended for reuse should (1) be treated to achieve biochemical oxygen demand (BOD) and total suspended solids levels of < 30 mg/L, during secondary or tertiary treatment and (2) receive additional disinfection by chlorination or other chemical disinfectants, UV radiation, ozonation, and membrane processing (Haering et al., 2009). Additionally, the U.S. Environmental Protection Agency (USEPA) has recommended concentration limits for 22 constitutes in reclaimed water for irrigation use (USEPA, 2004).

Even though drinking water is a small fraction of all water used, it is most critical due to its impact on human health. Because of this, many probably think of water quality primarily in reference to drinking water. A dominant concern regarding drinking water quality is disease transmittal. The World Health Organization (WHO) estimated that diarrheal diseases account for 4.1% of the total daily illnesses and is responsible for the deaths of 1.8 million people every year (WHO, 2005). The Centers for Disease Control and Prevention (CDC) reported over 400 waterborne-disease outbreaks and 469,000 cases associated with consumption of public drinking water from 1986 to 2006 in the United States (calculated based on data from the Surveillance Summaries for Waterborne Disease and Outbreaks). When water is contaminated with pathogens, some pathogens will survive for only a short time, while others (e.g., *Cryptosporidium*) may survive for months. When such pathogens are consumed in drinking water, a person may become seriously ill. In the early 1850s, many London residents were infected by cholera, which was spread when drinking water was contaminated by leaking sewers (Crittenden et al., 2005). In response, the Metropolitan Water Act was passed; it was one of the first instances of governmental regulation of water quality, requiring the filtration of all drinking water supplied to the area. In 1993, an outbreak of a parasitic disease caused by *Cryptosporidium* in Milwaukee, Wisconsin, became the largest water quality-related disease in current U.S. history (Corso et al., 2003). Approximately 403,000 residents fell ill with stomach cramps, fever, diarrhea, and dehydration, resulting in over 100 deaths and $96 million in damage. The state of Wisconsin made numerous changes to water treatment and testing procedures, which included increasing the frequency of testing of turbidity in drinking water (Wisconsin Department of Natural Resources, 2008). Many examples of disease outbreaks have been linked to water quality. Therefore, numerous water quality regulations and standards have been developed and enforced in developed countries. Currently, the USEPA sets standards for approximately 90 contaminants in drinking water.

As water users, industries have relatively less concern for water quality, but industries such as mining, oil production, chemical factories, pulp and paper mills, and auto and computer factories generate huge volumes of solid waste and wastewater containing toxic chemicals. Industrial wastewater and urban and agricultural runoff are polluting water resources, especially in developing countries. Worldwide, 884 million people do not have access to safe drinking water. Approximately 90% of the groundwater under China's cities is contaminated; 70% of India's rivers and lakes are unsafe for drinking or bathing; 75% of people in Latin America and the Caribbean suffer from chronic dehydration because they do not have access to safe drinking water; and all of Africa's 677 major lakes are now threatened to varying degrees by unsustainable use and pollution (Sterling and Vintinner, 2008). The United States has spent much more money to protect and preserve water quality and has promulgated more water quality regulations than any other country in the world. Despite this legislation, water quality is still a serious problem in the United States today. Based on the Assessment Total Maximum Daily Load (TMDL) Tracking and Implementation System (ATTAINS) that provides information reported by the states under the Clean Water Act requirements, 49.5% of rivers and streams; 66% of lakes, reservoirs, and ponds; 63.6% of bays and estuaries; 38% of coastal shoreline; 82.3% of ocean and near coastal waters; and 36.3% of wetlands are impaired and do not meet the criteria adopted by the states to protect designated uses (Table 1.1).

TABLE 1.1
Water Quality Conditions Reported by the USEPA (2009)

	Good[a]	Threatened	Impaired
Water Type		%	
Rivers and streams	49.8	0.7	49.5
Lakes, reservoirs, and ponds	33.7	0.3	66.0
Bays and estuaries	36.3	0.1	63.6
Coastal shoreline	62.0	0.0	38.0
Ocean and near coast	17.7	0.0	82.3
Wetlands	63.6	0.1	36.3

[a] Good—the designated use is met; threatened—the designated use is currently met but water quality conditions appear to be declining; impaired—the designated use is not met.

The causes of impairment of U.S. waters include chemical contaminants (such as PCBs, metals, and oxygen-depleting substances), physical conditions (such as elevated temperature, excessive siltation, or alterations of habitat), and biological contaminants (such as bacteria and noxious aquatic weeds). In general, pathogens—mainly fecal coliform and *E. coli*—have topped the list of causes for water impairment with 10,625 cases reported nationally. Mercury, other metals (e.g., lead and arsenic), and nutrients (mainly phosphorus) are the next three major causes (Table 1.2) with pathogens leading the causes of impairment of rivers and streams while mercury is the primary impairment for lakes, reservoirs, ponds, and coastal waters (Table 1.3). Organic enrichment, originating from nutrient and sewage discharges, is the primary cause for impairment of wetlands, bays, and estuaries (Table 1.4). Polychlorinated biphenyls (PCBs), a group of organochlorine compounds, are the second major cause of water impairment for lakes, reservoirs, ponds, and ocean water. Although PCB use was banned in the United States in 1977, they continue to be released into water bodies through runoff from landfills and discharges from waste chemicals.

These mentioned water quality contaminants and others have led to federal and state laws to protect U.S. waters. While these laws are continuously evolving, many current water quality regulations are linked to the 1972 Federal Water Pollution Control Act (or the Clean Water Act). Today, we are still challenged with developing and implementing components of this Act, including the Total Maximum Daily Loads (TMDLs) and water quality criteria (see Chapters 2 and 3), and achieving the level of water quality envisioned for U.S. waters.

1.2 WHAT IS WATER QUALITY MONITORING?

Water quality monitoring is the practice of assessing the chemical, physical, and biological characteristics of water in streams, lakes, estuaries, and coastal waters and groundwater relative to set standards and providing information on whether these waters are adequate for specific uses such as drinking, swimming, irrigation, and ecosystem services. The objectives of water quality monitoring often include (1) identifying specific water quality problems that affect the health of humans and ecosystems, (2) determining long-term trends in water quality, (3) documenting effects of pollution prevention or remediation, and (4) providing evidence for regulation compliance and legal disputes.

Thousands of federal, state, interstate, local agencies, universities, private organizations, companies, and citizen volunteers in the United States are involved in water quality monitoring. The USEPA provides funding for many of these water quality monitoring programs performed by states and tribes in addition to its own monitoring. The United States Geological Service (USGS) conducts the National Stream Quality Accounting Network (NASQAN) that monitors water quality at various locations on rivers and performs reconnaissance of emerging contaminants. Other federal agencies involved in water quality monitoring are the U.S. Fish and Wildlife Service, the National Oceanic and Atmospheric Administration (NOAA), the U.S. Army Corps of Engineers, and the United States Department of Agriculture (USDA). Some water bodies such as those in south Florida are intensively monitored while others are rarely sampled. Water in south Florida has been sampled and analyzed for many years by many agencies such as the South Florida Water Management

TABLE 1.2
Causes of Impairment for Waters Listed in the Assessment Total Maximum Daily Load (TMDL) Tracking and Implementation System (ATTAINS)

Cause of Impairment Group Name	Number of Causes of Impairment Reported
Pathogens	10,625
Mercury	8,864
Metals (other than mercury)	7,485
Nutrients	6,826
Organic enrichment/oxygen depletion	6,399
Sediment	6,293
Polychlorinated biphenyls (PCBs)	6,206
pH/acidity/caustic conditions	3,811
Cause unknown—impaired biota	3,266
Turbidity	3,036
Temperature	3,009
Pesticides	1,798
Salinity/total dissolved solids/Cl/SO_4	1,731
Cause unknown	1,253
Noxious aquatic plants	981
Habitat alterations	702
Dioxins	542
Algal growth	539
Toxic organics	459
Ammonia	356
Toxic inorganics	341
Total toxics	318
Other cause	222
Oil and grease	155
Taste, color, and odor	114
Flow alteration(s)	108
Trash	57
Fish consumption advisory	56
Radiation	44
Chlorine	34
Nuisance exotic species	29
Cause unknown—fish kills	12
Nuisance native species	6

Source: Modified from USEPA. 2009. U.S. Environmental Protection Agency, Washington, DC. http://www.epa.gov/waters/ir/index.html (accessed November 20, 2009).

District (SFWMD), USGS, NOAA, Everglades National Park, Biscayne National Park, Miccosukee Indian Tribe of Florida, Miami-Dade Environmental Resources Management (DERM), University of Florida (UF), and Florida International University (FIU). Currently, SFWMD has 1,800 monitoring stations on rivers, channels, wetlands, and lakes through 16 counties in South Florida and conducts over

TABLE 1.3
Top Ten Causes of Impairment for Specific Waters Reported by the USEPA (2009)

Rank	Rivers and Streams	Lakes, Reservoirs, and Ponds	Bays and Estuaries	Coastal Shoreline	Ocean and Near Coastal	Wetlands
1	Pathogens	Mercury	Organic enrichment	Mercury	Mercury	Organic enrichment
2	Sediment	Polychlorinated biphenyls (PCBs)	Polychlorinated biphenyls (PCBs)	Pathogens	Polychlorinated biphenyls (PCBs)	Mercury
3	Nutrients	Nutrients	Pathogens	Metals (other than mercury)	Organic enrichment	Metals (other than mercury)
4	Organic enrichment	Organic enrichment	Mercury	Turbidity	Pesticides	Habitat alterations
5	Habitat alterations	Metals (other than mercury)	Nuisance aquatic species	Pesticides	Pathogens	Nutrients
6	Polychlorinated biphenyls (PCBs)	Algal growth	Metals (other than mercury)	Others	Dioxins	Pathogens
7	Metals (other than mercury)	Turbidity	Pesticides	Nutrients	Nuisance exotic species	Flow alteration(s)
8	Flow alteration(s)	Sediment	Toxic organics	Cause unknown	Unknown toxic	Boron and other inorganic
9	Mercury	Nuisance exotic species	Nutrients	Algal growth	Metals (other than mercury)	Sediment
10	Temperature	Temperature	Others	Organic enrichment	Toxic organics	Unknown toxic

TABLE 1.4
Top Ten Probable Sources of Impairment Reported by the USEPA (2009)

Rank	Rivers and Streams	Lakes, Reservoirs, and Ponds	Bays and Estuaries	Coastal Shoreline	Ocean and Near Coastal	Wetlands
1	Agriculture	Atmospheric deposition	Municipal discharges/ sewage	Unspecified nonpoint source	Unknown	Agriculture
2	Unknown	Unknown	Atmospheric deposition	Natural/wildlife	Atmospheric deposition	Unknown
3	Atmospheric deposition	Agriculture	Unknown	Urban-related runoff/ stormwater	Municipal discharges/ sewage	Atmospheric deposition
4	Hydromodification	Natural/wildlife	Natural/wildlife	Municipal discharges/ sewage	Recreational boating and marinas	Industrial
5	Natural/wildlife	Hydromodification	Industrial	Industrial	Hydromodification	Natural/wildlife
6	Unspecified nonpoint source	Unspecified nonpoint source	Agriculture	Other	Recreation and tourism (nonboating)	Hydromodification
7	Municipal discharges/ sewage	Other	Hydromodification	Commercial harbor and port activities	Unspecified nonpoint source	Mining
8	Habitat alterations (not hydromodification)	Legacy/historical pollutants	Other	Recreation and tourism (nonboating)	Urban-related runoff/ stormwater	Other
9	Urban-related runoff/ stormwater	Urban-related runoff/ stormwater	Habitat alterations (not hydromodification)	Agriculture	Construction	Unspecified nonpoint source
10	Mining	Municipal discharges/ sewage	Urban-related runoff/ stormwater	Unknown	Industrial	Construction

300,000 analyses annually for nutrients, physical parameters, inorganics, pesticides, and mercury with an annual budget of $18 million for water quality monitoring. More information on south Florida water quality monitoring programs is provided in Chapter 15 and the attached DVDs.

1.3 WHAT IS THE PURPOSE OF THIS BOOK?

The purpose of this book is to present the latest information and methodologies for water quality policy, regulation, monitoring, field measurement, laboratory analysis, and data analysis. It was written as a handbook or manual for anyone whose work involves water quality regardless of their educational background, including water managers, teachers, scientists, chemists, biologists, ecologists, college students, extension agents, environmental consultants, environmental engineers, environmental health officials, environmental regulators, and others. Oftentimes, individuals occupy positions in their chosen fields that may require some understanding of basic water quality science for which they previously did not receive professional training. Although some professionals may have received instruction in the water sciences, they may not be current with the science or may need a refresher. Hopefully, this book, which was developed based on five workshops on "Water Quality Concepts, Sampling, and Analysis" organized over the last 4 years, will provide useful information for both situations. One key component of water quality education that is often missing from an environmental professional's set of skills is the knowledge and ability to collect and analyze water samples and to evaluate a water quality sampling report. To address this issue, a team of extension specialists and agents at the University of Florida (UF), Institute of Food and Agricultural Sciences (IFAS), Tropical Research and Education Center (TREC), Homestead, Florida, developed and administered training workshops. The goal of these workshops was to transfer information and skills to the attendees so that they could integrate this information into their current professional activities. Workshops were very successful with pre- and posttests showing an increase in knowledge of 28%–73%. One of the workshops was videotaped and is included in this book (see Chapter 15). Detail information on some of these workshops was also reported by Li et al. (2006).

This book gathers essential information for developing or understanding water quality monitoring programs. In contrast to other water quality books, this handbook assembles multidisciplinary, integrated knowledge essential for a water quality monitoring program including water quality concept, policy, and regulatory development and criteria (Chapters 1, 2, and 3); water quality monitoring planning and sampling techniques for surface water, groundwater, and pore water (Chapters 4, 5, 6, and 7); field and laboratory measurements (Chapters 8, 9, and 10); emerging water quality contaminants (Chapter 11); and statistical methods and data uncertainty with water quality data (Chapters 12 and 13). Chapter 14 presents an example of how to conduct a surface water quality monitoring program, and Chapter 15 is a description of the video presentations for the attached DVDs. The book includes two DVDs with videos of lectures, hands-on field and laboratory activities, and field tours. This is the first book that provides video-presented methodology for water quality measurements and analysis. In summary, the book will provide a very practical and applied perspective to water quality from data collection and laboratory analyses to data interpretation so that the audience can use the book as a guide in assessing, developing, and implementing complete water quality programs or components of a complete water quality program.

REFERENCES

Corso, P.S., M.H. Kramer, K.A. Blair, D.G. Addiss, J.P. Davis, and A.C. Haddix. 2003. Cost of illness in the 1993 waterborne *Cryptosporidium* outbreak, Milwaukee, Wisconsin. *Emerging Infectious Diseases* 9: 426–431.

Crittenden, J.C., R.R. Trussell, D.W. Hand, K.J. Howe, and G. Tchobanoglous. 2005. *Water Treatment: Principles and Design*. Second Edition, New Jersey: Wiley.

Haering, K.C., G.K. Evanylo, B. Benham, and M. Goatley. 2009. Water Reuse: Using Reclaimed Water for Irrigation. Virginia Cooperative Extension publication 452–014. http://pubs.ext.vt.edu/452/452-014/452-014.pdf (accessed November 20, 2009).

Jesperson, K. 2009. Drinking Water History. http://www.nesc.wvu.edu/old_website/ndwc/ndwc_DWH_1.html (accessed November 2, 2009).

Li, Y.C., T. Olczyk, and K. Migliaccio. 2006. County faculty in-service training for water sampling and chemical analysis. *Proc. Fla. State Hort. Soc.* 119: 249–254.

Sterling, E., and E. Vintinner. 2008. How much is left? An overview of the crisis. In *Water Consciousness—How We All Have to Change to Protect Our Most Critical Resource*, ed. T. Lohan, 15–25. San Francisco: AlterNet Books.

USEPA (U.S. Environmental Protection Agency). 2000. The History of Drinking Water Treatment. Office of Water. Washington, DC. http://www.epa.gov/ogwdw/consumer/pdf/hist.pdf (accessed November 20, 2009).

USEPA. 2004. Guidelines for water reuse. EPA 625-R-04-108. U.S. Environmental Protection Agency, Washington, DC. http://epa.gov/nrmrl/pubs/625r04108/625r04108.pdf (accessed November 20, 2009).

USEPA. 2009. Water Quality Assessment and Total Maximum Daily Loads Information (ATTAINS). U.S. Environmental Protection Agency, Washington, DC. http://www.epa.gov/waters/ir/index.html (accessed November 20, 2009).

WHO (World Health Organization). 2005. World Health Report 2005. Geneva: World Health Organization. http://www.who.int/whr/2005/whr2005_en.pdf (accessed November 20, 2009).

Wisconsin Department of Natural Resources. 2008. A Risk to Our Drinking Water. http://www.dnr.state.wi.us/org/water/dwg/crypto.htm (accessed November 20, 2009).

2 Water Quality Regulations and Policy Development

Kati W. Migliaccio and Mary Jane Angelo

CONTENTS

2.1 INTRODUCTION

Civilization has developed and sustained throughout history near adequate water supplies. As populations increased, protection of water became crucial to sustain societal needs and to prevent waterborne illnesses. Protection and treatment of drinking water supplies were reported to occur as early as 4000 BC and primarily focused on aesthetic issues. This was followed centuries later by the discovery that unseeable (using the naked eye) water contaminates were contributing to waterborne illness, as determined by John Snow and Louis Pasteur in the 1800s (USEPA, 2000c). Thus, water resource protection has evolved over time based on public need, societal values, and scientific discovery.

Water managers will continue to encounter the same challenges of maintaining adequate water supply to meet the demands placed on them by civilization. In addition, modern society faces new challenges due to decades of disposing of domestic and industrial wastes into oceans, rivers, lakes, and streams. As a result of such practices, many waterbodies are contaminated with organic and inorganic pollutants, which can render waterbodies inappropriate for their designated uses. While drinking water remains a dominant concern for all people, other designated uses (e.g., aquatic habitat for fish and wildlife, swimming and recreation, and agricultural supply) are also recognized as valuable by society. These uses are generally protected in modern

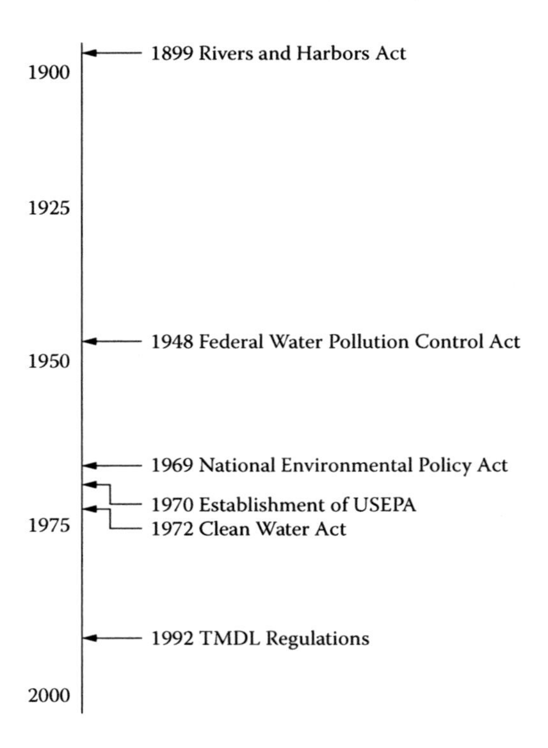

FIGURE 2.1 Summary of key water quality legislative events.

society by governments. The U.S. government's role in protection of U.S. waters has evolved over time starting with state-enforced programs and later with more comprehensive federally enforced programs. The most dramatic changes (to date) in water quality policy in the United States occurred during the last half of the 20th century. Initial U.S. legislation protecting water quality focused on protecting waterways for commercial navigation. Over time, the public outcry over ever-increasing water contamination from domestic and industrial pollutants resulted in an evolution of the law to address this type of water resource contamination. Consequently, present legislative emphasis goes well beyond protecting navigability of specific waterways and extends to protection and/or restoration of water bodies on a watershed or ecosystem level to achieve a specific pollutant criteria standard (see Chapter 3 for more information on water quality criteria and standards). The evolution of water quality protection in the United States is discussed further in this chapter, examining major legislation and the driving factors behind the movement to protect water resources (Figure 2.1).

2.2 UNITED STATES WATER QUALITY LEGISLATIVE HISTORY

2.2.1 Rivers and Harbors Act

Protection of U.S. surface waters through legislation started at the beginning of the 20th century with the Rivers and Harbors Act (RHA) of 1899 (33 USC. 403;

Chapter 425, March 3, 1899; 30 Stat. 1151). This Act required Congressional approval for the construction of any bridge, dam, dike, or causeway over or in navigable waters. A caveat to this law was that such structures could be authorized by state legislatures to be erected if the waters are totally within one state and the plan is approved by the Chief of Engineers and the Secretary of Army (33 USC. 401). The RHA included a provision (known as the Refuse Act) that addressed the dumping of refuse matter into waterways (Downing et al., 2003). This Act also addresses the building of wharfs, piers, jetties, and other structures within navigable waters, and the excavation or fill within navigable waters. The federal government was responsible for permitting and enforcement under this law. Although this Act primarily addressed protecting commercial navigation by regulating the construction of structures and the dumping of refuse into waterways that were used for navigational purposes, a secondary benefit of the Act was reducing environmental contamination and degradation. However, few of the environmental policies set forth by the RHA and the Refuse Act were actively enforced. After the passage of the RHA, little environmental legislation was passed for the next 50 years.

2.2.2 Federal Water Pollution Control Act

The first major legislation that was passed to directly address water pollution was the 1948 Federal Water Pollution Control Act (WPCA). The Act was passed in response to polluted waters within the United States resulting from the industrial growth and urban growth fueled by World War II. The original 1948 statute authorized the development of programs to eliminate or reduce pollution to improve the sanitary condition of surface water and groundwater. This Act placed responsibility for controlling water pollution on the states and primarily focused on the treatment of sewage wastes (Deason et al., 2001). Thus, early water protection efforts were directed toward point sources of pollution. Point source pollution refers to pollution from a discrete conveyance, such as a pipe, ditch, or ship. This law differed from previous environmental legislation as it had an objective of restoring waters that were polluted (Ferrey, 2004). This law, however, was not strong and lacked enforcements standards or punishment for violators (Milazzo, 2006).

2.2.3 National Environmental Policy Act

Water quality, as well as all environmental issues, began to receive greater and more notorious attention in the late 1960s as a result of growing public awareness of water quality decline due to high profile events such as the highly polluted Cuyahoga River in Cleveland, Ohio, catching fire in 1969 and the publication of Rachel Carson's book—*Silent Spring* (Carson, 1962). During this period, Americans were also concerned with the impact of massive federal roadway projects and the construction of hydroelectric dams. These elements set the stage for a shift in public awareness and concern toward greater protection of water resources.

Public concerns were eventually translated into the political response of passing laws (Ferrey, 2004). Specifically, public outcry for federal intervention lead to the passage of the 1969 National Environmental Policy Act (NEPA; 42 USC. 4321 et seq.).

This marked a change in water quality protection from primarily being managed by state and municipal authorities to having federal requirements through an established U.S. national policy for environmental enhancement. The 1969 Act also established the President's Council on Environmental Quality (CEQ; Houck, 1999). NEPA requires federal agencies to consider, through the development of an Environmental Impact Statement (EIS), the environmental effects their actions may have before finalizing a project proposal. Water quality concerns are among the wide array of environmental issues that must be considered under NEPA. NEPA is considered to be a primarily procedural statute. It does not necessarily protect against environmental harm, but instead ensures that investigation is made and options are considered regarding environmental impacts of a particular project (Ferrey, 2004).

Shortly following the passage of NEPA, President Nixon proposed the need for a separate regulatory agency to oversee enforcement of environmental policy and with congressional approval the U.S. Environmental Protection Agency (USEPA) was established. The birth of the USEPA in 1970 was a time to envision the following missions:

- The establishment and enforcement of environmental protection standards consistent with national environmental goals
- The conduct of research on the adverse effects of pollution and on methods and equipment for controlling it; the gathering of information on pollution; and the use of this information in strengthening environmental protection programs and recommending policy changes
- Assisting others, through grants, technical assistance, and other means, in arresting pollution of the environment
- Assisting the Council on Environmental Quality in developing and recommending to the President new policies for the protection of the environment (USEPA, 1992)

Since its inception, USEPA has been responsible for implementing a number of regulatory and nonregulatory programs designed to protect and improve water quality. Most of these programs were created by the Clean Water Act or the Safe Drinking Water Act.

2.2.4 Clean Water Act

The establishment of the USEPA was followed by the 1972 Federal Water Pollution Control Act. This Act is commonly referred to as the Clean Water Act (CWA). The CWA identified the goal of restoring surface waters considering their chemical, physical, and biological integrity and sought to eliminate point source discharges into waters of the United States by 1985 (Milazzo, 2006). One of the most significant features of the CWA was the establishment of the National Pollutant Discharge Elimination System (NPDES) program. This program requires permits for any discharge of a pollutant from a point source into waters of the United States. The permitting agency (USEPA, or states that have been delegated the authority to implement the program) must ensure that permitted discharges meet two different standards: technology-based standards and water quality-based standards. Technology-based standards are established on an industry-wide basis to ensure that polluters are employing the best available technology (or in the case of conventional pollutants,

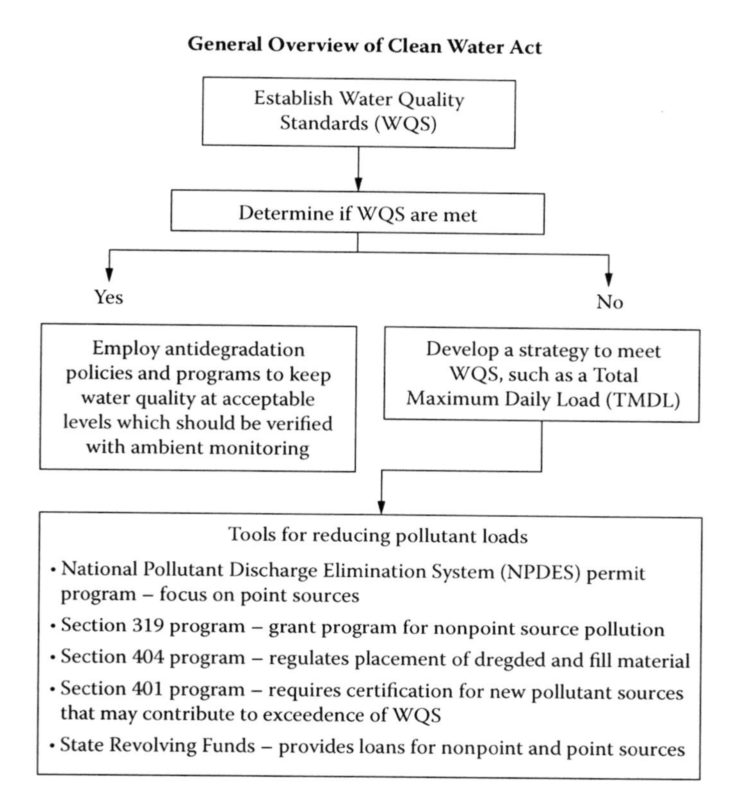

FIGURE 2.2 Flow chart of key elements in the Clean Water Act. (From USEPA, 2008. Watershed Academy Web: Introduction to the Clean Water Act. http://www.epa.gov/watertrain/cwa/.)

the best conventional technology) to treat their discharges. USEPA establishes these standards by surveying available technologies and determining what level of treatment is achievable for each pollutant if the best of these technologies is employed. USEPA adopts these levels, which are referred to as effluent limitation guidelines, by rule (Battle and Lipeles, 1998).

The second type of standards that permitting agencies are charged with are ensuring that dischargers meet water quality-based standards. Under the CWA, states are directed to establish water quality standards for each water body within their jurisdiction. Water quality standards consist of two different elements: (1) designated use and (2) water quality criteria. Each state determines the designated use of each water body within the state. For example, states may determine a particular water body should be designated for drinking water, for shellfish harvesting, for fishing and swimming, for agricultural use, or for industrial use. Then, numerical or, in some cases, narrative criteria are established for particular pollutants to protect such uses. In theory, all NPDES permits must ensure that these water quality criteria are met, and therefore by extension, the designated uses are protected. Water quality standards were intended to serve as a backstop to technology-based standards, to protect designated uses in situations where technology-based standards were not sufficient to protect a designated use of a particular water body (Battle and Lipeles, 1998). Unfortunately, the water quality standard component of the NPDES program was often overlooked and not implemented or enforced.

USEPA and the states implemented and enforced the NPDES program throughout the 1970s, 1980s, and 1990s, basing effluent limitations in permits primarily on the industry-wide technology-based standards, but largely ignoring state water quality standards. The failure to implement water quality standards was due, in large part, to the fact that a key component of the CWA, the Total Maximum Daily Load (TMDL) program, which was designed to implement water quality standards, was yet to be established. Although the CWA directed states to adopt TMDLs for all water quality-limited water body segments starting in the late 1970s, by the mid-1990s, TMDLs still had not been established for the vast majority of impaired waterbodies in the United States. This failure to establish and implement TMDLs led to a rash of lawsuits in the mid-1990s where more than 35 states and environmental groups sued USEPA, alleging that it failed to comply with the provisions of the CWA that mandated it either approve state-established TMDLs or establish federal TMDLs (Copeland, 2005). These lawsuits resulted in 22 states with court orders or consent decrees to establish TMDLs within an agreed upon time frame or USEPA will establish TMDLs (as of 2008). These states are Alabama, Alaska, Arkansas, California, Delaware, District of Columbia, Florida, Georgia, Iowa, Kansas, Louisiana, Mississippi, Missouri, Montana, New Mexico, Ohio, Oregon, Pennsylvania, Tennessee, Virginia, Washington, and West Virginia. Currently, the TMDL program is administered pursuant to USEPA's 1992 TMDL regulations (specifically, Part 130 of Title 40 of the C. F. R., Section 130.7). A more comprehensive review of the TMDL program and the law and policy behind its implementation is provided by Houck (1999) and DeBusk (2001).

As the states began to adopt and implement TMDLs, the TMDL program has taken center stage in most states because, unlike the NPDES program, the TMDL program addresses both point and nonpoint sources of pollutants and considers a watershed approach to allocating pollutant loads. The USEPA defines a TMDL as "the sum of allocated loads of pollutants set at a level necessary to implement the applicable water quality standards, including waste load allocations from point sources and load allocations from nonpoint sources and natural background conditions." (Nonpoint sources are sources that are diffuse or without a single point of origin, such as runoff from agriculture, urban, and construction.) A TMDL must contain a margin of safety and a consideration of "seasonal variations" (USEPA, 2007). In other words, a TMDL can be described as the amount of a particular pollutant that a particular water body can assimilate without resulting in a violation of a water quality standard. The TMDL is sometimes expressed as an equation:

$$\text{TMDL} = \text{WLA} + \text{LA} + \text{MOS} \tag{2.1}$$

where WLA is the waste load allocation from point sources, LA is the load allocation from nonpoint sources and natural background concentrations, and MOS is the margin of safety. MOS is used to account for uncertainties and variability in estimating

WLA and LA. Often, the MOS is considered to be a percentage (10%–15%) of WLA and LA. Others have considered conservative estimations of WLA and LA and thus described the estimated MOS as implicit due to these conservative assumptions.

The CWA requires states, territories, and authorized tribes to complete several tasks as part of their TMDL programs. These tasks are to (1) submit a list of waters that are impaired and/or threatened by pollutants (often referred to as the 303(d) list); (2) establish priority ranking of the listed waterbodies, taking into account the severity of pollution and the designated uses of the water; (3) identify waters targeted for TMDL development; and (4) develop and implement TMDLs.

TMDLs have been approved by USEPA in certain water bodies in certain states for dissolved oxygen, mercury, metals, nutrients, organics, pathogens, pesticides, pH, sediment, and temperature. Each state is in the process of assessing waterbodies and developing TMDLs (as needed). The methods employed to complete the TMDL process vary by state. Information on each approved TMDL is published by the USEPA on their Web site (www.epa.gov).

Once TMDLs are established by states and approved by USEPA, the next challenge is the allocation of TMDLs among all point and nonpoint source dischargers and the implementation of the TMDLs. For point source discharges, TMDLs will be allocated and implemented through the NPDES permitting program and may require pollution reductions beyond what would be required using only technology-based standards. For nonpoint sources, which include urban, suburban, and agricultural discharges that are not addressed by the NPDES permitting program, the allocation and implementation of TMDLs will be much more challenging. In most places, it is likely that a multifaceted watershed-based approach will be needed. Components of such a multifaceted approach will most likely have to include, among other things, some or all of the following pollution reduction approaches: state regulation of urban, suburban, and agricultural run-off; adoption of best management practices to reduce pollutant loadings in stormwater and agricultural discharges; retrofitting existing urban areas to treat stormwater; land acquisition programs to protect riparian areas that provide the function of filtering pollutants from run-off; wetland and water body restoration programs; and public education.

Another component of related to TMDLs that is currently receiving attention is the mandate to develop water quality standards or water quality numeric criteria. Water quality criterion are linked to TMDLs as they are, in general, concentrations levels that should not be exceeded to meet designated uses, and these concentrations can be translated into loads to be used in developing and implementing TMDLs. Detailed information on water quality standards is provided in Chapter 3.

2.2.5 Other Water-Related Legislation

While the CWA is probably the best known water law in the United States, other legislative acts also protect water resources. Two examples are the Coastal Zone Management Act and the Safe Drinking Water Act (SDWA). The Coastal Zone Management Act of 1972 sought to preserve, protect, develop, and restore resources in the coastal zone of the United States. The SDWA of 1974 was created to protect and ensure the safety of public drinking water supplies (Ferrey, 2004). Under the SDWA, USEPA establishes Maximum Contaminant Levels (MCLs) designed to protect human health. Public drinking water supplies must comply with these standards. Many different government regulations and policies help to protect our water resources. Some of these influence water resources directly, such as through the CWA and the SDWA, and others influence water resources indirectly, such as the Clean Air Act, which can be used to indirectly influence water quality by limiting the amount of airborne pollutants that may ultimately be deposited into waterbodies.

2.3 FUTURE DIRECTION OF WATER QUALITY POLICY

As more information is known and public awareness of water quality issues increases, better designed government programs are being developed and enforced to preserve water resources. Protecting and conserving water supplies is likely to be an ever increasing societal concern due to the increase in competing water uses (e.g., growing population, energy production, agriculture, etc.), limited water supplies, and the desire to protect the ecological integrity of water resources.

While there is growing need to preserve and restore water resources, there are also costs associated with achieving this goal. A good example of this is the Comprehensive Everglades Restoration Plan (CERP), which was originally estimated (1998) to cost US$7.8 billion with an additional annual US$182 million for operation, maintenance, and monitoring. These funds are to be supplied by the federal government and the state of Florida. Obviously, allocation of funds to one project implies that other projects will not be funded. As more “CERP” type projects are identified, a balance will have to be reached where the water resource protection desires and the costs to achieve them are publicly acceptable. Science and technology are available that can prevent or mitigate most water pollution problems; however, their implementation may not be feasible due to costs, conflicting interests, and/or public acceptance.

2.4 SUMMARY

Water quality protection of U.S. surface waters has received much attention in the 20th century and continues to be an important political issue. The formation of the USEPA in 1970 and the passage of the CWA set a path to protecting and restoring U.S. water resources that still continues. The primary drivers behind the legislation and its enforcement have been citizen activism and legal action. Water protection in the 21st century will have available the best technology and science that researchers can offer; however, the costs of successfully implementing this may be great and potential benefits will need to be evaluated by society.

REFERENCES

Battle, J.B., and M.I. Lipeles. 1998. *Water Pollution*. 3rd ed. Cincinnati, OH: Anderson Publishing Co.

Carson, R. 1962. *Silent Spring*. Boston, MA: Houghton Mifflin.

Copeland, C. 2005. Clean Water Act and Total Maximum Daily Loads (TMDLs) of pollutants. CRS Report for Congress. Order Code 97-831 ENR. http://digital.library.unt.edu/govdocs/crs/permalink/meta-crs-10107:1.

Deason, J.P., T.M. Schad, and G.W. Sherk. 2001. Water policy in the United States: A perspective. *Water Policy* 3:175–192.

DeBusk, W.F. 2001. Overview of the Total Maximum Daily Load (TMDL) Program. SL188, Soil and Water Sciences Department, Florida Cooperative Extension Service, Florida Cooperative Extension Service, Institute of Food and Agricultural Sciences. University of Florida. http://edis.ifas.ufl.edu/SS400.

Downing, D.M., C. Winer, and L.D. Wood. 2003. Navigating through Clean Water Act jurisdiction: a legal review. *Wetlands* 23(3):475–493

Ferrey, S. 2004. *Environmental Law*. 3rd ed. Aspen Law & Business. New York.

Houck, O.A. 1999. *The Clean Water Act TMDL Program: Law, Policy, and Implementation*. Environmental Law Institute: Washington, D.C.

Milazzo, P.C. 2006. *Unlikely Environmentalists: Congress and Clean Water, 1945–1972*. Lawrence, KS: University Press of Kansas.

USEPA (U. S. Environmental Protection Agency). 1992. The Guardian: Origins of the EPA. EPA Historical Publication-1. http://www.epa.gov/history/publications/print/origins.htm.

USEPA. 2000a. Ambient water quality criteria recommendations, rivers and streams in nutrient ecoregion IX. Office of Water. EPA 822-B-00-019.

USEPA. 2000b. Ambient water quality criteria recommendations, rivers and streams in nutrient ecoregion XII. Office of Water. EPA 822-B-00-021.

USEPA. 2000c. The History of Drinking Water Treatment. Office of Water. EPA-816-F-00-006.

USEPA. 2001. Fact sheet: Ecoregional nutrient criteria. Office of Water. EPA-822-F-01-010. http://www.epa.gov/waterscience/criteria/nutrient/ecoregions/files/jan03frnfs.pdf.

USEPA. 2007. Overview of current Total Maximum Daily Load—TMDL—program and regulations. http://www.epa.gov/owow/tmdl/overviewfs.html.

USEPA. 2008. Watershed Academy Web: Introduction to the Clean Water Act. http://www.epa.gov/watertrain/cwa/.

3 Water Quality Standards
Designated Uses and Numeric Criteria Development

Brian E. Haggard and J. Thad Scott

CONTENTS

3.1 INTRODUCTION

What are water quality standards? Why is it necessary to have them? And how do we establish and enforce them? We answer these questions in great detail in this chapter. In short, water quality standards are numeric values or narrative descriptions of water quality parameters that are meant to sustain the designated uses of a water body. Therefore, water quality standards involve not only the actual criteria associated with water quality parameters, but how certain levels of those parameters negatively affect the use of that water for human and/or ecological purposes.

The Federal Water Pollution Control Act (FWPCA) of 1956 directed states in the United States to develop water quality standards for interstate waters. This protection was expanded to all surface waters in the 1972 amendments to FWPCA, which also established the National Pollution Discharge Elimination System (NPDES). Collectively, this legislation is generally referred to as the Clean Water Act (CWA). According to the 1972 CWA, point source pollution discharges into U.S. surface waters were required to obtain and follow an NPDES permit that includes technology-based limits on certain pollutants. Water quality-based limits were used as a guideline for standardizing technologies across diverse geographic and socioeconomic areas. Amendments to the CWA in the 1970s and 1980s required that states develop numeric criteria for pollutants if the presence of the pollutant was likely to affect the water body's use. In response to these amendments, the United States Environmental Protection Agency (USEPA) developed guidelines to assist states in developing water quality standards (designated uses and associated numeric criteria) and published them in the *Water Quality Standards Handbook*. The handbook was finalized in its second edition (USEPA, 1994), and a Web-based version was published in 2007 (USEPA, 2007). According to the Web-based version, the handbook is "intended to serve as a 'living document,' subject to future revisions as the USEPA water quality program moves forward, and to reflect the needs and experiences of USEPA and the States." Thus, water quality standards are dynamic rather than static and require periodic updates that consider changes to designated uses and new technologies that could improve water quality. In fact, states are required to review and update their water quality standards every 3 years. For this reason, a comprehensive review of current standards in this text is not warranted. Rather, practitioners should be aware of the general process by which water quality standards are developed. Our intention is to outline this general process while providing some specific examples from individual states.

This chapter includes a summary of the major designated use categories for surface waters in the United States and provides examples of how individual states specifically define these categories. Major water quality parameters used in assessing whether a water body is of sufficient quality for its designated uses and development of numeric criteria for those parameters are also discussed. Finally, ongoing effort to develop numeric criteria for nutrients and the general procedures being used nationwide in the United States to establish nutrient water quality standards are presented.

3.2 DESIGNATED USES AND USE ATTAINABILITY ASSESSMENTS

In establishing water quality standards, states are first required to identify and describe how surface waters are used and what traits (i.e., water quality parameters) should be managed to protect the use or uses. The "designated uses" of a water body can be numerous and quite diverse (USEPA, 1994, 2007), but most are often grouped into four general categories: (1) agricultural and industrial water supply, (2) recreation, (3) public water supply, and (4) aquatic life. Agricultural and industrial water supplies are those waters that are the source of crop irrigation, livestock drinking water, or process water in industrial activities. Recreational waters are those where human activities involve either complete immersion, such as swimming, diving, and water-skiing, or incomplete immersion, such as boating, fishing, or wading. Public water supplies are those waters that are the source of human drinking water. Finally, the aquatic life use designation includes waters that support the growth and reproduction of wildlife species. Surface waters in the United States must be of sufficient quality to support each of these uses, and water quality criteria are developed to provide a clear and measurable indicator of whether or not a designated use is attained.

USEPA defines attainable uses of a water body as "the uses that can be achieved when [technology-based] effluent limits are imposed on point source dischargers and when cost-effective and reasonable best management practices are imposed on nonpoint source dischargers" (USEPA, 1994, 2007). States are not required to designate each of these use categories to all waters. However, states must demonstrate that use attainment for categories that are excluded is not possible or unreasonable, and cannot be obtained in the future. States may also remove specific uses already assigned to waters if the specific use is demonstrated to be unattainable currently or in the future.

The potential for attainment of all designated uses, but most often the aquatic life use, is determined by conducting a Use Attainability Assessment (UAA; USEPA, 1994, 2007). UAAs involve two phases: (1) a water body survey and assessment and (2) a socioeconomic assessment. The water body survey and assessment is a comprehensive analysis of the physical, chemical, and biological characteristics of a water body that influence the ability of that water body to support aquatic life use. Guidance documents are available to states describing USEPA's preferred methods for water body surveys and assessments in streams and rivers (USEPA, 1983), lakes (USEPA, 1984a), and estuaries (USEPA, 1984b). In recent years, USEPA and states have increasingly recognized the need to quantify specific ecosystem attributes that may influence use attainability in certain waters. For example, water impoundment reservoirs exhibit well-known spatial variability in their physical, chemical, and biological attributes (Thornton et al., 1990; Straškraba et al., 1993), and some states are exploring the need to identify specific reservoir "zones" in which assessment methodologies might differ (Brooks et al., 2009). USEPA permits states to use alternative approaches to assessing specific waters if the methods used are "scientifically and technically supportable" (USEPA, 1994, 2007). Results of water body survey and assessments are used to identify the levels of water quality parameters that would be required to achieve the desired aquatic life use. The cost for improving point source treatment technology and implementing nonpoint source best management practices are then considered along with regional socioeconomic data to determine whether to proceed with improvements, or eliminate or lower the aquatic life use designation (USEPA, 1994, 2007).

3.3 WATER QUALITY STANDARDS

Once the designated uses of a water body have been established, quantitative water quality criteria are established to protect those uses. In cases where multiple uses require different levels of the same parameter, the most sensitive value is used as the standard. USEPA has provided federal guidance to states on water quality criteria for specific pollutants and toxic substances (USEPA, 1986). In the following pages, water quality criteria for the four aforementioned designated uses are summarized, with the majority of information dedicated to the aquatic life use designation. Examples of state standards are provided in each category from states in USEPA Region 6 (which includes Arkansas, Louisiana, New Mexico, Oklahoma, Texas, and 66 Tribes).

3.3.1 Industrial and Agricultural Water

Few states list specific water quality criteria for industrial and agricultural water uses. This is most likely because most waters used in industrial and agricultural practices also have other designated uses (such as aquatic life use) with more sensitive water quality criteria. However, some parameters are occasionally listed for protection. For example, the state of Oklahoma lists region-specific criteria for sulfate, chloride, and total dissolved solids for agricultural water use (see Table 3.1 for details). More pressing for many states is the availability of water for agricultural and industrial practices. Many environmental concerns outside of water quality, such as aquifer depletion and minimum stream flows, drive environmental regulations on industrial water and agricultural water.

3.3.2 Contact Recreation

All states have adopted water quality criteria to protect the health of humans coming into contact with waters during recreational activities. USEPA permits states to classify waters as either primary contact recreation or secondary contact recreation. Primary contact recreation is defined as any activity that involves the potential ingestion or complete immersion in water. This would include swimming, water-skiing, diving, and surfing. Secondary contact recreation is defined as activities that result in incomplete immersion, such as wading and boating. States may conduct a UAA to demonstrate that specific waters cannot naturally support contact recreation criteria.

Water quality criteria for contact recreation uses generally involve monitoring indicator organisms to assess the risk of encountering fecal contamination from humans and other warm-blooded animals. Total coliforms and fecal coliforms were historically used as indicator organisms in freshwater and brackish waters. However, recent research has indicated that *Escherichia coli* counts in freshwaters and entercocci counts in brackish waters may provide more reliable indications of fecal contamination (Jin et al., 2004). As a result, most states have adopted *E. coli* and enterococci counts for contact recreation criteria in freshwater and brackish water, respectively, but some still use fecal coliform counts in assessments until monitoring data can be expanded.

3.3.3 Drinking Water Supply

State water policies do not often include a great deal of detail regarding drinking water quality standards. Most state regulations include narrative information stating that toxic substances should not be present in quantities alone or in combination that can be toxic to humans. Although some states do have numeric criteria for many known carcinogens and some other compounds that are potentially toxic for humans, drinking water quality standards actually fall under federal jurisdiction, as described in the Safe Drinking Water Act (SDWA) of 1974. Under SDWA authority, USEPA established national primary drinking water standards for microbiological contaminants, inorganic and organic chemical contaminants, and disinfectants and disinfection by-products. Drinking water standards are most often expressed in the form of a Maximum Contaminant Level (MCL).

Standards for inorganic and organic chemical contaminants are too numerous to list but may be easily obtained through USEPA documents and the USEPA Web site

TABLE 3.1
Summary of Aquatic Life Use Designations by States in USEPA Region 6 and Examples of Associated Numeric Water Quality Criteria for These Designations

State	Aquatic Life Use Category	Temperature (°C)	Dissolved Oxygen (mg/L)	pH	Turbidity (NTU)
Arkansas[1]	All fisheries			6–9[7]	
	Trout	20	6.0		10[8]
	Lakes and reservoirs	32	5.0		25[8]
	Streams	29–32[3]	2.0[6]–6.0[3]		10–75[8]
Louisiana[1]	All freshwaters	32.2	5.0	6–9[7]	
	Lakes and reservoir	1.7 AA[4]			25
	Streams and rivers	2.8 AA[4]			25 or 150[9]
New Mexico[1,2]	All aquatic life				10 AA[4] or 20% AA[4]
	Coldwater aquatic life	20	6.0	6.6–8.8[7]	
	Marginal cold water aquatic life	25	6.0	6.6–9.0[7]	
	Warm water aquatic life	32.2	5.0	6.6–9.0[7]	
	Marginal warm water aquatic life	32.2[5]	5.0	6.6–9.0[7]	
Oklahoma[1]	Fish and wildlife propagation	2.8 AA[4]		6.5–9.0[7]	50
	Lakes	1.7 AA[4]	5.0[6]		25
	Warm water aquatic community	28.9	6.0[6]		10
	Cool water aquatic community	20	6.0[6]		10
	Trout fishery		3.0[6]		
	Habitat limited				
Texas[1]	All aquatic life	29.5–35.0[3]		6.5–9.0[3,7]	Narrative
	Exceptional		4.0[6]		
	High		3.0[6]		
	Intermediate		3.0[6]		
	Limited		2.0[6]		

[1] State standards summarized by USEPA at www.epa.gov/waterscience/standards/.
[2] Separate standards for classified water bodies (basin-specific) and nonclassified water bodies that are considered either ephemeral, intermittent, or perennial streams.
[3] Regionally dependent.
[4] AA = above ambient.
[5] May be exceeded; evaluated on a case-by-case basis.
[6] 24-h minimum during critical period (generally summer and early fall).
[7] Measurements must be within the defined range.
[8] Excluding stormflow conditions.
[9] Selected rivers.

(USEPA, 2001). The contaminants are from many diverse sources including industrial and agricultural wastes and runoff from urban areas. These contaminants can cause a variety of adverse effects in humans, including kidney, liver, and nervous system disorders and increased risk of various cancers.

Another contaminant monitored in drinking water is microorganisms. Microorganisms such as *Cryptosporidium*, *Giardia*, *E. coli*, and other enterococci bacteria enter waters from human and animal wastes. These organisms, or other organisms with which they are commonly associated, can cause severe gastrointestinal disorders when ingested. USEPA (2009) requires that disinfection of public drinking water remove 99.9% of *Cryptosporidium* and *Giardia*. Any positive detection in tests for *E. coli* is considered a violation of the MCL (USEPA, 2001).

Disinfectants are used by municipal water authorities to remove the threat of microbiological contamination of waters. However, these disinfection compounds, which include chlorine, chlorine dioxide, and chloramines, are themselves harmful for human consumption. Disinfection compounds can cause eye/nose and stomach irritation and even anemia and nervous system disorders in children and young adults. These disinfection compounds can undergo chemical reactions with organic matter in water to form various disinfection by-products such as chlorite, haloacetic acids, and trihalomethanes. These compounds are even more harmful, causing liver, kidney, and nervous system disorders, and increased risk of various cancers. Therefore, USEPA limits the levels of both disinfectants and disinfection by-products in finished drinking water (USEPA, 2001).

3.3.4 Aquatic Life Use

The designation of aquatic life uses is different among states due to the wide range of geomorphologic features and regional climate patterns in the United States. Aquatic life use is intended to protect the propagation of fish and wildlife and aquatic biodiversity. Of course, virtually all waters support some level of biological growth and reproduction. Thus, the aquatic life use designation is highly subjective and requires that states assess the attainability of uses by individual water bodies. Table 3.1 demonstrates the diversity of approaches taken by states in USEPA Region 6 to categorize aquatic life use. This is most often done for "classified segments" that are usually divided into individual lakes, reservoirs and wetlands, and stream or river reaches confined to a given geographic area. Toxic compounds including organics, metals, and other inorganics usually have specific numeric criteria developed by acute and chronic toxicity tests using model organisms such as the zooplankton *Daphnia magna* and *Ceriodaphnia dubia* (APHA, 2005). Other water quality parameters have numeric criteria necessary to support various degrees of aquatic life use that may be regionally or system specific. Some of these variables include water temperature, pH, dissolved oxygen, and turbidity.

Most states require that dischargers do not increase the temperature of receiving waters above a threshold that could impact aquatic life. Streams supporting trout populations, such as some streams in Arkansas, Oklahoma, and New Mexico in USEPA Region 6 (Table 3.1), require that maximum stream temperature not exceed 20°C. Other states have regionally dependent temperature criteria (e.g., Texas), use temperature criteria for waters categorized by type (e.g., Arkansas and Louisiana), or apply temperature criteria based on type of habitat certain waters provide (e.g., New Mexico and Oklahoma). Numeric criteria for pH are usually expressed as a range of values that measurements should be within, and most states apply a broad range to all waters (Table 3.1). However, some states such as New Mexico have modified these ranges according to the type of habitat certain waters provide (Table 3.1).

Dissolved oxygen criteria vary substantially by state according to each state's aquatic life use designations (Table 3.1). Some states use single measurements while others use 24-h average or minimum concentrations for comparison to numeric criteria. Similar to temperature criteria, streams supporting trout populations receive the highest degree of protection for dissolved oxygen. Otherwise, most states assign numeric dissolved oxygen criteria based on the type of water body (e.g., Arkansas), the type of habitat certain waters provide (e.g., New Mexico and Oklahoma), or by the level of biological diversity that certain waters support (e.g., Texas).

Dissolved oxygen standards were originally developed in order to address the introduction of untreated sewage wastes to surface waters. Breakdown of this organic matter in streams and lakes caused these systems to become anoxic and not support aquatic life use. In the last two decades, scientists and policy makers have increasingly recognized the role of nutrients in driving dissolved oxygen dynamics in water bodies (Carpenter et al., 1998). Increased nutrients derived from anthropogenic sources can cause algal blooms that create an oxygen demand. And there is growing recognition that nutrients derived from point and nonpoint sources may be causing a dramatic loss of ecosystem services historically provided by streams, lakes, and rivers (Tilman et al., 2001; Mulholland et al., 2009). Although states regulate standards for dissolved oxygen, very few states have adopted numeric criteria for nutrients such as nitrogen and phosphorus. However, most states are now either developing or planning the development of water quality nutrient criteria. In the remainder of this chapter, we discuss the process of nutrient criteria development and the successes and pitfalls many states have encountered.

3.4 NUTRIENT CRITERIA DEVELOPMENT

Historically, narrative criteria for nutrients have been used by many states to protect designated uses. For example, Arkansas's Regulation 2 states that "Materials stimulating algal growth shall not be present in concentrations sufficient to cause objectionable algal densities or other nuisance aquatic vegetation or otherwise impair any designated use of the water body." These narrative criteria were intended to protect the designated uses of the water bodies, but the exceedance of narrative criteria requires subjective determinations. States have often assessed variables that would indicate potential impairments due to nutrient enrichment; these variables include water clarity (i.e., turbidity), periphyton and or phytoplankton production (i.e., chlorophyll-a, mg/m^2, or concentration, μg/L), dissolved oxygen (i.e., saturation, concentration, and diurnal changes), pH fluctuations, aquatic life communities (i.e., fish and macroinvertebrate community structure), and other biological metrics based on aquatic life. Turbidity, dissolved oxygen, and pH were often used to indirectly assess designated use attainability related to nutrients because these variables change in response to changing nutrient availability and because they have numeric criteria defined for aquatic life use.

The CWA has long required that numeric criteria for water quality indicators, even nutrients, be developed and adopted at the state (and tribal) level, although the establishment of these numeric nutrient criteria was not put into action until the 1998 Clean Water Action Plan (CWAP). In 1998, the USEPA also released its National Strategy for the Development of Regional Nutrient Criteria and subsequently provided technical guidance manuals to assist the states in the development of nutrient criteria (USEPA, 2000a, 2000b). The technical documents presented three general approaches from which nutrient criteria can be developed, including (1) the use of frequency distributions of nutrient concentrations from selected reference streams or all available data, (2) the use of predictive relations between nutrients and selected response variables (e.g., chlorophyll-a, dissolved oxygen, aquatic life community structure), and (3) the modification of established nutrient and algal thresholds as defined in the literature (e.g., nuisance periphytic biomass defined as chlorophyll-a ranging from 100 to 150 mg/m^2; Welch et al., 1989). In 2001, the Federal Register had noticed that states were strongly recommended to develop a plan to adopt numeric nutrient criteria and to submit this outline to USEPA. Subsequently, the USEPA published recommended numeric water quality criteria for nutrients and chlorophyll-a concentration in selected water body types (e.g., streams, rivers, lakes, and reservoirs) for Level III eco-regions. However, USEPA's recommendations were intended as starting points for states and tribes, as these entities have the ability to develop numeric criteria specific to physical, chemical, and biological conditions within their own jurisdictions.

The intent of many states and tribes is to translate existing narrative nutrient criteria into numeric criteria that protect the designated uses of its water bodies. Although the methods and procedures used to translate narrative criteria into numeric values vary across states, there are several commonalities. Some of these methods are further discussed in the following sections, and specific examples of criteria are given from the states within USEPA Region VI.

3.4.1 Frequency Distributions and Gradients of Nutrient Concentrations

USEPA recommends using frequency distribution on median data from water bodies (Figure 3.1) across large spatial scales such as states, eco-regions, or large basins to develop numeric criteria. The use of frequency distributions to evaluate numeric nutrient criteria requires that multiple data points for individual water bodies be reduced down to often a single value, for example, the median concentration. This way each water body has the same influence of the frequency distribution of observed nutrient concentrations across the region of interest. Thus, the median database is less likely to be influenced by one individual site or concentration outliers than a database using all observations. For more basic information data distributions and statistical evaluations, see Chapter 13.

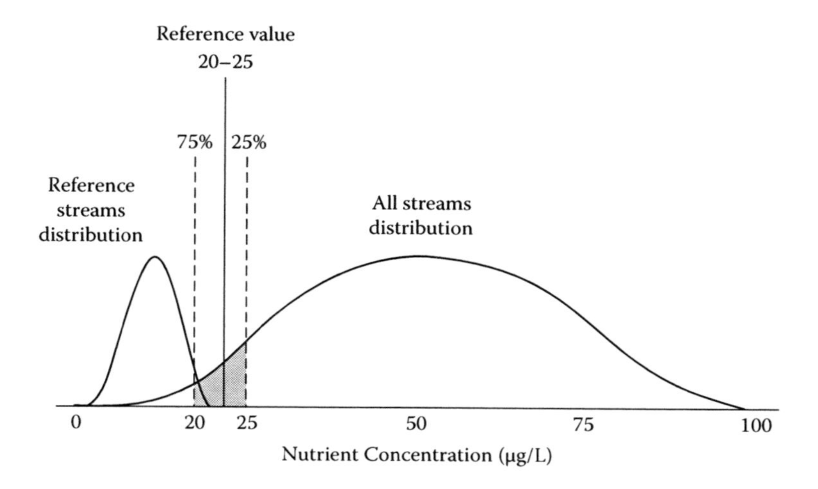

FIGURE 3.1 Example of the estimation of the 75th percentile nutrient concentration from selected reference streams and the 25th percentile concentrations from all streams; the suggested or recommended criterion would be either or in between these nutrient concentrations per USEPA guidance. (Modified from USEPA [U.S. Environmental Protection Agency]. 2000b. Nutrient Criteria Technical Guidance Manual—Rivers and Streams. EPA-822-B-00-002, Washington, D.C.)

USEPA recommended two specific methods for frequency distribution of medians. The first method involves the assessment of water bodies representing reference conditions, from which the 75th percentile of the median nutrient concentration should serve as the reference condition (Figure 3.1). The reference condition approach uses sites that are relatively undeveloped (minimal human influence) within its watershed, suggesting that these sites represent the natural, physical, chemical, and biological conditions of the region. For example, the Oklahoma Water Resources Board used the 75th percentile of flow-weighted total phosphorus (TP) concentrations of relatively undeveloped streams from the USGS Hydrologic Benchmark Network and its National Water Quality Assessment Program (Clark et al., 2000) to establish its TP criterion at 0.037 mg/L for scenic rivers across the state.

The second USEPA recommended method involves evaluating the frequency distribution across all water bodies (including reference conditions), and using the 5th to 25th percentile as the reference condition as this distribution of data includes impaired water bodies. However, much of the discussion and recommendations that have followed since these guidance materials were released has focused on the 25th percentile of nutrient data from all water bodies (example in Figure 3.1). The general thoughts behind these two techniques are that the 75th percentile from the reference population and the 25th percentile from the general population would be relatively similar in values. However, it is not always appropriate to assume that this reference-to-general population relationship for all nutrients exists (Suplee et al., 2007). The USEPA suggests that the frequency distribution approach represents one of the components of nutrient criteria development, that is, the reference conditions, and that this information should be used in conjunction with additional information to establish nutrient criteria.

An alternative method that may be used to establish reference nutrient concentrations for watersheds or specific regions considers the nutrient concentration gradient as human activity increases or land use changes. For example, Dodds and Oakes (2004) developed multiple regression models using nutrient concentrations (i.e., log-means in this example) as the dependent variable and percent land use classes (i.e., crop and urban) as the independent variable, where the intercept of these models represent reference conditions under the absence of human activities. This approach produces concentration values that are sometimes similar to that attained from the 75th percentile method and other times much different depending upon the specific region that is compared (Dodds and Oakes, 2004).

Establishing reference conditions allows for some variability among values to account for local or regional landscape characteristics such as geology, soils, and climate, among other factors. The concept of eco-regions has been used to integrate these landscape attributes, where eco-regions are developed displaying similarities across these factors (Omernik, 1987). The USEPA has developed numeric nutrient criteria recommendations across the 14 Level III eco-regions as defined across the United States for many water bodies (e.g., lakes and reservoirs, Table 3.2; and

TABLE 3.2
Recommended USEPA Criteria for Total Phosphorus (TP), Total Nitrogen (TN), Chlorophyll *a* (Chl *a*), and Secchi Depth for Lakes and Reservoirs in Each Aggregate Ecoregion

Parameter	TP (μg/L)	TN (mg/L)	Chl *a* (μg/L)	Secchi Depth (m)
Agg Ecoregion I[1]	—	—	—	—
Agg Ecoregion II	9	0.10	1.90	4.50
Agg Ecoregion III	17	0.40	3.40	2.70
Agg Ecoregion IV	20	0.44	2.00 S[2]	2.00
Agg Ecoregion V	33	0.56	2.30 S[2]	1.30
Agg Ecoregion VI	38	0.78	8.59 S[2]	1.36
Agg Ecoregion VII	15	0.66	2.63	3.33
Agg Ecoregion VIII	8	0.24	2.43	4.93
Agg Ecoregion IX	20	0.36	4.93	1.53
Agg Ecoregion X[1]	—	—	—	—
Agg Ecoregion XI	8	0.46	2.79 S[2]	2.86
Agg Ecoregion XII	10	0.52	2.60	2.10
Agg Ecoregion XIII	18	1.27	12.38 T[3]	0.79
Agg Ecoregion XIV	8	0.32	2.90	4.50

[1] Under development.
[2] Chlorophyll *a* measured by spectrophotometric method with acid correction.
[3] Chlorophyll *a*, *b*, and *c* measured by trichromatic method.
Source: http://www.epa.gov/waterscience/criteria/nutrient/ecoregions/files/sumtable.pdf

TABLE 3.3
Recommended USEPA Criteria for Total Phosphorus (TP), Total Nitrogen (TN), Chlorophyll *a* (Chl *a*), and Turbidity for Rivers and Streams in Each Aggregate Ecoregion

Parameter	TP (μg/L)	TN (mg/L)	Chl *a* (μg/L)	Turbidity (FTU/NTU)
Agg Ecoregion I	47	0.31	1.80	4.25
Agg Ecoregion II	10	0.12	1.08	1.30 N[4]
Agg Ecoregion III	22	0.38	1.78	2.34
Agg Ecoregion IV	23	0.56	2.40	4.21
Agg Ecoregion V	67	0.88	3.00	7.83
Agg Ecoregion VI	76	2.18	2.70	6.36
Agg Ecoregion VII	33	0.54	1.50	1.70 N[4]
Agg Ecoregion VIII	10	0.38	0.63	1.30
Agg Ecoregion IX	37	0.69	0.93 S[3]	5.70
Agg Ecoregion X	128[2]	0.76	2.10 S[3]	17.50
Agg Ecoregion XI	10	0.31	1.61 S[3]	2.30 N[4]
Agg Ecoregion XII	40	0.90	0.40 S[3]	1.90 N[4]
Agg Ecoregion XIII[1]	—	—	—	—
Agg Ecoregion XIV	31	0.71	3.75 S[3]	3.04

[1] Under development.
[2] This value appears inordinately high and may either be a statistical anomaly or may reflect a unique condition. Further regional investigation is indicated to determine the source(s) (i.e., measurement error, notational error, statistical anomaly, natural enriched conditions, or cultural impacts).
[3] Chlorophyll *a* measured by spectrophotometric method with acid correction.
[4] NTU; unit of measure for turbidity.
Source: http://www.epa.gov/waterscience/criteria/nutrient/ecoregions/files/sumtable.pdf

streams and rivers, Table 3.3). The basis for having a single nutrient criterion per eco-region assumes that the concentrations and sources of nutrients are different across eco-regions, and that the expression of water quality impairment relative to specific nutrient concentrations is different across eco-regions. However, nutrient concentrations in water bodies often show a gradient across the land use composition of different watersheds, especially in streams and rivers (Migliaccio et al., 2007). Wickham et al. (2005) showed that the variance in nutrient concentrations among land use was three to six times greater than that across eco-regions. Thus, numeric criteria may be developed considering the increase in median (or mean, geomean, etc.) nutrient concentrations along a land use gradient representing increased human activity.

3.4.2 Relations between Nutrients and Biological Response Variables

If the intent of numeric nutrient criteria is to sustain aquatic life as the designated beneficial use, then it is necessary to document and understand the relation between nutrient concentration and biological responses. Eutrophic conditions generally relates to water bodies with high primary productivity, although this relationship is less clear in streams where tree canopies attenuate light (Dodds, 2006, 2007). Lake studies have focused on the positive linear relationship between nutrient concentrations and chlorophyll-a (Dillon and Rigler, 1974; Carlson, 1977; Kratzner and Brezonik, 1981), and numeric criteria in lakes and reservoirs have been developed by some states in USEPA Region VI. For example, a site-specific criteria for chlorophyll-a has been developed for Beaver Lake in northwest Arkansas (i.e., 8 μg/L), and reservoir-specific criteria for phosphorus have been developed in northeastern Oklahoma (e.g., 17 μg/L in Lake Eucha and 0.014 μg/L in Lake Spavinaw).

Although algal growth in flowing water bodies may impair designated beneficial uses, the relation between nutrients and biological responses in these systems is not well understood (e.g., Lohman et al., 1992; Biggs, 1996; Lohman and Jones, 1999). Nutrients often explain less of the variability in chlorophyll-a in streams compared to lakes, and maybe reservoirs. Several studies have shown that chlorophyll-a concentrations in flowing waters are correlated to dissolved (Biggs, 2000) or total nutrients (e.g., Welch et al. 1989; Dodds et al., 1997, 1998). Welch et al. (1988) suggested that nuisance biomass levels might be present in flowing waters when benthic chlorophyll-a levels exceed 100 to 150 mg/m^2; this was based upon the dominance of filamentous algae (e.g., cladophora).

Another approach to assessing the nutrient status of streams involves quantifying the growth of algae on artificial substrate that is either enriched or not enriched with nutrients. The response ratio is the amount of algal growth on unenriched substrate divided by the amount of algal growth on enriched substrate (Matlock et al., 1999a, 1999b). This ratio theoretically ranges from zero to one, where a value of one suggests that ambient nutrient concentrations in the water body are sufficient to promote periphytic growth that equals that observed with nutrient enrichment on artificial substrates. When the ratio has been plotted as a function of nutrient concentration, it can display an exponential rise to maximum or an asymptotic relation (Popova et al., 2006). Kiesling et al. (2001) suggested that a ratio of 0.5 was the boundary between mesotrophic and eutrophic conditions, which corresponded to a dissolved phosphorus concentration of 40 μg/L in the North Bosque River in central Texas. However, many factors may influence the response ratio, including light availability and the concentration of nutrients in enrichment substrata (Matlock et al., 1999a).

Nutrient enrichment influences not only autotrophic response (e.g., periphyton biomass, chlorophyll-a) but also heterotrophic biomass and production, including heterotrophic microbes, macroinvertebrates, and fish (Dodds, 2006). Many streams are net heterotrophic due to light attenuation from forest canopy, and the relationship between nutrient enrichment and heterotrophic production is not well understood (Dodds, 2007). Recent studies have started evaluating how algal, heterotrophic microbes, macroinvertebrate, and fish communities change along a nutrient concentration gradient (e.g., Richardson et al., 2007; Wang et al., 2007; Weigel and Robertson, 2007; Scott et al., 2008; Stevenson et al., 2008; Evans-White et al., 2009).

There are various statistical techniques that states and tribes should use to evaluate biological response along a nutrient gradient, including correlation, simple linear and nonlinear regression, locally weighted scatterplot smoothing (LOWESS), change point, and classification analysis. Most recent nutrient criteria studies have used non-parametric statistical methods to detect nutrient thresholds (i.e., change points; Qian et al., 2003) in which a measurable change in biological variables occurs (Table 3.4). These studies indicate that the threshold nutrient concentrations that result in a biological response can be highly variable among watersheds, regions, and states.

TABLE 3.4
Recommended Stream Nutrient Criteria for Total Phosphorus (TP) and Total Nitrogen (TN) Based on Algae, Macroinvertebrate, Fish, Macrophyte, and Watershed Responses from Select Scientific Literature

Biological Response Variable	TP (µg/L)	TN (µg/L)	Method	Reference
International Data				
Algae	30	40	Break point regression	Dodds et al. (2002)
Pennsylvania				
Watershed[1]	70	2010	Frequency distribution	Sheeder and Evans (2004)
Florida				
Macroinvertebrates	13–19		Change point analysis	King and Richardson (2003)
Minnesota				
Algae	100	2700	Modeling	Carleton et al. (2009)
Mid-Atlantic Highlands				
Algae	10–12		Change point analysis	Stevenson et al. (2008)
Florida				
Macroinvertebrates	10–14		Change point analysis	Qian et al. (2003)
Illinois				
Algae	70		Change point analysis	Royer et al. (2008)
Iowa, Kansas, & Missouri				
Macroinvertebrates	50	930–1140	Change point analysis	Evans-White et al. (2009)
Florida				
Algae, macrophytes, macroinvertebrates	12–15		Change point analysis	Richardson et al. (2007)
Wisconsin				
Macroinvertebrates, fish	60	640	Change point analysis	Weigel & Robertson (2007)
Wisconsin				
Macroinvertebrates, fish	60–70	540–610	Change point analysis	Wang et al. (2006)
West Virginia				
Algae, macroinvertebrates		310–1800[2]	Change point analysis	Zheng et al. (2008)
Kentucky & Michigan				
Algae	30	1000	Cladophora presence	Stevenson et al. (2006)

[1] This study defined watershed as unimpaired and impaired to establish criteria.
[2] This study identified threshold values based on nitrate-nitrogen (NO_3-N) concentrations.

In order to establish numeric nutrient criteria, states and tribes should first indentify which biological response variables represent the aquatic life use they wish to protect. Next, the relationship between these biological variables and nutrients should be examined using sound statistical methods such as change point analysis. Finally, nutrient concentration thresholds where the biological responses suggest impairment should be used to establish criteria. The variability in the nutrient concentrations established as environmental thresholds shows that one numeric value cannot work for all circumstances and the development of numeric nutrient criteria will likely be basin or water body specific.

3.4.3 Weight of Evidence Approach

The establishment of numeric nutrient criteria by states and tribes should consider multiple approaches and available literature, including the use of frequency distribution and regression models to estimate reference conditions, the statistical relations between nutrients and select biological response variables, and studies published in the refereed journals as well as technical documents. USEPA has recommended that the states and tribes use multiple lines of evidence to establish nutrient criteria, and that these different lines of evidence be subjectively weighted based upon best professional judgment. More specifically, the USEPA (2000a, 2000b) has stated that "a weight of evidence approach that combines any or all three approaches will produce criteria of greater scientific validity." The relationship between nutrient and impairments in water bodies relative to designated uses are complex, and an approach that uses multiple lines of evidence is warranted in complex systems. The following paragraphs will demonstrate the use of the weight of evidence approach using example data from hypothetical watersheds, where an abundance of data may exist.

States and tribes might use data as shown in Figures 3.1 and 3.2 to establish nutrient criteria, where Figure 3.1 provides the frequency distribution of the nutrient of concern for available data within a large watershed or across region. The 75th percentile nutrient concentration of the reference streams and 25th percentile of all streams are assumed to reflect the range in concentrations representative of the reference conditions, which protect the aquatic life designation of these streams. In this example, the range in nutrient concentrations is 20 to 25 μg/L and states and tribes would consider this as a recommended starting point, which could further be defined to establish numeric water quality standards. The states and tribes might operate under the understanding that streams with nutrient concentration below this reference range support its aquatic life use designation. However, it is likely that streams above this reference concentration range may also support designated uses as defined by the states.

From this point, the states and tribes might consider the biological response along an increasing gradient of the nutrient of concern as shown in the examples depicted in Figure 3.2. The next step would be evaluating biological responses (including but not limited to benthic and sestonic chlorophyll-a, heterotrophic microbial biomass/decomposition, and macroinvertebrate/fish community composition) to increasing nutrients with emphasis on identifying threshold responses. For example, nuisance levels of benthic algae occur when benthic chlorophyll-a exceeds 100 mg/m^2 (Welch et al., 1988; Dodds, 2006). Based on the example regression equations, the level

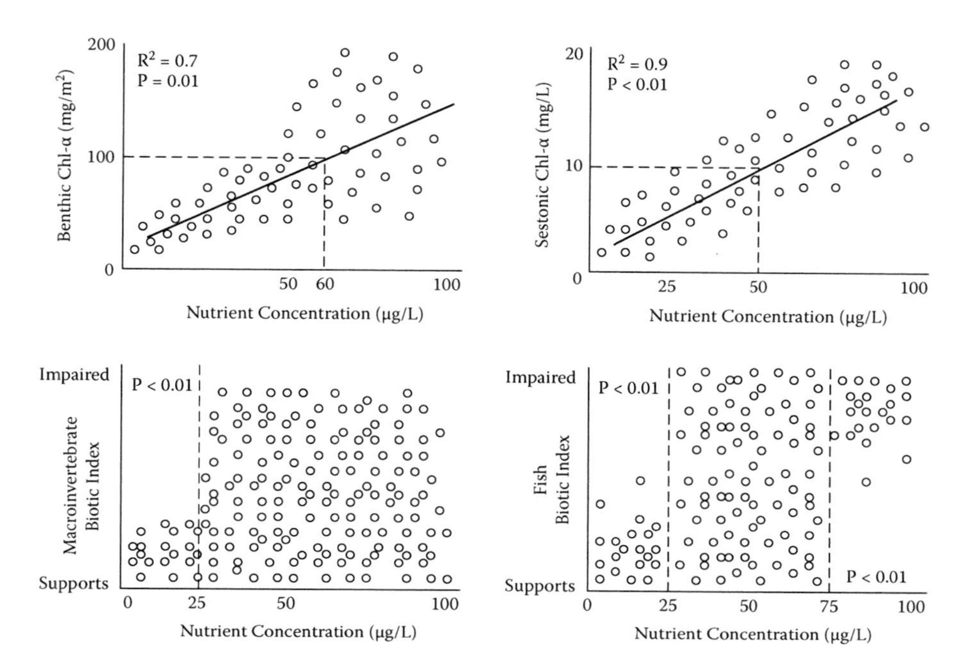

FIGURE 3.2 Examples showing the relation between nutrient concentrations and benthic chlorophyll-a (chl-α), sestonic chlorophyll-a, macroinvertebrate community structure, and fish community structure; these are the type of data that may be used in conjunction with percentile distributions of nutrient concentrations to establish nutrient criteria.

of benthic algae relates to a concentration of 60 μg/L of the nutrient of concern. However, sestonic chlorophyll-a concentrations greater than 10 μg/L suggest eutrophic conditions in large rivers and impoundments (USEPA, 2000b). Again, based on the example regression equations, this concentration of sestonic algae corresponds to a nutrient concentration of 50 μg/L of the nutrient of concern. States and tribes must decide on how much weight to put on these analyses compared to the range in reference conditions established by the percentile distributions.

Finally, this example considers watersheds or a region where substantial biological data exist, including macroinvertebrate and fish community structure and associated metric of biotic integrity. The states and tribes may then consider breakpoint in the biotic indices that exist along the gradient of increasing concentrations of the nutrient of concern (Figure 3.2). In this example, the biotic indices for macroinvertebrates and fish are defined in terms of supporting designated use and biological communities that would be considered impaired. These criteria would have to be established by the individual regulatory authorities for states and tribes based on natural history surveys of least-impacted water bodies. The example relation with macroinvertebrates shows that when concentrations are below 20 μg/L of the nutrient of concern, the biological community always supports its aquatic life use; however, the biological community may be impaired above this hypothetical concentration but not always. The example with fish communities shows a slightly different relation, where two thresholds exist.

First, the fish community always supports its designated use when concentrations are below 25 μg/L, whereas its designate use would be impaired most likely when concentration of the nutrient of concern exceeds 75 μg/L. Again, the states and tribes must consider how much emphasis to put on these numeric values from each source of information.

The role of nutrients in driving biological responses in aquatic systems, particularly streams, rivers, and reservoirs, remains poorly understood. Individual states and tribes must consider the relevant approaches (provided as only examples here) when developing numeric nutrient criteria to implement within its water quality standards. The defined designated use of individual water bodies should be used to guide its criteria development, and the weight of evidence approach allows states and tribes to place different levels of emphasis on the individual lines of evidence. Nutrient criteria development should be pursued when links between nutrients and biological response (such as in these examples) are better understood and scientifically documented. USEPA allows flexibility to states and tribes to establish numeric nutrient criteria to protect the designated uses, particularly aquatic life use, for water bodies within a geographical context such that one numeric value does not apply to all water bodies, watersheds, or regions. USEPA promotes the use of its regional technical assistance groups (RTAGs), including member states, tribes, and regional scientists, to further refine numeric nutrient criteria.

REFERENCES

APHA (American Public Health Association). 2005. *Standard Methods for the Examination of Water and Wastewater: Part 8000, Toxicity*. Washington, DC:APHA.

Biggs, B.J.F. 1996. Patterns in benthic algae of streams. In *Algal Ecology—Freshwater Benthic Ecosystems*, ed. R.J. Stevenson, M.L. Bothwell, and R.L. Lowe, 31–56. San Diego: Academic Press.

Biggs, B.J.F. 2000. Eutrophication of streams and rivers: Dissolved nutrient-chlorophyll relations for benthic algae. *Journal of the North American Benthological Society* 19: 17–31.

Brooks, B.W., J.T. Scott, M.G. Forbes, T.W. Valenti, J.K. Stanley, R.D. Doyle, K.E. Dean, J. Patek, R.M. Palachek, R.D. Taylor, and L. Koenig. 2009. Reservoir zonation and water quality. *Lakeline* 28(4): 39–43.

Carleton, J.N., R.A. Park, and J.S. Clough. 2009. Ecosystem modeling applied to nutrient criteria development in rivers. *Environmental Management* 44: 485–492.

Carlson, R.E. 1977. A trophic state index for lakes. *Limnology and Oceanography* 22(2): 361–369.

Carpenter, S.R., N.F. Caraco, D.L. Correll, R.W. Howarth, A.N. Sharpley, and V.H. Smith. 1998. Nonpoint source pollution of surface waters with nitrogen and phosphorus. *Ecological Applications* 8: 559–568.

Clark, G.M., D.K. Mueller, and M.A. Mast. 2000. Nutrient concentrations and yields in undeveloped stream basins of the United States. *Journal of the American Water Resources Association* 36(4): 849–860.

Dillon, P.J., and F.H. Rigler. 1974. The phosphorus chlorophyll relationship in lakes. *Limnology and Oceanography* 19(5): 767–773.

Dodds, W.K. 2006. Eutrophication and trophic state in rivers and streams. *Limnology and Oceanography* 51(1&2): 671–680.

Dodds, W.K. 2007. Trophic state, eutrophication, and nutrient criteria in streams. *Trends in Ecology and Evolution* 22: 669–676.

Dodds, W.K., and R.M. Oakes. 2004. A technique for establishing reference nutrient concentrations across watersheds affected by humans. *Limnology and Oceanography: Methods* 2: 333–341.

Dodds, W.K., J.R. Jones, and E.B. Welch. 1998. Suggested classification of stream trophic state distributions of temperate stream types by chlorophyll, total nitrogen and phosphorus. *Water Research* 32(5): 455–462.

Dodds, W.K., V.H. Smith, and K. Lohman. 2002. Nitrogen and phosphorus relationships to benthic algal biomass in temperate streams. *Canadian Journal of Fisheries and Aquatic Sciences* 59: 865–874.

Dodds, W.K., V.H. Smith, and B. Zander. 1997. Developing nutrient targets to control benthic chlorophyll levels in streams—a case study of the Clark Fork River. *Water Research* 31(7): 1738–1750.

Evans-White, M.A., W.K. Dodds, D.G. Huggins, and D.S. Baker. 2009. Thresholds in macroinvertebrate biodiversity and stoichiometry across water-quality gradients in Central Plains (USA) streams. *Journal of the North American Benthological Society* 28(4): 855–868.

Jin, G., A.J. Englande, H. Bradford, and H. Jeng. 2004. Comparison of *E. coli*, enterococci, and fecal coliform as indicators for brackish water quality assessment. *Water Environment Research* 76(3): 245–255.

Kiesling, R.L., A.M.S. McFarland, and L.M. Hauck. 2001. Nutrient targets for Lake Waco and North Bosque River, Developing Ecosystem Restoration Criteria. 52 p. Texas Institute for Applied Environmental Research, Tarleton State University, TR0107.

King, R.S., and C.J. Richardson. 2003. Integrating bioassessment and ecological risk assessment: An approach to developing numerical water-quality criteria. *Environmental Management* 31(6): 795–809.

Kratzner, C.R., and P.L. Brezonik. 1981. A Carlson type trophic state index for nitrogen in Florida lakes. *Water Resources Bulletin* 17(4): 713–715.

Lohman, K., and J.R. Jones. 1999. Nutrient—sestonic chlorophyll relationships in northern Ozark streams. *Canadian Journal of Fisheries and Aquatic Sciences* 56(1): 124–130.

Lohman, K., J.R. Jones, and B.D. Perkins. 1992. Effects of nutrient enrichment and flood frequency on periphyton biomass of northern Ozark streams. *Canadian Journal of Fisheries and Aquatic Sciences* 49(5): 1198–1205.

Matlock, M.D., D.E. Storm, M.D. Smolen, and M.E. Matlock. 1999a. Determining the lotic ecosystem nutrient and trophic status of three streams in eastern Oklahoma over two seasons. *Aquatic Ecosystem Health and Management* 2: 115–127.

Matlock, M.D., D.E. Storm, M.D. Smolen, M.E. Matlock, A.M.S. McFarland, and L.M. Hauck. 1999b. Development and application of a lotic ecosystem trophic status index. *Transactions of the American Society of Agricultural Engineers* 42(3): 651–656.

Migliaccio, K.W., B.E. Haggard, I. Chaubey, and M.D. Matlock. 2007. Linking watershed subbasin characteristics to water quality parameters in War Eagle Creek Watershed. *Transactions of the American Society of Agricultural and Biological Engineers* 50: 2007–2016.

Mulholland, P.J., A.M. Helton, G.C. Poole, R.O. Hall, S.K. Hamilton, B.J. Petersen, J.L. Tank, L.R. Ashkenas, L.W. Cooper, C.N. Dahm, W.K. Dodds, S.E.G. Findlay, S.V. Gregory, N.B. Grimm, S.L. Johnson, W.H. McDowell, J.L. Meyer, H.M. Valett, J.R. Webster, C.P. Aragno, J.J. Beaulieu, M.J. Bernot, A.J. Burgin, C.L. Crenshaw, L.T. Johnson, B.R. Neiderlehner, J.M. O'Brien, J.D. Potter, R.W. Sheibley, D.J. Sobota, and S.M. Thomas. 2009. Stream denitrification across biomes and its response to anthropogenic nitrate loading. *Nature* 452: 202–206.

Omernik, J.M. 1987. Ecoregions of the conterminous United States. Map (scale 1:7,500,000). *Annals of the Association of American Geographers* 77(1): 118–125.

Popova, Y.A., V.G. Keyworth, B.E. Haggard, D.E. Storm, R.A. Lynch, and M.E. Payton. 2006. Stream nutrient limitation and sediment interactions in the Eucha-Spavinaw Basin. *Journal of Soil and Water Conservation* 61(2): 105–115.

Qian, S.S., R.S. King, and C.J. Richardson. 2003. Two statistical methods for the detection of environmental thresholds. *Ecological Modeling* 166: 87–97.

Richardson, C.J., R.S. King, S.S. Qian, P. Vaithiyanathan, R.G. Qualls, and C.A. Stow. 2007. Estimating ecological thresholds for phosphorus in the Everglades. *Environmental Science and Technology* 41: 8084–8091.

Royer, T.V., M.B. David, L.E. Gentry, C.A. Mitchell, K.M. Starks, T. Heatherly II, and M.R. Whiles. 2008. Assessment of chlorophyll-a as a criterion for establishing nutrient standards in the streams and rivers of Illinois. *Journal of Environmental Quality* 37: 437–447.

Scott, J.T., J.A. Back, J.M. Taylor, and R.S. King. 2008. Does nutrient enrichment decouple algal–bacterial production in periphyton? *Journal of the North American Benthological Society* 27: 332–344.

Sheeder, S.A., and B.M. Evans. 2004. Estimating nutrient and sediment threshold criteria for biological impairment in Pennsylvania watersheds. *Journal of the American Water Resources Association* 40(4): 881–888.

Stevenson, R.J., B.H. Hill, A.T. Herlihy, L.L. Yuan, and S.B. Norton. 2008. Algae-P relationships, thresholds, and frequency distributions guide nutrient criterion development. *Journal of the North American Benthological Society* 27(3): 783–799.

Stevenson, R.J., S.T. Rier, C.M. Riseng, R.E. Schultz, and M.J. Wiley. 2006. Comparing effects of nutrients on algal biomass in streams in two regions with different disturbance regimes and with applications for developing nutrient criteria. *Hydrobiologia* 561: 149–165.

Straškraba, M., J.G. Tundisi, and A. Duncan, Eds. 1993. *Developments in Hydrobiology: Comparative Reservoir Limnology and Water Quality Management.* Dordrecht, The Netherland: Kluwer Academic Publishers.

Suplee, M.W., A. Varghese, and J. Cleland. 2007. Developing nutrient criteria for streams: An evaluation of the frequency distribution method. *Journal of the American Water Resources Association* 43(2): 453–472.

Thornton, K.W., F.E. Payne, and B.L. Kimmel. 1990. *Reservoir Limnology: Ecological Perspectives.* New York: John Wiley & Sons, Inc.

Tilman, D., J. Fargione, B. Wolff, C. D'Antonio, A. Dobson, R. Howarth, D. Schindler, W.H. Schlesinger, D. Simberloff, and D. Swackhamer. 2001. Forecasting agriculturally driven global environmental change. *Science* 292: 281–284.

USEPA (U.S. Environmental Protection Agency). 1983. Technical Support Manual: Waterbody Surveys and Assessments for Conducting Use Attainability Analyses. EPA-440486037, Washington, D.C.

USEPA. 1984a. Technical Support Manual: Waterbody Surveys and Assessments for Conducting Use Attainability Analyses—Volume 2, Estuarine Systems. EPA-440486038, Washington, D.C.

USEPA. 1984b. Technical Support Manual: Waterbody Surveys and Assessments for Conducting Use Attainability Analyses—Volume 3, Lake Systems. EPA-440486039, Washington, D.C.

USEPA. 1986. Quality Criteria for Water. EPA-440/5-86-001, Washington, D.C.

USEPA. 1994. Water Quality Standards Handbook: Second Edition. EPA-823-B-94-005a, Washington, D.C.

USEPA. 2000a. Nutrient Criteria Technical Guidance Manual—Lakes and Reservoirs. EPA-822-B-00-001, Washington, D.C.

USEPA. 2000b. Nutrient Criteria Technical Guidance Manual—Rivers and Streams. EPA-822-B-00-002, Washington, D.C.

USEPA. 2001. National Primary Drinking Water Standards. EPA-816-F-01-007, Washington, D.C.

USEPA. 2007. Water Quality Standards Handbook: Web Edition. http://www.epa.gov/waterscience/standards/handbook/index.html, Washington, D.C.

USEPA. 2009. National Primary Drinking Water Regulations. EPA-816-F-09-004, Washington, D.C.

Wang, L., D.M. Robertson, and P.J. Garrison. 2006. Linkages between nutrients and assemblages of macroinvertebrates and fish in wadeable streams: Implication to nutrient criteria development. *Environmental Management* 39: 194–212.

Weigel, B.M., and D.M. Robertson. 2007. Indentifying biotic integrity and water chemistry relations in nonwadeable rivers of Wisconsin: Toward the development of nutrient criteria. *Environmental Management* 40: 691–708.

Welch, E.B., R.R. Horner, and C.R. Patmont. 1989. Prediction of nuisance periphytic biomass: A management approach. *Water Research* 23:401–405.

Welch, E.B., J.M. Jacob, R.R. Horner, and M.R. Seely. 1988. Nuisance biomass levels of periphytic algae in streams. *Hydrobiologia* 157: 161–168.

Wickham, J.D., K.H. Riitters, T.G. Wade, and K.B. Jones. 2005. Evaluating the relative roles of ecological regions and land-cover composition for guiding establishment of nutrient criteria. *Landscape Ecology* 20: 791–798.

Zheng, L., J. Gerritsen, J. Beckman, J. Ludwig, and S. Wilkes. 2008. Land use, geology, enrichment and stream biota in the Eastern Ridge and Valley Ecoregion: Implication for nutrient criteria development. *Journal of the American Water Resources Association* 44(6): 1521–1536.

4 Project Planning and Quality System Implementation for Water Quality Sampling Programs

Delia Ivanoff

CONTENTS

4.1 INTRODUCTION

Reliable, cost-effective, and defensible environmental decisions depend highly on the quality of results gathered during the monitoring and data collection processes. An effective water quality monitoring program is not just about collecting samples, generating results in the laboratory, and tabulating them for data reporting and analysis purposes; it takes careful planning, proper training and communication, performance tracking, and adaptive implementation to help ensure that collected data are useful for the project. The U.S. Environmental Protection Agency (USEPA) and various state environmental regulatory agencies set standards, provide guidance, and disseminate information to maintain scientific data integrity and maintain consistency.

This chapter presents the different aspects of a project planning process related to water quality monitoring. It includes a discussion on the project life cycle, implementation of a quality system, formulation of data quality objectives, preparation and implementation of a project plan, and preparation and implementation of standard operating procedures (SOPs).

4.2 WATER QUALITY PROJECT CYCLE

A successful water quality monitoring program is described by a life cycle similar to that illustrated in Figure 4.1. Project planning should be documented in the form of a project plan. The project plan includes the specific data quality objectives, the specific questions and constraints for answering the objectives, and the strategies and methodologies for implementation in the form of a written document assembled by the project team. Further information about preparation of this plan is discussed in Section 4.4.

The next stage of a monitoring project is implementation. During implementation, the project team performs sample collection, laboratory analyses, and data review. Documentation and coordination among the project team members are critical parts of a successful implementation phase. Each data point collected must be traceable in terms of its life history from sample collection to reporting; this is possible through documentation. A description of the equipment used for sample collection, the method used, the date and time of collection, the location, the sampling depth, and field observations are among the critical information that is documented as part of the sampling process. Similarly, details about sample preparation and laboratory analysis should be recorded in sufficient details to allow for tracing the history of data generated through that process.

Once data have been collected and assembled, they must undergo a review and validation process. This phase evaluates results against the project data quality objectives and performance criteria. Data validation is usually performed by personnel that are familiar not only with the project objectives but also with field and laboratory report content and quality indicators. The ultimate product of data validation

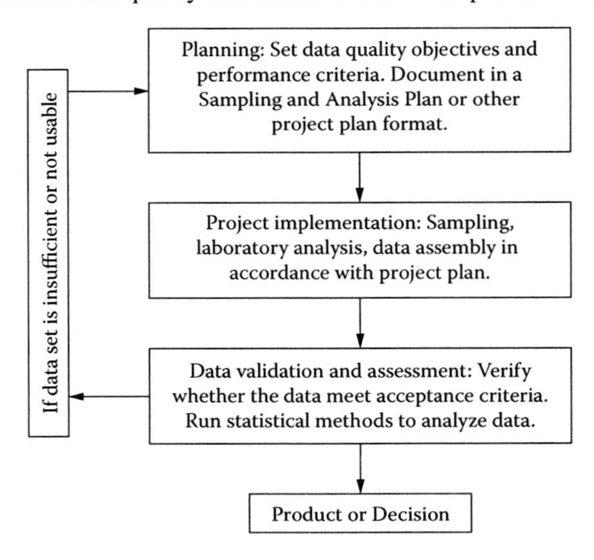

FIGURE 4.1 Project life cycle for a water quality monitoring program.

is a set of data with documented quality and known level of usability. When the generated data set is incomplete or unusable, the project team should review the project plan including the objectives and methodologies and determine if any adjustment, for example, revised methodology or collecting additional data, is necessary. Oftentimes, the project team foregoes this step due to resource limitation or lack of knowledge. When data are interpreted and used as-is, without review and validation, this could result in costly mistakes, wasted effort, and/or health risks.

4.3 QUALITY SYSTEM IN WATER QUALITY PROGRAMS

Quality system is a structured and documented system describing the policies, objectives, principles, organizational authority, responsibilities, accountability, and implementation plan of an organization for ensuring the quality in its work processes, products, items, and services (ASQ, 2004). It provides the framework for planning, implementing, and assessing activities leading to the generation of results.

Quality assurance (QA) is described as a management function that rests on the documentation and establishment of quality control protocols and on the evaluation and summarization of their outcomes. A good QA program for a water quality monitoring project covers the entire project life cycle including planning, sample collection, laboratory analysis, data assessment, and reporting results. *Quality control* (QC) is the overall system of technical activities whose purpose is to measure and control the quality of the process that led to generation of results, that is, field sampling or laboratory analysis, so that the results meet the requirements for the project and under certain industry standards.

Effective implementation of a quality system is a worthwhile investment of time and effort. When quality and traceability of work and product are maintained, there is greater reliability on and defensibility of the data, lesser risks in the back end of the process, and usually less costs since there will be less resampling, reanalysis, or data loss. In many cases, erroneous results or wrong decisions could be more costly economically and environmentally than the implementation of a quality system.

4.4 DEFINING DATA QUALITY OBJECTIVES

Data quality objectives (DQOs) are qualitative and quantitative statements of the overall level of uncertainty that a decision-maker will accept in results or decisions based on environmental data (USEPA, 2001). The effectiveness of a monitoring program depends largely on carefully designed data quality objectives. A well-planned project that is centered on the DQOs results in the following:

- Focused data requirements and optimized design for data collection, therefore helping ensure collection of the right type and amount of data for the project at an optimum budget
- Well documented procedures and requirements for data collection and evaluation
- Clearly developed analysis plans with sound, comprehensive QA project plans
- Early identification of the sampling design and data collection processes

The USEPA developed a seven-step process in formulating DQOs for environmental monitoring projects (USEPA, 2006):

Step 1. *State the problem.* Define the problem that necessitates the study; identify the planning team, examine budget, and schedule.
Step 2. *Identify the goal of the study.* State how environmental data will be used in meeting objectives and solving the problem, identify study questions, and define alternative outcomes.
Step 3. *Identify information inputs.* Identify data and information needed to answer study questions. Identify the type of data needed.
Step 4. *Define the boundaries of the study.* Specify the target population and characteristics of interest, define spatial and temporal limits, and scale of inference. Identify constraints to data collection.
Step 5. *Develop the analytic approach.* Define the parameter of interest, specify the type of inference, and develop the logic for drawing conclusions from findings.
Step 6. *Specify performance or acceptance criteria.* Specify probability limits for false rejection and false acceptance decision errors. Develop performance criteria for new data being collected or acceptable criteria for existing data being considered for use. The DQOs are typically expressed as tolerable limits on the probability or chance (risk) of the collected data leading you to making an erroneous decision.
Step 7. *Develop the plan for obtaining data.* Select the resource-effective sampling and analysis plan that meets the performance criteria. Data collection design is developed to meet the set quantitative and qualitative criteria.

4.5 SAMPLING AND ANALYSIS PLANS

Project plans come in various names such as sampling plan, sampling and analysis plan, monitoring plan, or quality assurance project plan. Overall, these documents include details about the project, the project organizational roles and responsibilities, the DQOs, the location, the scope of sampling and analysis, the methodologies that will be used, the quality assurance and quality control requirements and procedures, data handling and analysis procedures, acceptance criteria, contingency plan for the project, and the project budget. In some cases, organizations tend to combine the preceding information with standard operating procedures. The goal is to provide details about the project so that all project requirements are clear and consistent among project personnel.

4.6 STANDARD OPERATING PROCEDURES

Standard Operating Procedures (SOPs) are written instructions that document routine or repetitive activities to guide an activity or a set of activities. In water quality programs, they serve as training guides for procedures and assist with maintaining consistency and integrity in the data collection process (ASTM, 2004). SOPs include technical and administrative details and are usually linked to other documentation such as quality plans and/or project plans, quality manuals, and standard or published methods. Individual SOPs may be written for each key task or activity, for example, equipment maintenance, purchase of supplies and reagents, surface water collection from a structure, or instrument calibration (FDEP, 2008).

Oftentimes, although organizations comply with regulatory and/or clientele requirements to have an established set of SOPs, their implementation is not always effective. For SOPs to be effectively implemented, they have to be

1. Organized to ensure ease and efficiency in use
2. Specific to the purpose for which it is developed
3. Written in simple language and format that is easy to follow
4. Detailed enough so that someone with a basic understanding of the field can successfully reproduce the activity or procedure when unsupervised
5. Reviewed by one or more individuals with appropriate training and experience with the process
6. Maintained and kept current

Recommended SOP Format: An example SOP for a water quality monitoring program is presented in Chapter 14. Format and structure are important to the users of the procedure. The document must be organized and easy to follow in order to be effectively used. Breaking the procedures into logical, small discussion topics is easier to follow than long paragraphs of information. A typical format and list of topics that should be included in field sampling SOPs are presented next. However, note that in some cases, the format is dictated by regulatory preference or guidance, such as laboratory procedures for an organization's laboratory accreditation.

Title Page: The title should clearly identify the activity or procedure; an SOP reference number; date of issue, revision, and effective dates; the name of the applicable organization to which the SOP applies; and the signatures and signature dates of SOP authors and approving authorities.

Table of Contents: A table of contents is needed for quick reference and particularly useful if the SOP is long or has multiple sections.

Scope and Application: This section should clearly state the scope of the SOP and its applicability. For example, the SOP should clearly state whether it was written specifically for collecting surface water samples or for collecting groundwater samples. It may also be useful to specify when the procedures are not applicable, for example, this SOP is not applicable when sampling at depths exceeding a specified depth.

Equipment, Instrumentation, and Supplies: A list of all the required equipment, instrumentation, and supplies should be included with sufficient details so that any qualified person could locate and use the proper items or replace them with the proper type and construction if necessary. If multiple units are available for major equipment and instrumentation, serial numbers or some other form of unique identification should be included in the SOP. Proper calibration and maintenance information for equipment should be included in the SOP, or reference materials should be provided through identifiable and/or linked documentation such as the technical manuals.

Step-by-Step Procedures: A good SOP should include sequential procedures that are easy to follow. It should also include any tips, caution statements, diagrams, and techniques to get the task done properly, efficiently, and consistently. Numbered list of steps or steps that are subdivided into logical categories and subcategories are easier to follow and understand than long paragraphs of instructions. Flow charts and photos are also very useful and provide visual breaks for long sections describing specific steps. When appropriate, SOPs can also refer to or link to other published methodologies. However, SOPs that refer or link their content to various references are not always effectively used and thus should be used judiciously.

Quality Control and Quality Assurance Section: QC requirements and activities, as described in Section 4.6, should also be part of the SOP. Describe the preparation of appropriate QC procedures (self-checks, such as calibrations, recounting, reidentification) and QC material (such as blanks-rinsate, trip, field, or method; replicates; splits; spikes; and performance evaluation samples) that are required to demonstrate successful performance of the method. Specific criteria for each should be included. Describe the frequency of required calibration and QC checks and discuss the rationale for decisions. Describe the limits/criteria for QC data/results and actions required when QC data exceed QC limits. Describe the procedures for reporting QC data and results.

Reference Section: Documents or procedures that interface with the SOP should be fully referenced (including version), including related SOPs, published literature, or methods manuals. Citations cannot substitute for the description of the method being followed in the organization. Attach any that are not readily available.

4.7 FIELD SAMPLING QUALITY CONTROL

As discussed in Section 4.3, quality control provides a way to evaluate the quality of a data set generated through a sampling or analysis process. Quality control measures may indicate one or more of the following data quality indicators:

Accuracy—The degree of agreement between an observed value and an accepted reference value; includes a combination of random error (precision) and systematic error (bias) components.

Precision—The degree to which a set of observations or measurements of the same property, obtained under similar conditions, conform to themselves; usually expressed in terms of variance as relative standard deviation or relative percent difference.

Bias—The systematic or persistent distortion of a measurement process that causes errors in one direction (above or below the true value or mean). Commonly reported as % Bias or % Recovery.

Representativeness—The degree to which sample data accurately and precisely describe the characteristics of a population of samples, parameter variations at a sampling point, or environmental condition.

The following is a list of commonly used quality control measures. The frequency and number of QC samples depend on the quality goals for the project or what is required in regulatory procedures. At a minimum, there should be an indication of the absence or presence of contamination (blanks), an indication of precision in sampling (replicates), and an indication that the field test meter has been properly calibrated (calibration verification standards).

Equipment blank. Measures background level in sampling equipment and containers, or the entire sampling train. It is usually acquired by processing analyte-free water through all sample collection equipment or sampling train then collecting the rinsate into sample containers. This blank is processed, preserved, and handled in the same manner as the field samples.

Field blank. Measures background level from sample processing environment, handling, and transport. A field blank is prepared by pouring analyte-free water directly into the sample container on site, preserved, and kept open until sample collection is completed for the routine sample site. Just like the equipment blank, field blanks are processed, preserved, and handled in the same manner as the field samples.

Trip blank. Measures background level or cross contamination resulting from sample handling and transport and is commonly required only for volatile organic carbon (VOC) collection. A trip blank is prepared before the trip by filling precleaned sample vials with analytical-grade water, and the vial remains unopened as it is transported back to the laboratory along with the field samples. In most cases, one trip blank is required per cooler of VOC samples.

Replicate sample. Replicate sample results are used to assess sampling precision or field variability. Replicate samples are two or more samples collected from the same source on the same day. Replicate sample results provide a measurement of sampler variability and inherent heterogeneity of the field samples.

Split samples. Split samples are identical samples taken from the same bulk sample using a splitter or partitioned in separate containers in repeated succession. These samples are submitted to different laboratories to determine comparability in laboratory analytical performance. It should be noted that submitting split field samples with unknown concentration to two laboratories can create a dilemma, as it would be difficult to determine which laboratory is producing the expected results. This can be resolved by submitting to multiple laboratories or having some reference values, for example, spike solution of known concentration or historical values.

Calibration and checks for field testing. When conducting field testing, it is critical that the instrument is properly calibrated. The calibration process may vary depending on the type of testing and the instrumentation used, but the basic principle is the same: set the instrument to baseline (usually at zero setting) then at target response for known calibration solution or material. Once the instrument is calibrated, it is a good practice (and required by some regulatory agencies) that an initial calibration (ICV) check be performed prior to measurement in the field. The ICV verifies if the calibration is acceptable or not. After a certain time period has lapsed, it is also good to check if the instrument response has drifted or not by running a continuing calibration verification standard (CCV).

The acceptable limits for blank and QC checks are provided by the regulatory agency or generated based on historical results. Target limits for equipment blank are commonly set at less than the method detection limit for most analytes. Most commonly, QC control limits are generated as the mean recovered results ± 3 standard deviations, and the warning limits as the mean recovered results ± 2 standard deviations. It is also prudent to track QC results so that the sampling personnel will readily see any trends or biases that may need to be addressed. When anomalous QC results are obtained, the sampling personnel should begin systematic troubleshooting to determine the root cause of the problem and initiate corrective action if necessary. Simply fixing the recovery for one sampling event does not guarantee that the problem is solved, unless the root cause is resolved. For example, if the results indicate that blanks are positive (blanks greater than method detection limits), then the sampling personnel should determine the source of the problem. It could be the water source, the equipment, the processing environment, or improper decontamination.

There will be times when resolving QC issues is not timely for the sampling event and resampling might not be an option. In this case, data should be qualified according to regulatory requirements or, if not under regulatory programs, some type of qualifier must be attached to the data so that present or future data users are aware of potential limitations in interpreting the data. For example, if it is clear that the field meter response was drifting during the sampling event, the sampling personnel should document the observation. During data validation, this note would be attached as a qualifier (also called *flag* or *remarks*) to the data.

4.8 DOCUMENTATION AND RECORDKEEPING

Documentation is a vital part of a monitoring program. Thorough and clear documentation provide for traceability and reconstruction of generated data. Aside from quality system documentation, every step of the data collection process requires some form of record. In sample collection, the following should be documented:

Decontamination
Equipment identification
Maintenance and troubleshooting logs
Calibration logs
Dates of use
Methods used
Quality control results

Unique sample identification
Location of collection
Date and time of collection
Any other pertinent descriptive field information about the sample collected (e.g., site condition, strong odor, sample color, presence of potential contaminants)
Names of individuals in the sampling team and what roles each performed during the sampling event
List of required analyses and corresponding preservatives used
Any communication with the laboratory pertinent to the sample collection

It is a general practice to use a chain of custody (COC) form for sample collection. COC is defined as an unbroken trail of accountability that ensures the physical security of samples, data, and records (USEPA, 2001). Most of the items listed earlier can be included in the COC. Additional details can be recorded in the field notebook or using additional sheets. It is important that these documents be linked or compiled together for traceability. Handwritten notes must be made using indelible and waterproof ink. All original notes and document must be kept and protected from weather elements and secured for future reference. A thorough documentation of field observations assists with addressing data problems later or for responding to audits or assessments. For example, if one is looking at anomalous results and field notes indicate presence of a large amount of decaying vegetation, the data user can then make a determination whether or not to use or exclude that data.

If at any point, field records have to be corrected, the original notes should remain intact with a notation about the correction. Obliteration of entries or use of correction fluid is not acceptable and should be avoided. A single cross-out with initials of the person making the correction and date, with the correct entry written next to it or attached, is oftentimes sufficient. There may be cases when a more thorough explanation about the corrected entries is needed for defensibility and data integrity reasons. For example, if an individual was making corrections on sampling location after the fact, ample explanation should be provided on why the correction is being made and supporting facts that justify the corrections should be provided.

In this modern age of computers and electronic devices, companies are taking advantage of using electronic documentation. Many regulatory agencies have been including provisions in their rules and guidance documents for electronic documentation. Computerized field notebooks are becoming more popular, especially for sampling groups that do repetitive and large amounts of sampling. Electronic documentation helps reduce the need to transcribe field information in the laboratory, which means less chance for errors. As with paper documentation, it is essential to have all the necessary information, electronic back-up, and a secured system to maintain integrity and traceability of records.

4.9 QUALITY SYSTEMS ASSESSMENT AND AUDITING

To monitor performance of work in accordance with the quality system and project documents, periodic review of the different activities should be done. This could be in the form of an audit, an assessment, or a management review. Many agencies require that internal systems audit be done at least annually; a more frequent audit or assessment may be necessary depending on the quality of results being generated. The objective is to verify, by thorough review of objective evidence, that applicable elements of the quality system are appropriate and have been developed, documented, and effectively implemented. The level of assessment or audit depends on the objective and the risk for a particular project, entity, or process. This could range from a simple review of data sets or individual interviews to extensive review of instrument output and bench sheets or complete systems audit. Technical audits are conducted to provide a systematic independent technical examination of a project to determine if a data collection activity is being conducted as planned and producing data and information of the type and quality specified in the sampling and analysis plan. A review or audit checklist is recommended so that all necessary requirements can be verified in a work list format. Findings should be communicated with the personnel conducting the affected activities, and an action plan should be developed to correct any deficiencies or vulnerabilities in the quality system implementation.

REFERENCES

American Society for Quality (ASQ). 2004. ANSI/ASQC E4-2004, Specifications and Guidelines for Quality Systems for Environmental Data Collection and Environmental Technology Programs. Milwaukee, WI.

American Society for Testing and Materials (ASTM). 2004. Standard Guide for Documenting the Standard Operating Procedures Used for the Analysis of Water. ASTM D 5172-91, American Society for Testing and Materials, West Conshohocken, PA.

Florida Department of Environmental Protection (FDEP). 2008. Standard Operating Procedures for Field Activities. DEP-SOP-001/01. Bureau of Assessment and Restoration Support Standards and Assessment Section, Florida Department of Environmental Protection, Tallahassee, FL.

USEPA (U.S. Environmental Protection Agency). 2001. EPA Requirements for Quality Assurance Project Plans (QA/R-5), EPA/240/B-01/003, Office of Environmental Information, U.S. Environmental Protection Agency, Washington, D.C.

USEPA. 2006. Guidance on systematic planning using the data quality objectives process (QA/G-4), EPA/240/B-06/001, Office of Environmental Information, U.S. Environmental Protection Agency, Washington, D.C.

5 Surface Water Quality Sampling in Streams and Canals

Kati W. Migliaccio, Daren Harmel, and Peter C. Smiley, Jr.

CONTENTS

5.1 INTRODUCTION

Surface water sampling and water quality assessments have greatly evolved in the United States since the establishment of the Clean Water Act in the 1970s. Traditionally, water quality referred to only the chemical characteristics of the water and its toxicological properties related to drinking water or aquatic life uses, but now water quality includes physical, chemical, and biological characteristics. Surface water sampling reflects the changing views of water quality and the emerging variety of sampling goals. Surface water sampling projects range from very simplistic to highly complex and differ based on spatial extent, response variables analyzed, flow conditions sampled, and technology used. This chapter provides up-to-date information on designing and implementing surface water quality sampling projects in wadeable streams and in wadeable canals constructed to improve drainage characteristics. This chapter not only focuses on procedures for determining chemical quality but also provides introductory concepts related to evaluation of the physical and biological aspects of water quality.

5.2 DEFINING PROJECT GOAL

The first and often the most difficult step in designing a surface water sampling project is clearly identifying the project goal. This step must be accomplished because the goal ultimately determines sampling methods, sampling frequency, and response variables. Typical sampling goals include but are not limited to the following:

1. Meeting total maximum daily load (TMDL) monitoring requirements
2. Quantifying the performance of Best Management Practices (BMPs)
3. Identifying source(s) of constituent(s) of concern
4. Quantifying reference or background constituent concentrations
5. Determining if the quality of water in a water body is meeting designated uses

These few examples highlight that water sampling goals can range from simple descriptive assessments intended to document existing conditions to complex studies intended to determine watershed level impacts of specific practices. Thus, it is critical to recognize that different goals result in different sampling intensities and sampling designs (Smiley et al., 2009).

The primary limitation in design and implementation of any sampling project is resource availability or constraints. As such, the success of a surface water sampling project meeting its established goal is typically determined by its ability to accurately characterize water quality conditions with available resources (Harmel et al., 2006a). Others have reported the difficulty and importance of achieving this balance (e.g., Preston et al., 1992; Shih et al., 1994; Tate et al., 1999; Agouridis and Edwards, 2003; Harmel et al., 2003; King et al., 2005; Harmel and King, 2005). As with most multifaceted objectives, optimization is required to ensure that resources are effectively and efficiently allocated (Abtew and Powell, 2004; Miller, 2005). Factors that affect resource allocation should be carefully considered in project design and implementation and include factors such as site selection, equipment maintenance, personnel requirements, discharge measurement, water chemistry sampling methodology, and physical and biological assessment methodologies.

5.3 SITE SELECTION

Selection of the initial set of potential sampling sites is primarily based on the goal of the sampling project. Data collected from sampling sites should answer a question or hypothesis proposed by the sampling goal. For example, a sampling goal may be to quantify the reference or background concentration of a constituent in a watershed. The question that needs to be answered would be "What is the concentration of a particular constituent in the most downstream location (or outlet) of selected subbasins in the watershed that are minimally or not impacted by human influences?" Constituent data collected from sampling sites located at the most downstream location of the minimally impacted subbasins that represented reference conditions would be able to answer this question. Another sampling goal might be to determine if water quality in a water body was meeting designated uses. The question that needs to be answered would be: "How do water quality measurements collected from a stream compare to water quality criteria associated with the designated uses of that stream?" Data from sampling sites located strategically in the stream would be needed so that sampling sites were placed where concentrations were thought to be the greatest (and variable) due to point source inputs, tributary influences, or other known features of the system. Data collected would be compared to water quality criteria to determine if designated use standards were being met. Thus, site selection requires identification of goals and some understanding of the system processes, so appropriate sampling sites are selected and the collected data can answer the hypothesis or water quality question being posed.

Site selection also depends on the scale associated with the sampling goal. The greater heterogeneity inherent with watershed scale studies compared to small field-scale studies requires consideration of multiple point and nonpoint constituent sources and how these sources may impact the appropriateness of particular sites considering the sampling goal. This is especially important when evaluating water quality near effluent discharges from Waste Water Treatment Plants (WWTPs; Migliaccio et al., 2007; Carey and Migliaccio, 2009) and immediately downstream of tributary confluences. Site selection should also consider potential stream modification that may occur naturally or due to anthropogenic activity. Modifications often include widening or shifting of the channel and erosion of embankment. Hence, sites should be selected in areas that are anticipated to remain stable and where signs of impending modification are not present, but this may not always be possible as stream systems naturally meander over time. For these sites, the physical dynamics of the stream should be considered for site selection. Sites should also be located (if possible) at existing flow gauges or hydraulic control structures with an available historical flow record and established stage–discharge relationship because of the difficulty of establishing accurate stage–discharge relationships (discussed in Section 5.6).

Potential sites may be identified using aerial photographs, detailed maps, or personal knowledge of the area (Benson and Dalrymple, 1984). Technology and Internet tools have progressed, and now this task may also be completed using GIS software or even ready-to-use tools such as Google Earth. Once potential sites have been selected, field visits are needed to evaluate site characteristics to ensure that each site will support project goals and optimize available resources. The accessibility and integrity of each potential sampling site should also be carefully considered. Site characteristics such as personnel and vehicle access (especially in wet conditions), flood likelihood, adjacent land ownership, and vandalism potential must be considered in site selection. Sampling projects with long-term goals should also consider the probability of significant construction or land modification near the sampling site, especially if these activities significantly alter flow and constituent transport conditions. If significant impact is expected from such activities, alternative sites should be selected. Sites should also be selected to minimize travel expenses (USDA, 1996). Travel costs to and from sampling sites can be substantial, especially as distances increase, requiring more personnel time and increased transportation expense. Trips to distant sampling locations can be especially difficult and costly in wet periods when frequent trips are required to collect samples from automated sampling systems (Harmel et al., 2006a). Similarly, sample preservation and related quality assurance measures associated with sample transport become more difficult as time between retrieval and analysis increases. First-hand visitation of sites and assessment of these challenges will likely result in a reduction in the amount of feasible sites from the original list of potential sites.

Once the physical location of a site has been selected, the spatial extent at which the evaluation will be conducted at each site should be determined. The spatial extent of a site will differ depending on what types of variables are being measured. Water chemistry measurements and discharge measurements are collected from one spot or at multiple spots along one transect, and thus the site is traditionally a point. In contrast, a site for the collection of physical habitat variables and biological sampling consists of multiple sampling locations within a defined reach of a specific length. The differences in what constitutes a site relates to the differences in the sampling protocols for each type of response variable. Physical habitat variables and the biota differ spatially within even short distances and, to ensure adequate characterization of these variables, sampling must be conducted in longer length sites than that typically used for water chemistry or discharge measurements. Water chemistry and discharge at one location are assumed to be representative of the combined flux (discharge and constituent) from upstream contributing areas.

The considerations that must be included in defining a site for sampling physical habitats or biota variables are further examined by considering the following example related to fish sampling. A common method for determining site lengths for fish sampling involves measuring the wetted width of the stream and then multiplying the mean base flow wet width by a constant to determine the site length to sample (Lyons, 1992). Recommended constants range from 14 times (Patton et al., 2000) to 120 times (Paller, 1995) the wetted stream widths, and they differ among response variables of interest, ecoregions, and sampling protocols. They even differ among national assessments of water quality. The USGS National Water Quality Assessment uses 20 times wetted width with a minimum of 150 m long and a maximum of 300 m long in wadeable streams (Lazorchak et al., 1998). In contrast, the USEPA Environmental Monitoring and Assessment program uses 40 times wetted width with a minimum of 150 m (Moulton et al., 2002). This proportional distance method of establishing site lengths is intended to ensure the accurate determination of species composition within a site without oversampling or undersampling. However, it results in variable site lengths among different-sized streams where small streams have shorter sites than large streams. Thus, if water quality comparisons among streams, sampling periods, or different categories of streams are important, then fixed length sites would be preferable to standardize sampling efforts and ensure comparability (Smiley et al., 2009).

5.4 EQUIPMENT MAINTENANCE

It is inevitable that equipment malfunction will occur, which makes proper installation and operation extremely important to limit errors and missing data. Sampling equipment should be maintained per the manufacturer's instructions and usually includes calibration, cleaning, and general maintenance. Failure to adequately maintain automated samplers is by far the greatest cause for malfunction and missed data. The important maintenance requirements for automated samplers include battery level (under load), solar panel output, sample lines (clogs or holes), stage/flow recorder accuracy, and desiccant capacity (Harmel et al., 2006a). All of these should be checked weekly or every other week at a minimum. A maintenance log should be kept and include a schedule for maintenance checks throughout the life of the project to ensure that maintenance is not overlooked. Each maintenance check and calibration procedure should be recorded in the maintenance log. This process should be included as part of the quality assurance/quality control plan (see Chapters 4 and 14).

Data collected and stored on-site with loggers should regularly be downloaded to minimize data loss due to power supply problems, lightning, vandalism, and other mishaps. Real-time data transfer and communication with remote sites can minimize data loss and unnecessary site visits by notifying personnel when problems occur and when samples have been collected. We emphasize that real-time equipment does not eliminate the need for frequent site visits to conduct maintenance and collect samples. Backup equipment should be purchased in anticipation of equipment malfunction or failure to allow for quick replacement and minimal data loss.

Surface water sampling equipment, such as automatic samplers, stage and/or velocity devices, data loggers, and power supplies, should be installed with consideration to environmental factors (i.e., plants, animals, vandals, and weather) and protected from damage that would alter sample integrity. All equipment should be housed in locked, sturdy structures positioned above the highest expected flow elevation to ensure accessibility during high flows (Haan et al., 1994; USEPA, 1997).

5.5 PERSONNEL REQUIREMENTS

The most critical component in a surface water sampling is the personnel. Personnel should be trained during project initiation to perform the necessary tasks and should receive refresher training throughout the project. It is best to have two-person teams dedicated to particular sampling aspects to minimize safety concerns and to ensure continuous staffing. It is best not to have multiple people collecting data separate from each other to limit personnel biases and/or technique inconsistencies.

Water sampling requires dedicated personnel and long hours of work often outside of conventional working hours; thus, project design should include estimation of personnel hours. Different sampling strategies (such as manual sampling versus automatic sampling) require different personnel needs and result in different costs. In addition, personnel will often need to travel to sites during adverse weather conditions; therefore, the time needed to complete tasks may be longer than if conditions were optimum.

5.6 DISCHARGE MEASUREMENT

The primary hydrologic component measured in surface water sampling projects is discharge. Discharge refers to the flow rate of water as measured in units of volume per time, such as gallons per minute (gpm) or cubic meters per second (m^3 s^{-1}). Collection of appropriate discharge data is essential to adequately characterize water quality in terms of (1) offsite constituent loss, (2) downstream constituent transport, and (3) channel erosion and deposition. In addition, discharge data and associated constituent concentrations are needed to determine constituent mass flux (or loads) and to differentiate among transport mechanisms (Harmel et al., 2006a).

5.6.1 CONTINUOUS DISCHARGE MEASUREMENT

Continuous discharge values are measured using monitoring instrumentation that is installed in the field and automatically recorded by a logging device. Such sites, often referred to as *gauging stations*, are often ideal water quality sampling locations. This is because colocation of sites allows for load calculations (from the product of water quality concentrations and discharge). Discharge measurements are typically stored in the field and collected using remote or on-site data transfer methods. The most common continuous discharge measurement method utilizes the stage–discharge relationship (also known as a rating curve). The basic premise of this method is that stage (water surface level or water depth) is continuously recorded and used to estimate the discharge based on an established mathematical relationship depicted by the rating curve. Frequent adjustments to this relationship are necessary in unstable channels to minimize the uncertainty in discharge data.

An established stage–discharge relationship accompanies precalibrated hydrologic structures such as flumes or weirs. For small watershed sites, precalibrated flow control structures are useful because they provide reliable and accurate flow data (Slade, 2004). However, weirs and flumes are expensive to purchase and install, and may result in flow ponding that impacts sediment transport (USDA, 1996). Another disadvantage is that precalibrated structures are limited in the discharge they support, which restricts their use as watershed size increases. Improper installation resulting in incomplete flow capture, unlevel alignment, or inappropriate approach channel characteristics can introduce considerable uncertainty (Rantz, 1982).

For sites where precalibrated structures are not feasible, a stage–discharge relationship can be established from a series of instantaneous stage and discharge measurements (discussed subsequently). Since stage and discharge data must be collected for the range of expected values, developing a stage–discharge relationship is a time consuming, long-term task.

Once a stage–discharge relationship has been established, sensors are used to continuously measure the stage. Common sensor types include bubblers, pressure transducers, noncontact sensors, and floats (Buchanan and Somers, 1982; USDA, 1996). Bubblers and pressure transducers are submerged sensors that estimate water stage by sensing the pressure head created by water depth. Noncontact sensors are suspended above the water surface and use ultrasonic or radar technology to measure surface water level (Costa et al., 2006). Float sensors float on the water surface and, in conjunction with a stage recorder, produce a graphical or electronic data record of stage. To increase the accuracy of stage measurement, stage sensors should be installed in a stilling well for protection and creation of a uniform water surface. Sensors must also be calibrated using a surveyed point for establishing correct stage measurements (Brakensiek et al., 1979; Haan et al., 1994). Installation of a permanent staff gauge is also recommended (USDA, 1996).

Although stage–discharge relationships are most often used for continuous discharge measurement, instream velocity meters and stage sensors are also appropriate alternatives. These instruments use multiple velocity measurements and corresponding stage data with cross-sectional survey data to determine the cross-sectional flow area and discharge (Harmel et al., 2006a). These devices may be preferred for sites where rapid morphological change or bidirectional flow occurs or where a structure is not feasible (e.g., waterways).

Continuous discharge may also be determined using Manning's equation (Maidment, 1993; Haan et al., 1994). Manning's equation estimates flow velocity using physical features of the system: channel roughness, slope, and cross-sectional geometry. Cross-sectional survey data are used with the velocity estimate to determine discharge. This method, however, introduces substantial uncertainty into discharge data as it was developed for uniform flow and because accurate channel roughness coefficients are difficult to select (Maidment, 1993). Thus, Manning's equation should only be used as a last alternative when other methods are not feasible for estimation of continuous discharge data.

5.6.2 Noncontinuous (Instantaneous) Discharge Measurement

Noncontinuous discharge measurements are collected using portable equipment that is not permanently installed in the field. Instantaneous discharge is often determined with the area–velocity method using velocity measurements and cross-sectional flow areas. The area–velocity method requires that water depth and velocity be measured in multiple vertical sections (each with no more than 5% to 10% of total flow) perpendicular to water flow in order to calculate mean velocity and cross-sectional area for each section (Buchanan and Somers, 1976; Figure 5.1). Velocity within the vertical profile of the stream is not uniform; thus, it is recommended that velocities be collected from 0.2 and 0.8 depths and averaged if the depth is greater than 0.6 m. Otherwise, velocity measurements are taken at the 0.6 depth to determine the mean velocity for the area–velocity method (Chow et al., 1988). The total discharge (Q) is the sum of discharges for each section and is calculated as

$$Q = \sum_{i=1}^{n} \left((X_{i+1} - X_i)(U_i Y_i + U_{i+1} Y_{i+1}) / 2 \right) \tag{5.1}$$

where n is the number of sections, X is the distance measurement on the horizontal scale, U is the velocity, and Y is the depth measured on a vertical scale.

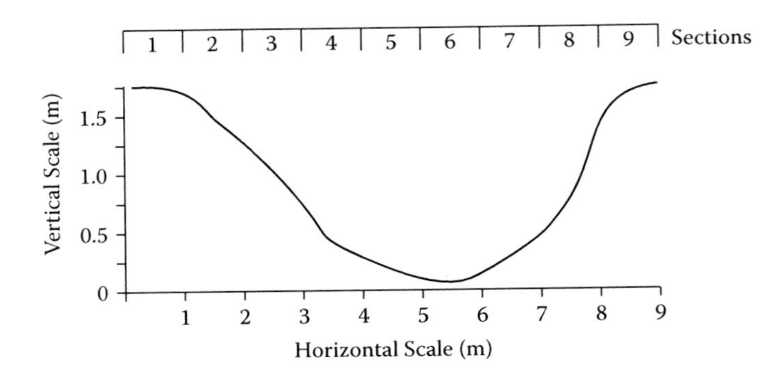

FIGURE 5.1 The stream bottom (the line) is presented with vertical and horizontal scale axis; each vertical section is measured for depth, width, and velocity; and the resulting measurements are used to calculate flow (discharge) using the area–velocity method.

Several portable devices are available to measure water velocity. Velocity meters or current meters may use revolving cups that spin at a rate proportional to the velocity, or they may use Doppler, electromagnetic, or radar technology to determine flow velocity (Rantz, 1982; Morlock, 1996).

Another instantaneous discharge method involves dye or tracer studies that can be used where stream geometry prevents the application of the area–velocity method. Dye and tracer studies are useful for understanding advection and longitudinal dispersion, groundwater inflow, and hydrologic retention properties of a stream (Lee, 1995; Harvey and Wagner, 2000).

The drift method is another noncontinuous discharge measurement alternative that might be considered to calculate discharge if safety concerns or costs prevent the area–velocity method. This method consists of watching and timing a floating device as it travels a specified distance in stream flow. This method is the least accurate and should be considered the least preferable option. When using this method, the floating device should be selected so that it floats with the majority of its volume underwater to prevent wind influences on velocity measurement. Drift velocity measurements should be collected by positioning the floating device in the middle of the stream and reducing the velocity value determined by approximately 60%. The discharge can then be determined as the product of the cross-section and the velocity (Burton and Pitt, 2002).

5.7 WATER CHEMISTRY SAMPLING

Water chemistry sampling refers to the activity of collecting a sample of water from a selected water body that represents the constituent concentration of that water body with the intent of analyzing the sample for selected constituent concentrations. Water chemistry sampling may also refer to the collection of a water characteristic (such as temperature, conductivity, or dissolved oxygen) in situ using portable sensors. Sampling project design should consider key elements (processes) of surface water systems that might hinder collection of a representative sample and attempt to minimize potential biases. The following sections briefly describe key sample collection considerations.

5.7.1 Sample Collection Point

Although the location of sample collection can result in considerable differences in constituent concentrations (Martin et al., 1992; Ging, 1999), this issue is commonly ignored in many sampling projects, especially those using automated samplers with a single intake. It is generally assumed that dissolved constituents can be adequately sampled at a single location within the flow cross-section for field-scale sites and small streams unless immediately downstream of significant point sources (Martin et al., 1992; Ging, 1999; Slade, 2004). However, recent research indicates that dissolved constituents may exhibit horizontal and/or vertical variability within the cross-section (Harmel, unpublished data). The assumption of well-mixed conditions can be evaluated by measuring pH, temperature, conductivity, and dissolved oxygen throughout a stream cross-section. If collected measurements differ by less than 5% throughout the cross-section, then a single measurement point at the centroid of flow adequately represents the cross-section (Wilde and Radtke, 2005). The degree to which constituents are distributed within the flow cross-section is a major component in the uncertainty associated with sample collection at a single location (see Chapter 12 regarding measurement uncertainty).

In sites that exhibit considerable concentration variability within the flow cross-section, sampling methods should be adjusted to account for this variability. Multiple grab samples can be collected, but integrated techniques (discussed in Section 5.7.3) are the preferred manual sampling technique. For automated sampling, the recommended methodology involves development of a relationship between concentrations at the sampler intake and mean concentrations as determined by integrated sampling at a range of discharges (e.g., Ging, 1999). With such a relationship, concentrations at the intake can be adjusted to represent mean concentrations for the total cross-section.

In contrast to dissolved constituents, sediment and sediment-bound constituent concentrations almost always vary substantially within the flow cross-section. For field-scale and small watershed sampling locations, a single sample intake may be adequate for sediment sampling because of well-mixed conditions and shallow water depths. Sediment sampling of coarse sediment or sediment in larger streams requires integrated sampling to adequately capture sediment concentration variability (Harmel et al., 2006a). Water quality sampling for sediment and sediment-associated constituents is more difficult than sampling for other water chemistry variables because of temporal and spatial variability in transport. Sediment particles <62 μm in diameter are generally homogeneously distributed throughout a channel's cross-section (Vanoni, 1975; Edwards and Glysson, 1999). However, particles >62 μm typically exhibit vertical and horizontal concentration gradients with the greatest concentration near the stream bed, and concentrations decrease with increasing distance from the bed. Therefore, the point at which a sample is collected will influence sediment concentration data. Samples for sediment sizes >62 μm should be collected with integrated techniques or should be adjusted to represent mean cross-sectional concentrations. Sediment sizes between 62 to 2,000 μm should be sampled using isokinetic techniques. Isokinetic sampling refers to collection of a water-sediment mixture in such a way that there is no change in velocity as the sample leaves the ambient flow and enters the sampler intake. Equipment for taking isokinetic samplers has been developed and evaluated by the Federal Interagency Sedimentation Project (Edwards and Glysson, 1999). Other techniques are available for sampling sediment concentrations, such as acoustic backscatter and optical backscatter (Wren et al., 2000; Gray, 2005). While these alternatives collect more detailed data in a temporal sense and may produce more accurate data, standardized methods and applicability to various conditions are not well established (Gray and Glysson, 2003; Gray, 2005).

While the previous discussion focused on collection of water samples for analysis in the laboratory, other technologies are available for in situ estimation of water quality parameters. Technological advances have resulted in greater application of portable sensors for measuring water parameters in situ. These devices are commonly used for measurements of physicochemical variables (pH, water temperature, dissolved oxygen, conductivity), but advancements in biosensor development have resulted in a broader application of such portable sensors (Glasgow et al., 2004). Sensors range from simple pen-type instruments capable of on-the-spot measurements of pH lacking data storage capabilities to multiparameter meters designed for continuous measurements of a dozen or more water chemistry parameters including those typically measured in the laboratory (i.e., chlorophyll, blue-green algae, rhodamine, nitrate, ammonium, and chlorine). Many commercial vendors enable multiparameter meters to be custom fitted with selected probes to meet specific research needs. Chapter 8 discusses field measurement methods and equipment in greater detail.

5.7.2 Flow Regime Considerations

Water samples may be collected during base flow and/or storm flow conditions, depending on the goal of the sampling project, and it should be expected that substantial differences in concentrations will occur between these discharge regimes due to differing source contributions.

5.7.2.1 Base Flow

Base flow is defined as the discharge derived from the seepage of groundwater in combination with upstream water through-flow without significant direct contribution from surface runoff resulting from precipitation. In small streams and canals, point source discharge can be an important base flow component.

Base flow water sampling is necessary at intermittent and perennial flow sites to quantify the contributions of point sources, tile drainage, shallow subsurface return flow, and constituent release from instream processes. Whereas base flow sampling is often needed in watershed-scale sampling projects, it can be unnecessary in field-scale projects that are usually conducted in runoff-dominated ephemeral sites (Harmel et al., 2006a). Since point sources may impact base flow concentrations, samples should be taken a distance from effluent discharges where flow is expected to be well mixed. Base flow samples collected at a single point in well-mixed flow, usually in the centroid of flow, are typically assumed to accurately represent true mean cross-sectional concentrations (Martin et al., 1992; Ging, 1999; Slade, 2004).

5.7.2.2 Storm Event Sampling

Storm event sampling refers to sampling when discharge conditions are influenced by surface runoff due to a precipitation event. Such events are typically defined by a specific amount of precipitation occurring after a specific interval of no precipitation. Depending on the site location and sampling goal, the definition of a storm event will differ. The National Pollutant Discharge Elimination System (NPDES) Storm Water Sampling Guidance Document (USEPA, 1992) defines a storm event as the occurrence of 2.54 mm of accumulated precipitation after 72 h of preceding dry weather. The guidance also suggests that "where feasible, the depth of rain and duration of the event should not vary by more than 50% from the average depth and duration." It is important to note that this definition is specific to storm water sampling to meet permit requirements and that increased discharge containing surface runoff contributions may or may not occur following this amount of rainfall. Adjustments from this protocol may be implemented depending on project sampling objectives.

Another strategy for defining and identifying storm event flows is to evaluate discharge records (Institute of Hydrology, 1980a, 1980b). This method separates base flows and storm flows in a dataset using the principle that if 70% or greater of total discharge for a day is composed of base flow, then it is considered base flow. Otherwise, the discharge regime is characterized as a storm event flow. The limitation to this method is that it characterizes discharge measurements as base flow or storm flow after the data have been collected. Whatever method is selected, it is important to establish the storm definition at the beginning of the project to ensure that adequate samples are collected to characterize base flow and storm flow as required by the project goals.

Storm flow is often the focus of sampling projects because of its increased transport ability for recently washed off and resuspended constituents that have been attenuated by instream processes. Increased constituent transport in storm events can occur early in storms, and this phenomenon is often referred to as *first flush*. Alternatively, constituent concentrations can peak with peak flow, increase throughout storm events, or remain relatively uniform as storm and base flow contributions interplay. In any case, storm flows should be sampled throughout events to ensure proper water chemistry characterization (Huber, 1993). The unique circumstances of storm event sampling and the resulting strategies for implementing water quality sampling have only recently been described (e.g., McFarland and Hauck, 2001; Harmel et al., 2003; Haggard et al., 2003; King and Harmel, 2003; Behrens et al., 2004; Harmel et al., 2006a). Characterization of storm water chemistry is much more difficult than base flow because storm runoff often occurs with little advance warning, outside conventional work hours, and under adverse weather conditions (USEPA, 1997).

5.7.3 Sampling Equipment Considerations

Surface water samples from small streams and canals are collected manually or by an automated sampler permanently or semipermanently installed at the sampling location and programmed to collect water samples at desired intervals. Each method has its advantages and disadvantages.

5.7.3.1 Manual Sampling

Manual grab sampling at a single location in the flow cross-section at each sampling site is the standard method for collection of base flow samples. Grab sampling has several advantages. It is relatively safe, simple, and inexpensive, can be performed at any location, and is not subject to equipment theft or vandalism. Unfortunately, grab sampling provides limited information on temporal variability of constituent concentrations unless frequent samples are collected. Manual sampling methods also introduce human errors due to sampling variability that might occur. Manual grab sampling may, however, be the only alternative if the capital costs of purchasing and maintaining automatic sampling equipment exceeds available resources (Burton and Pitt, 2002).

Integrated sampling is an alternative manual technique that collects subsamples throughout the flow cross-section to accurately determine mean cross-sectional constituent concentrations. Integrated sampling typically utilizes the USGS equal-width increment or equal-discharge increment procedures (Wells et al., 1990; USGS, 1999). With these procedures, multiple depth-integrated samples are obtained across the stream cross-section and have been shown to produce accurate concentration measurements even in large streams. However, these techniques require substantial personnel time, especially for multiple sites, and can be difficult for sample collection throughout the range of observed discharges. If human entry into the stream or canal is necessary for sample collection, personnel safety must be the utmost priority. If entry is safe, then streambed and bank disturbances should be limited to limit constituent resuspension. Similarly, water samples should be collected upstream from the point of entry.

5.7.3.2 Automatic Sampling

Automated samplers offer several advantages to manual sampling. A major advantage of automated samplers is their ability to use consistent sampling procedure and simultaneously collect samples at multiple sites. Automatic samplers are advantageous for distant, hard-to-reach (e.g., steep inclines), and dangerous sites due to wildlife (e.g., alligators) or storm conditions (e.g., high flow, lightning). Automatic samplers are particularly useful for storm event sampling because of their ability to sample throughout runoff events of various durations and magnitudes. However, automated samplers also have some disadvantages such as their single sample intake and impossibility of keeping the intake in the centroid of flow (Harmel et al., 2006a). In addition, automatic samplers require frequent maintenance (Burton and Pitt, 2002; Harmel et al., 2006a).

Mechanical samplers, such as the rotating slot sampler and the multislot divisor sampler, may be practical alternatives to electronic automatic samplers, but they commonly have a discharge rate limitation at which they can effectively sample. Mechanical samplers collect flow-weighted samples and estimate flow volume, allowing for the calculation of event mean concentrations and mass loads. The rotating slot sampler requires minimal maintenance, no electrical power, and collects a single flow-proportional runoff sample (Parsons, 1954, 1955; Edwards et al., 1976). Others have modified this design for their specific application needs (Bonta, 1999; 2002; Malone et al., 2003). The multislot divisor has also been applied by many investigators for collecting surface runoff from fields or slopes (Geib, 1933; Sheridan et al., 1996; Franklin et al., 2001; Pinson et al., 2003).

5.8 PHYSICAL HABITAT AND BIOLOGICAL ASSESSMENTS

Stream ecosystems are complex, dynamic systems and often physical, biological, and chemical variables are interrelated. Thus, understanding these relationships can help with developing watershed management recommendations. Physical habitat and biological assessments are an integral part of water quality monitoring programs in the United States as all states have a monitoring program that includes physical habitat, biological, and traditional water chemistry and discharge assessments (USEPA, 2002). Many stream sampling protocols consist of both descriptions of physical habitat and biological sampling in the same protocol (USEPA, 2002; Somerville and Pruitt, 2004). These joint assessments are necessary as the physical and chemical characteristics of streams are needed to interpret the observed biological responses (see also Chapter 3, Section 3.4.2). More than 400 habitat and biological sampling protocols have been developed for a wide range of studies (Johnson et al., 2001; NRCS, 2001; USEPA, 2002; Somerville and Pruitt, 2004). However, few if any sampling protocols are universally accepted (Frissell et al., 2001). In this section, an introduction to key concepts and commonly used sampling methods for assessing physical habitat and biological characteristics of streams is provided.

5.8.1 Physical Habitat Assessments

Physical variables, such as channel cross-section area, discharge, number of riparian trees, and percent sand substrate, are often referred to by ecologists as physical habitat variables because they serve as descriptors of the space or the "habitat" that fishes, insects, aquatic plants, and other stream organisms occupy. Types of physical variables include measurements of watershed characteristics (i.e., watershed size, shape, land use), riparian habitat (i.e., riparian width, canopy cover, tree species composition), geomorphology (i.e., sinuosity, gradient, cross-section area), and instream habitat (i.e., wet width, water depth, water velocity, substrate types). Watershed habitat variables are typically evaluated through examination of topographic maps or use of geographic information systems to analyze electronic aerial photos, topographic maps, and other information sources. Alternatively, measurement of riparian habitat, geomorphology, and instream habitat requires field work.

Transect-based and visual-based sampling methods represent the two general approaches used for the measurement of riparian habitat, geomorphology, and instream habitat. Transect-based sampling methods consist of measuring selected habitat variables at predetermined points along multiple transects within a site. Transect-based habitat sampling methods result in quantitative habitat data, reduce observer bias, and ensure comparability among sampling sites, time periods, and categories of interest (Simonson et al., 1994; Wang et al., 1996). Transect-based sampling methods are a must when quantitative information on water depth, water velocity, riparian characteristics, channel size, and adjacent land use is needed. Visual-based habitat methods involve the estimation rather than the actual measurement of a site's habitat variables. Visual habitat methods are frequently used because they require less equipment and time than quantitative habitat measurements. Visual-based habitat methods are dependent on observer training and skills, thus scores differ among observers (Somerville and Pruitt, 2004). Visual-based habitat sampling protocols are useful for descriptive purposes and for preliminary habitat assessments intended to assist with site selection. Additionally, visual-based methods are often developed to result in the calculation of a habitat quality index, i.e., USEPA Rapid Bioassessment Habitat Index (Barbour et al., 1999), that provides an easily interpretable value representative of the physical habitat quality of the site.

The frequency of measuring physical habitat variables also depends on the goal of the surface water quality program and the type of habitat variable. For example, if the objective of the surface water project is to characterize the water chemistry conditions within the watershed and physical habitat is just being measured to provide accurate descriptions of the study sites, one-time measurements of selected habitat variables may be appropriate. However, if the objective of the surface water project is to evaluate the potential impacts of urbanization on water chemistry, geomorphology, and instream habitat, physical habitat variables may need to be measured more than once. Watershed variables, such as watershed size and shape, are not likely to change for the duration of the monitoring period and one-time measuring of these variables is appropriate. However, watershed land use, riparian, and geomorphology variables can fluctuate annually and may require annual measurements at a minimum. Many instream habitat variables exhibit seasonal and even daily fluctuations and, thus, these variables need to be measured more often than once a year. Additionally, monitoring projects with objectives to link changes in instream habitat to biological characteristics need to ensure that habitat and biological sampling events are conducted concurrently.

5.8.2 Biological Assessments

There are different types of biological assessments that can be used to assess stream water quality, and these different approaches are capable of assessing individual, population, or community responses. For example, one can sample animals or plants from different sites and measure the amount of contaminants within their tissues. Additionally, one could collect water from selected sites and conduct controlled laboratory bioassays that investigate the survival of laboratory animals in collected water. A common approach used in the United States and Europe is to evaluate the community characteristics of aquatic organisms from different sites (Cairns and Pratt, 1993). Also, community assessments are cheaper than conducting toxicological bioassays and in some cases cheaper than measuring water chemistry variables in the laboratory (Yoder and Rankin, 1995). We will focus on descriptions of community assessments in this section.

In general, community assessments involve the use of standardized sampling protocols to collect all species of aquatic organisms from a sampling site with the intent of calculating different metrics that describe the diversity, abundance, and species composition of the sampled communities. Sampling protocols for a wide range of stream organisms have been established by state agencies in the United

States that include algae, aquatic plants, plankton, aquatic macroinvertebrates, fish, amphibians, waterfowl, and other vertebrates (USEPA, 2002). Forty-nine of 50 states sample aquatic macroinvertebrates as part of their stream bioassessments and 35 states sample fishes (USEPA, 2002). Additionally, 33 states evaluate both aquatic macroinvertebrates and fishes as part of their stream bioassessments of water quality (USEPA, 2002).

Numerous sampling methods exist for aquatic macroinvertebrates, but the most common method used by state agencies in the United States is the dipnet (Figure 5.2).

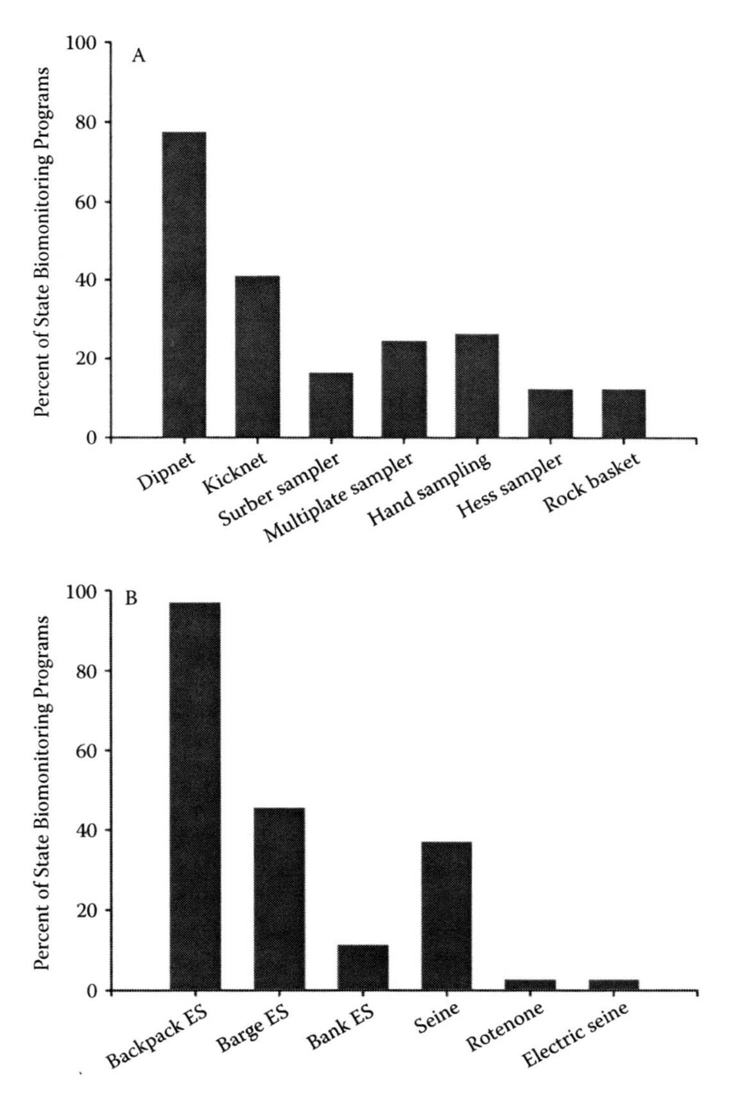

FIGURE 5.2 Percent of use of different macroinvertebrate (A) and fish sampling techniques (B) by state biomonitoring programs in the United States (ES—electroshocker).

Essentially, this is a net with a wooden handle. The advantage of the dipnet is its capability of sampling many different types of microhabitats (i.e., pools, riffles, runs, etc.) found within a site. The disadvantage of the dipnet is that it is considered a qualitative sampling technique because one cannot calculate density (number of organisms per square meter) from dipnet samples. However, despite this flaw, the dipnet sampling methodology can be modified to standardize for sampling effort and to obtain comparable estimates of macroinvertebrate abundance. Another sampling method consists of using a surber sample, that is, a portable, stream bottom sampler consisting of two folding frames with netting that are affixed at right angles (Figure 5.3). For sampling, the horizontal frame is positioned in the substrate, and silt and rocks are stirred up so that the current transports bottom organisms into the net. The surber sampler has been most frequently used in quantitative biomonitoring studies evaluating the effects of different disturbances in gravel bottom streams (Resh and McElravy, 1993). However, the surber sampler is most effective in shallow riffles, and this explains the lack of its adoption in statewide biomonitoring programs that require multihabitat assessments of macroinvertebrate communities. Additionally, it is frequently recommended that multiple sampling techniques be used for sampling macroinvertebrates due to the diversity of the taxa involved (Karr and Chu, 1999), and many state biomonitoring programs in the United States have incorporated this recommendation (USEPA, 2002).

Capture of stream fishes typically involves direct current electrofishing; backpack-mounted electrofishers are the most commonly used in wadeable streams (Figure 5.2).

FIGURE 5.3 Use of dipnet (A) and surber sampler (B) to collect macroinvertebrates from agricultural drainage ditches in Ohio.

Techniques for use of backpack-mounted electrofishers involve two or more people where one person carries and operates the backpack electrofisher while the other(s) dipnet the stunned fishes. The effectiveness of electrofishing has resulted in less emphasis on the use of multiple sampling techniques for fishes. However, no single sampling gear is effective in the capture of all types and sizes of organisms; thus, the use of multiple sampling techniques for fishes is also a recommended practice (Moulton et al., 2002; Smiley et al., 2009). Seining is the most frequently used non-electrofishing gear and can be a supplementary sampling method for electrofishing.

5.9 SUMMARY

This chapter focuses on procedures for determining chemical quality in wadeable streams and canals but also provides introductory concepts related to physical and biological aspects of water quality. Surface water sampling reflects the changing views of water quality and the emerging variety of sampling projects and goals. The first step in designing a surface water sampling project is to identify its goal and the resources available to achieve that goal. Whether the project and its goals are very simplistic or highly complex, project design should include initial planning that considers (1) the spatial extent of sampling, (2) the response variables to be analyzed, (3) the flow conditions to be sampled, and (4) the type of technology to be used. Then sampling project components (site locations, discharge measurement, sampling methodology, and personnel and equipment requirements) should be evaluated and implemented to achieve the project goal with the available resources. This chapter does not provide an exhaustive guide but rather concepts that should be integrated into a surface water quality sampling project. Readers interested in further details regarding selected topics should refer to the recommended readings section in Table 5.1.

TABLE 5.1
Suggested References for Further Information on Selected Topics

Topic	References
Discharge measurements—theory and procedures	Brakensiek et al., 1979; Buchanan and Somers, 1976, 1982; Kennedy, 1984; Carter and Davidian, 1989; Chow et al., 1988; Haan et al., 1994; Maidment, 1993; Burton and Pitt, 2002
Weirs and flumes	Bos, 1976; Brakensiek et al., 1979; USDIBR, 2001
Water chemistry sampling	Dissmeyer, 1994; USDA, 1996; USEPA, 1997; Harmel et al., 2006a
Manual water chemistry sampling	Wells et al., 1990; USGS, 1999
Water chemistry quality assurance/quality control	Dissmeyer, 1994; USDA, 1996; USEPA, 1997; Harmel et al., 2006b
Sediment sampling	Edwards and Glysson, 1999
Storm event sampling	Wells et al., 1990; USGS, 1999; Harmel et al., 2003, 2006a
Habitat and biological sampling	Merritt and Cummins, 1996; Murphy and Willis, 1996; Bain and Stevenson, 1999; Barbour et al., 1999; Johnson et al., 2001; NRCS, 2001; Ohio EPA, 2002; Moulton et al., 2002; Somerville and Pruitt, 2004; Hauer and Lambertini, 2006

REFERENCES

Abtew, W., and B. Powell. 2004. Water quality sampling schemes for variable-flow canals at remote sites. *J. Amer. Water Resources Assoc.* 40(5): 1197–1204.

Agouridis, C. T., and D. R. Edwards. 2003. The development of relationships between constituent concentrations and generic hydrologic variables. *Trans. ASAE* 46(2): 245–256.

Bain, M. B., and N. J. Stevenson. 1999. Aquatic habitat assessment: Common methods. American Fisheries Society, Bethesda, MD.

Barbour, M. T., J. Gerritsen, B. D. Snyder, and J. B. Stribling. 1999. Rapid Bioassessment Protocols for Use in Streams and Wadeable Rivers: Periphyton, Benthic Macroinvertebrates and Fish, Second Edition. EPA 841-B-99-002. U.S. Environmental Protection Agency; Office of Water; Washington, D.C.

Behrens, B., J. H. Riddle, and J. Gillespie. 2004. Tips to improve wet weather monitoring. *American Public Works Assoc. Reporter* Sept. 2004 (online). Also presented by K. F. Holbrook as "Maximizing reliability of storm water monitoring data" at the North American Surface Water Quality Conference, Palm Desert, Cal., July 29, 2004.

Benson, M. A., and T. Dalrymple. 1984. General Field and Office Procedures for Indirect Discharge Measurements, USGS, Washington, DC.

Bonta, J. V. 1999. Water sampler and flow measurement for runoff containing large sediment particles. *Trans. ASAE* 42(1): 107–114.

Bonta, J. V. 2002. Modification and performance of the Coshocton wheel with the modified drop-box weir. *J. Soil Water Cons.* 57(6): 364–373.

Bos, M. G., ed. 1976. Discharge measurement structures. Publication No. 20. Wageningen, The Netherlands: International Institute for Land Reclamation and Improvement.

Brakensiek, D. L., H. B. Osborn, and W. J. Rawls, coordinators. 1979. Field Manual for Research in Agricultural Hydrology. Agriculture Handbook No. 224. Washington, D.C.: USDA.

Buchanan, T. J., and W. P. Somers. 1976. Chapter A8: Discharge measurements at gaging stations. Techniques of Water-Resources Investigations of the U.S. Geological Survey, Book 3. Washington, D.C.: USGS.

Buchanan, T. J., and W. P. Somers. 1982. Chapter A7: Stage measurement at gaging stations. Techniques of Water-Resources Investigations of the U.S. Geological Survey, Book 3. Washington, D.C.: USGS.

Burton, G. A., Jr., and R. E. Pitt. 2002. *Stormwater Effects Handbook*. Boca Raton, FL: CRC Press.

Cairns, J. C., Jr., and J. R. Pratt. 1993. A history of biological monitoring using benthic macroinvertebrates. Pages 10–27 in D. M. Rosenberg and V. H. Resh (eds.), *Freshwater Biomonitoring and Benthic Macroinvertebrates*. Chapman and Hall.

Carey, R. O., and K. W. Migliaccio. 2009. Contribution of wastewater treatment effluents to nutrient dynamics in aquatic systems: A review. Environmental Management 44(2):205–217. DOI 10.1007/s00267-009-9309-5.

Carter, R. W., and J. Davidian. 1989. Chapter A6: General procedure for gaging streams. Techniques of Water-Resources Investigations of the U.S. Geological Survey, Book 3. Washington, D.C.: USGS.

Chow, V. T., D. R. Maidment, and L. W. Mays. 1988. *Applied Hydrology*. New York: McGraw-Hill.

Costa, J. E., R. T. Cheng, F. P. Haeni, N. Melcher, K. R. Spicer, E. Hayes, W. Plant, K. Hayes, C. Teague, and D. Barrick. 2006. Use of radars to monitor stream discharge by noncontact methods. *Water Resources Research* 42:1–14. doi:10.1029/2005WR004430.

Dissmeyer, G. E. 1994. Evaluating the effectiveness of forestry best management practices in meeting water quality goals or standards. Misc. Publication No. 1520. Atlanta, Ga.: USDA Forest Service, Southern Region.

Edwards, T. K., and G. D. Glysson. 1999. Chapter C2: Field methods for measurement of fluvial sediment. Techniques of Water-Resources Investigations of the U.S. Geological Survey, Book 3. Washington, D.C.: USGS.

Edwards, W. M., H. E. Frank, T. E. King, and D. R. Gallwitz. 1976. Runoff sampling: Coshocton vane proportional sampler. ARS-NC-50. Washington, D.C.: USDA-ARS.

Franklin, D. H., M. L. Cabrera, J. L. Steiner, D. M. Endale, and W. P. Miller. 2001. Evaluation of percent flow captured by a small in-field runoff collector. *Trans. ASAE* 44(3): 551–554.

Frissell, C. A., N. L. Poff, and M. E. Jensen. 2001. Assessment of biotic patterns in freshwater ecosystems. Pages 390–403 in M. E. Jensen and P. S. Bourgeron (eds), *A Guidebook for Integrated Ecological Assessments*. Springer: New York.

Geib, H. V. 1933. A new type of installation for measuring soil and water losses from control plots. *J. American Soc. Agron.* 25: 429–440.

Ging, P. 1999. Water-quality assessment of south-central Texas: Comparison of water quality in surface-water samples collected manually and by automated samplers. USGS Fact Sheet FS-172-99. Washington, D.C.: USGS.

Glasgow, H. B., J. M. Burkholder, R. E. Reed, A. J. Lewitus, and J. E. Kleinman. 2004. Real-time remote monitoring of water quality: A review of current applications, and advancements in sensor, telemetry, and computing technologies. *J. Experimental Marine Biol. Ecology* 300(1–2):409–448.

Gray, J. R., ed. 2005. *Proceedings of the Federal Interagency Sediment Monitoring Instrument and Analysis Research Workshop*. USGS Circular 1276. Reston, VA: USGS).

Gray, J. R., and G. D. Glysson, eds. 2003. *Proceedings of the Federal Interagency Workshop on Turbidity and Other Sediment Surrogates*. [USGS Circular 1250]. Reston, VA: USGS).

Haan, C. T., B. J. Barfield, and J. C. Hayes. 1994. *Design Hydrology and Sedimentology for Small Catchments*. New York: Academic Press.

Haggard, B. E., T. S. Soerens, W. R. Green, and R. P. Richards. 2003. Using regression methods to estimate stream phosphorus loads at the Illinois River, Arkansas. *Applied Eng. Agric.* 19(2): 187–194.

Harmel, R. D., and K. W. King. 2005. Uncertainty in measured sediment and nutrient flux in runoff from small agricultural watersheds. *Trans. ASAE* 48(5): 1713–1721.

Harmel R. D., K. W. King, and R. M. Slade. 2003. Automated storm water sampling on small watersheds. *Applied Eng. Agric.* 19(6): 667–674.

Harmel, R. D., K. W. King, B. E. Haggard, D. G. Wren, and J. M. Sheridan. 2006a. Practical guidance for discharge and water quality data collection on small watersheds. *Trans. ASABE* 49(4): 937–948.

Harmel, R. D., R. J. Cooper, R. M. Slade, R. L. Haney, and J. G. Arnold. 2006b. Cumulative uncertainty in measured streamflow and water quality data for small watersheds. *Trans. ASABE* 49(3): 689–701.

Hauer, F. R., and G. A. Lamberti. 2006. *Methods in Stream Ecology*. Academic Press, Boston, MA.

Harvey, J. W., and B. J. Wagner. 2000. Chapter 1: Quantifying hydrologic interactions between streams and their subsurface hyporheic zones. In *Streams and Ground Waters*. Jones, J.B., and P.J. Mulholland., Ed. Academic Press: San Diego, CA.

Huber, W. C. 1993. Chapter 14: Contaminant transport in surface water. *Handbook of Hydrology* (ed., David R. Maidment) New York: McGraw-Hill, Inc.

Institute of Hydrology. 1980a. Low flow studies. Report No. 1. Wallingford, U.K.

Institute of Hydrology. 1980b. Low flow studies. Report No. 3. Wallingford, U.K.

Johnson, D. H., N. Pimman, E. Wilder, J. A. Sliver, R. W. Plotnikoff, B. C. Mason, K. K. Jones, P. Roger, T. A. O'Neill, and C. Barrett. 2001. Inventory and Monitoring of Salmon Habitat in the Pacific Northwest: Directory and Synthesis of Protocols for Management/Research and Volunteers in Washington, Oregon, Idaho, Montana, and British Columbia. Washington Department of Fish and Wildlife, Olympia, WA.

Karr, J. R., and E. W. Chu. 1999. *Restoring Life in Running Waters: Better Biological Monitoring*. Island Press: Washington, D.C.

Kennedy, E. J. 1984. Chapter A10: Discharge ratings at gaging stations. Techniques of Water-Resources Investigations of the U.S. Geological Survey, Book 3. Washington, D.C.: USGS.

King, K. W., and R. D. Harmel. 2003. Considerations in selecting a water quality sampling strategy. *Trans. ASAE* 46(1): 63–73.

King, K. W., R. D. Harmel, and N. R. Fausey. 2005. Development and sensitivity of a method to select time- and flow-paced storm event sampling intervals. *J. Soil Water Cons.* 60(6): 323–331.

Lazorchak, J. M., D. J. Klemm, and D.V. Peck. 1998. Environmental Monitoring and Assessment Program - Surface Waters: Field Operations and Methods for Measuring the Ecological Condition of Wadeable Streams. EPA/620/R-94/004F. U.S. Environmental Protection Agency, Washington, D.C.

Lee, K. K. 1995. Stream Velocity and Dispersion Characteristics Determined by Dye-Tracer Studies on Selected Stream Reaches in the Willamette River Basin, Oregon. U.S. Geological Survey Water-Resources Investigations Report 95-4078, 39 pp.

Lyons, J. 1992. The length of stream to sample with a towed electrofishing unit when fish species richness is estimated. *North Amer. J. Fisheries Management* 12:198–203.

Maidment, D. R., ed. 1993. *Handbook of Hydrology*. New York: McGraw-Hill.

Malone, R. W., J. V. Bonta, and D. R. Lightell. 2003. A low-cost composite water sampler for drip and stream flow. *Applied Eng. Agric.* 19(1): 59–61.

Martin, G. R., J. L. Smoot, and K. D. White. 1992. A comparison of surface-grab and cross-sectionally integrated streamwater-quality sampling methods. *Water Environ. Res.* 64(7): 866–876.

McFarland, A., and L. Hauck. 2001. Strategies for monitoring nonpoint-source runoff. TIAER Report No. 0115. Stephenville, Texas: Tarleton State University, Texas Institute for Applied Environmental Research.

Merritt, R. W., and K. W. Cummins. 1996. *An Introduction to the Aquatic Insects of North America*. Third edition. Kendall/Hunt, Dubuque, IA.

Migliaccio, K. W., B. E. Haggard, I. Chaubey, and M. D. Matlock. 2007. Water quality monitoring at the War Eagle Creek watershed, Beaver Reservoir, Arkansas. *Trans. ASABE* 50(6):2007–2016.

Miller, T. 2005. Monitoring in the 21st century to address our nation's water-resource questions. USGS Congressional Briefing Sheet. Washington, D.C.: USGS.

Morlock, S. E. 1996. Evaluation of acoustic Doppler current profiler measurement of river discharge. U.S. Geological Survey. Water Resources Investigation Report 95-4218, 37 pp.

Moulton, S. R., II, J. G. Kennen, R. M. Goldstein, and J. A. Hambrook. 2002. Revised Protocols for Sampling Algal, Invertebrate, and Fish Communities as Part of the National Water-Quality Assessment Program. Open-file report 02-150, United States Geological Survey, Reston, Virginia.

Murphy, B. R., and D. W. Willis. 1996. *Fisheries Techniques*. American Fisheries Society, Bethesda, MD.

Natural Resources Conservation Service (NRCS), 2001. Stream Corridor Inventory and Assessment Techniques: A Guide to Site, Project, and Landscape Approaches Suitable for Local Conservation Programs. Watershed Science Institute Technical Report, Natural Resources Conservation Service, Washington, D.C.

Ohio EPA. 2002. Field Evaluation Manual for Ohio's Primary Headwater Habitat Streams. Ohio EPA, Division of Surface Water, Columbus, Ohio.

Paller, M. H. 1995. Relationships among number of fish species sampled, reach length surveyed, and sampling effort in South Carolina Coastal Plain streams. *North Amer. J. Fisheries Mgmt.* 15:110–120.

Parsons, D. A. 1954. Coshocton-type runoff samplers: Laboratory investigations. SCS-TP-124. Washington, D.C.: USDA-SCS.

Parsons, D. A. 1955. Coshocton-type runoff samplers. ARS-41-2. Washington, D.C.: USDA-ARS.

Patton, T. M., W. A. Hubert, F. J. Rahel, and K. G. Gerow. 2000. Effort needed to estimate species richness in small streams on the Great Plains in Wyoming. *North Amer. J. Fisheries Mgmt.* 20: 394–398.

Pinson, W. T., D. C. Yoder, J. R. Buchanan, W. C. Wright, and J. B. Wilkerson. 2003. Design and evaluation of an improved flow divider for sampling runoff plots. *Applied Eng. Agric.* 20(4): 433–437.

Preston, S. D., V. J. Bierman, and S. E. Silliman. 1992. Impact of flow variability on error in estimation of tributary mass loads. *J. Environ. Eng.* 118(3): 402–419.

Rantz, S. E. 1982. Measurement and computation of streamflow. Vol. 1, Measurement of Stage and Discharge. U.S. Geological Survey Water Supply Paper, 2175, 284 pp.

Resh, V. H., and E. P. McElravy. 1993. Contemporary quantitative approaches to biomonitoring using benthic macroinvertebrates. Pages 159–194 in D. M. Rosenberg and V. H. Resh (eds.), *Freshwater Biomonitoring and Benthic Macroinvertebrates*. Chapman and Hall.

Sheridan, J. M., R. R. Lowrance, and H. H. Henry. 1996. Surface flow sampler for riparian studies. *Applied Eng. Agric.* 12(2): 183–188.

Shih, G., W. Abtew, and J. Obeysekera. 1994. Accuracy of nutrient runoff load calculations using time-composite sampling. *Trans. ASAE* 37(2): 419–429.

Simonson, T. D., J. Lyons, and P. D. Kanehl. 1994. Quantifying fish habitat in streams: Transect spacing, sample size, and a proposed framework. *North Amer. J. Fisheries Mgmt.* 14:607–615.

Slade, R. M. 2004. General methods, information, and sources for collecting and analyzing water-resources data. CD-ROM.

Smiley, P. C., Jr., F. D. Shields Jr., and S. S. Knight. 2009. Designing impact assessments for evaluating the ecological effects of conservation practices on streams in agricultural landscapes. *J. American Water Resources Assoc.* 45: 867–878.

Somerville, D. E., and B. A. Pruitt. 2004. Physical Stream Assessment: A Review of Selected Protocols for Use in the Clean Water Act Section 404 Program. Prepared for U. S. Environmental Protection Agency, Wetlands Division (Order # 3W-0503-NATX), Washington, D.C.

Tate, K. W., R. A. Dahlgren, M. J. Singer, B. Allen-Diaz, and E. R. Atwill. 1999. Timing, frequency of sampling affect accuracy of water-quality monitoring. *California Agric.* 53(6): 44–49.

USDA. 1996. Part 600: Introduction. In *National Water Quality Handbook*. Washington, D.C.: USDA-NRCS.

USDIBR. 2001. Water Measurement Manual. A Water Resources Technical Publication. U.S. Department of the Interior Bureau of Reclamation, Water Resources Research Laboratory. Washington, D.C. Web source: http://www.usbr.gov/pmts/hydraulics_lab/pubs/wmm/.

USEPA (U.S. Environmental Protection Agency). 1992. NPDES Storm Water Sampling Guidance Document. Office of Water. EPA 833-8-92-001. Web source: http://www.epa.gov/npdes/pubs/owm0093.pdf.

USEPA. 1997. Monitoring guidance for determining the effectiveness of nonpoint-source controls. EPA 841-B-96-004. Washington, D.C.: USEPA.

USEPA. 2002. Summary of biological assessment programs and biocriteria development for states, tribes, territories, and interstate commissions: Streams and wadeable rivers. U.S. Environmental Protection Agency, Office of Water, EPA-822-R-02-048, Washington, D.C.

USGS. 1999. Section A: National Field Manual for Collection of Water-Quality Data. Techniques of Water-Resources Investigations of the U.S. Geological Survey, Book 9. Washington, D.C.: USGS.

Vanoni, V. A., ed. 1975. *Sedimentation Engineering*. New York: ASCE.

Wang, L., T. D., Simonson, and J. Lyons. 1996. Accuracy and precision of selected stream habitat estimates. *North Amer. J. Fisheries Mgmt.* 16:340–347.

Wells, F. C., W. J. Gibbons, and M. E. Dorsey. 1990. Guidelines for collection and field analysis of water-quality samples from streams in Texas. USGS Open-File Report 90-127. Washington, D.C.: USGS.

Wilde, F. D., and D. B. Radtke. 2005. Chapter A6: Field measurements: General information and guidelines. Techniques of Water-Resources Investigations of the U.S. Geological Survey, Book 9. Washington, D.C.: USGS.

Wren, D. G., B. D. Barkdoll, R. A. Kuhnle, and R. W. Derrow. 2000. Field techniques for suspended-sediment measurement. *J. Hydraulic Eng.* 126(2): 97–104.

Yoder, C. O., and E. T. Rankin. 1995. Biological criteria program development and implementation in Ohio. Pages 109 to 144 in W. S. Davis and T. P. Simon (eds.), *Biological Assessment and Criteria*. Lewis Publishers, Boca Raton, FL.

6 Groundwater Sampling

Qingren Wang, Rafael Muñoz-Carpena, Adam Foster, and Kati W. Migliaccio

CONTENTS

6.1 INTRODUCTION

Groundwater is protected in most areas as it is a primary source of drinking water. In the United States, 50% of the population relies on groundwater supplies (Reilly et al., 2008). Groundwater sampling in the United States became commonplace in the 20th century as contaminated water resources became apparent and a growing public concern emerged to protect water resources. In response to this concern, the U.S. government mandated a study in which scientists identified six categories of groundwater contaminant sources (OTA, 1984):

- Category 1—sources designed to discharge substances (e.g., injection well)
- Category 2—sources designed to store, treat, and/or dispose of substances; discharge through unplanned release (e.g., landfills)
- Category 3—sources designed to retain substances during transport or transmission (e.g., pipelines)
- Category 4—sources discharging as consequence of other planned activities (e.g., pesticide application)
- Category 5—sources providing conduit or inducing discharge through altered flow patterns (e.g., construction excavation)
- Category 6—naturally occurring sources whose discharge is created and/or exacerbated by human activity (e.g., salt water intrusion).

To identify and mitigate sources of pollution for a particular groundwater quality concern, groundwater sampling must be conducted. Groundwater sampling refers to the sampling of water that is in an aquifer or groundwater table and underneath the ground surface for chemical, biological, or physical analyses. Sampling is typically conducted through some constructed feature that allows access to the groundwater, such as a groundwater well. However, the constructed feature may introduce complications that may hinder the extraction of a truly representative groundwater sample. To minimize this problem, special procedures should be followed for the construction of sampling wells, sample collection, sample handling, and quality control and assurance methods (QA/QC). This chapter provides general procedures for groundwater quality sampling. Groundwater sampling should be conducted only after a review has been completed on sampling regulations to ensure proper sample collection for legally and scientifically defensible results.

6.2 SPECIFYING THE OBJECTIVES FOR GROUNDWATER SAMPLING

Although the specific objectives vary with projects, the overall objective in most groundwater sampling projects is to obtain "representative" water samples from the subsurface aquifer under in situ conditions that provide high-quality reproducible data in order to guide decision making (Nielsen and Nielsen, 2007). "High-quality" data means data with sufficient accuracy, precision, and completeness to meet the project objectives. "Precision" refers to the repeatability of sampling and analysis, while "accuracy" is the degree of closeness of a measured value to the actual value. Both are required for high-quality data, and their obtainment depends, in part, on the correct choice of sampling equipment and the analytical procedures to be followed (Puls and Barcelona, 1996). Groundwater sampling can easily be compromised due to the manner in which the sample is collected (i.e., generally from a groundwater well). Any possible source of interference to meet the sampling objective should be avoided. As a simple example, if the nonpoint pollution is a major concern, any point pollution source, such as manufacturing units, mines, industrial or municipal waste landfills, tailing dumps, or drainage, adjacent to the sampling sites or possibly entering to the sites through any channel before or during the sampling process, can affect the result and interpretation and should be avoided or highlighted as critical concerns in the proposed procedure. Groundwater sampling objectives generally can be categorized as contaminant detection, contaminant evaluation, resource assessment, or remediation evaluation.

6.2.1 Contaminant Detection

The purpose of contaminant detection is to identify the presence and concentration of selected specific contaminant(s) of interest. Groundwater sampling often targets hazardous substances that might have been disposed of or stored improperly. Alternatively, groundwater sampling might target compounds that have ecological consequences (e.g., nutrients). Common contaminants include simple inorganic ions (e.g., nitrate [NO_3-N]), chlorides (from deicing salts), salt water intrusion, heavy metal ions (from industrial processes), and complex synthetic organic compounds (from cleaning fluids and pesticides; Patrick et al., 1987). As analysis of all contaminants is not economically feasible, probable contaminants or contaminants of particular concern for a given groundwater system should be targeted. A preliminary step that can reduce sampling analysis cost is to first conduct a screening analytical procedure, followed by more precise chemical measurements as warranted.

6.2.2 Contaminant Evaluation

Contaminant evaluation refers to the characterization of a known groundwater contaminant. This includes identifying the source, the three-dimensional extent and concentration of contamination, transport properties of the contaminant in the hydrogeologic location, and the number and types of receptors affected or that potentially will be affected. Measurable factors may be integrated into groundwater simulation models to predict potential transport, treatment, and/or attenuation scenarios. As groundwater contamination is a potential health hazard, groundwater contaminant evaluation should be conducted and reported by qualified professionals considering regulatory guidelines.

6.2.3 Resource Assessment

Resource assessment of groundwater includes sampling activities that characterize subsurface hydrological characteristics, such as aquifer boundaries, groundwater flow rate and direction, and aquifer interconnections. In addition, the parameters obtained for groundwater characteristics can be used to evaluate the relative risk related to documented or potential releases of contaminants (Thornton et al., 1997).

6.2.4 Remediation Evaluation

Remediation of groundwater refers to the implementation of some method (e.g., excavation, hydraulic control, bioremediation) to remove a contaminant or to limit its impact. Groundwater sampling for remediation evaluation provides information identifying the optimal remediation approach, and the consequence and effectiveness when such an approach has been implemented in the impacted area.

6.3 GENERAL CONSIDERATIONS IN GROUNDWATER SAMPLING

There are a large number of factors that can affect the collection of representative and reproducible samples that can lead to valid data. A detailed account and examples of groundwater sampling guidelines can be found in the U.S. Geological Survey (USGS) *National Field Manual for the Collection of Water-Quality Data* (USGS, 2006). Some of the most important are highlighted in subsequent sections.

6.3.1 Hydrological Characteristics of the Investigated Area

It is critical to understand the hydrologic system to properly design and install groundwater sampling and monitoring wells. Full understanding of the hydrologic characteristics (depicted in Figure 6.1) of the area of interest enables investigators to collect the most representative groundwater samples and to appropriately interpret the data. Factors that influence water movement underground, such as the location and size of groundwater pumps, areas of any artificial recharge, drainage and receptors, and water level fluctuations in the adjacent surface water systems, need to be identified. Figure 6.2 illustrates the groundwater movement with a horizontal flow and the "stagnant" water formation that may affect the water quality monitoring substantially. Often the hydrology of a location changes depending on the season; therefore, seasonal considerations should be included when assessing hydrologic characteristics.

6.3.2 Current and Historical Land Uses and Management

Land use has a significant influence on groundwater quality and quantity as modifications from natural land use conditions alter hydrologic processes (e.g., evapotranspiration, surface runoff) and contaminant sources. While groundwater quantity and quality in a region are largely determined by both natural characteristics (e.g., lithology, groundwater velocity, quantity and quality of recharge water, interaction between surface water and the aquifer) and human activities, anthropogenic land use activities

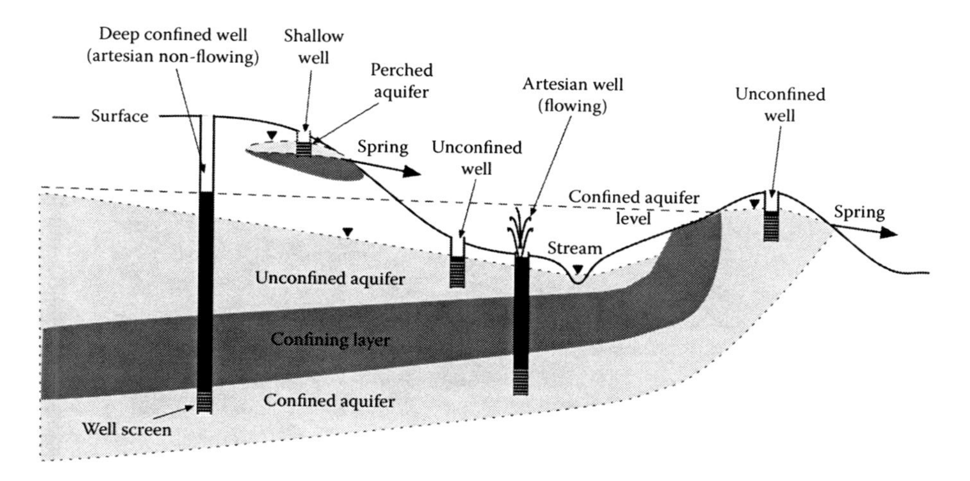

FIGURE 6.1 Illustration of aquifer types and wells. (Modified from Brassington, R. 1998. *Field Hydrogeology*. John Wiley & Sons, Chichester [England]).

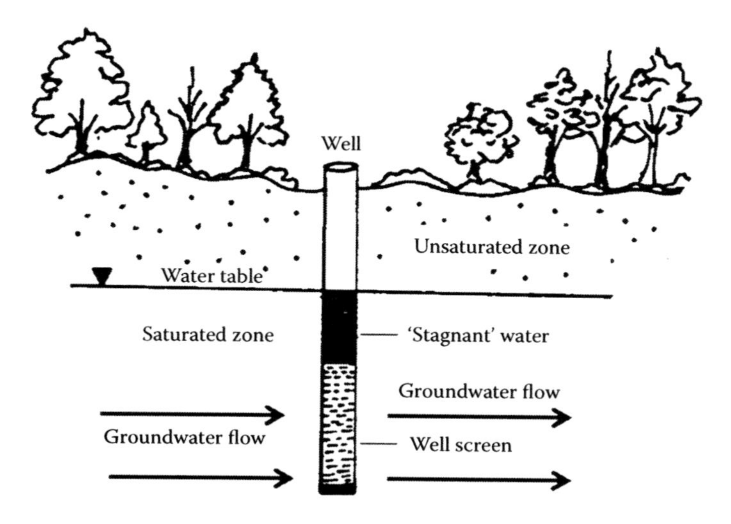

FIGURE 6.2 Groundwater movement through the well screen with a horizontal flow dominated and the stagnant water isolated.

are increasing, resulting in greater degradation of water resources (Helena et al., 2000). Leading sources of groundwater pollution as a result of anthropogenic activity are underground storage tanks, landfills, waste disposal injection wells, septic systems, agricultural wastes, and mining (Bedient et al., 1999). These and other sources of groundwater pollution have been investigated by many researchers with results linking groundwater quality degradation to land use activities (e.g., Pionke and Urban, 1985; Spalding and Exner, 1993; Eckhardt and Stackelberg, 1995; Kolpin, 1997).

One contaminant worthy of note that is linked to a particular land use is NO_3-N. Elevated NO_3 concentrations (greater than the USEPA standard of 10 mg L^{-1}) in water that is consumed can result in methemoglobinemia or "blue baby" syndrome in infants. A dominant source of NO_3–N in U.S. groundwater is from agricultural land use, particularly irrigated agriculture, and thus a direct land use groundwater quality correlation occurs (Power and Schepers, 1989). Many other researchers (i.e., Strebel et al., 1989; Ryker and Jones, 1995; Canter, 1996; Zhang et al., 1996) have reported similar conclusions as NO_3–N is a primary component of fertilizers and is easily transported by water and leached into groundwater.

While current land use is easy to quantify, historic land use is also needed to properly interpret groundwater quality data. This is due to the residence time of some pollutants that are stored or temporarily bound in the soil and released over time into groundwater. In addition, historic storage and wash areas, old septic systems, and other remnants of past land use may influence current groundwater quality. Therefore, knowledge of the current and historic land use should be quantified to ensure that sampling sites are selected and data are analyzed to meet the groundwater sampling objective.

6.3.3 Scale and Duration of the Project

The spatial and temporal extents of the groundwater quality monitoring program depend upon the objectives of the program. The questions to be answered by the program generally dictate the list of constituents, detection limits, scale, and duration of the program. Sampling duration may be influenced by many factors, the most likely being resource availability for the project and meeting project objectives. Knowledge or consideration of seasonal and annual variations in groundwater hydrology and quality is essential to estimating the required duration for sampling groundwater so that collected data will meet the project objective. Groundwater sampling features (such as wells) should be designed, installed, and protected with respect to the anticipated project duration to maintain consistence in data collection.

6.3.4 Stability and Potential Transfer in Land Use

Groundwater sampling sites are specific locations as most groundwater sampling is conducted using an installed groundwater monitoring well. Thus, to maintain sampling consistency and ensure data quality, sites should be selected that are expected to not change substantially during the duration of the project (unless the "change" is itself the monitoring interest). Changes, such as dramatic land use change and construction, could impact sampling results and prevent project objective obtainment. Up-gradient activities may also impact sampling and should be considered when selecting a site.

6.3.5 Representativeness of Groundwater Samples

Environmental factors influence groundwater and thus influence samples collected from groundwater wells. Environmental factors of particular concern, such as hydraulic variation, geochemical changes, and water chemistry, should be considered when conducting sampling programs and are further discussed in subsequent sections.

6.3.5.1 Factors Related to Well Hydraulics and Sampling Point Placement

Installation of a groundwater monitoring well provides a means to collect a groundwater sample for analysis, but it also creates a new feature in the system. The hydraulic characteristics within and surrounding this feature need to be assessed and considered to collect the most representative water quality sample. In particular, some understanding of groundwater directional movement should be quantified (Nielson and Nielson, 2007).

6.3.5.2 Factors Related to Geochemical Changes

Even under ideal conditions, water samples may be influenced by the sampling process due to the inherent nature of groundwater sampling. Factors that might influence geochemical changes include temperature changes, pressure changes, particulate matter, and aeration. Temperature and pressure changes occur when a groundwater sample is collected from a rather stable underground environment and moved to the ground surface where it is exposed to atmospheric conditions. Temperature and pressure differences can result in collected water samples that overestimate or underestimate true constituent values in groundwater (Nielson and Nielson, 2007). Particulate matter is another concern when sampling as sampling equipment may disturb settled materials and introduce particulates into a sample uncharacteristic of actual groundwater. Lastly, aeration is a factor and may occur within wells due to sampling device operation or on the ground surface as a sample is being collected.

6.3.6 Developing a Project-Specific Standard Operating Procedure (SOP)

Developing a project-specific standard operation procedure (SOP) is required prior to the groundwater quality monitoring. The SOP is a guideline that must be standardized, detailed, instructive, and practical (see Chapter 4). Under the guideline created by the Environmental Protection Agency (EPA), most states, if not all, have their own general SOP based on their specific requirements. A good example from the Florida Department of Environmental Protection (FDEP) is the FS 2200 Groundwater Sampling document (FDEP, 2008).

6.4 DESIGN AND INSTALLATION OF GROUNDWATER WELLS

Ideal groundwater sampling sites are located in the landscape of interest, a distance from surface water features and pump stations, easily assessable, and unlikely to experience substantial alteration or land use change. A typical groundwater monitoring well (Figure 6.3) consists of a 5-cm (2-inch) diameter well casing, well screen, bottom cap, filter pack, annular seal, and surface grout and protection (Figure 6.3 and 6.4a). The monitoring well should also include a lid or cap (Figure 6.3 and 6.4e–d) with a lock. A clear sign to easily locate the well is recommended, even if accurate geographic coordinates for the site are known. This is especially important in agricultural land use areas since agronomical operations may alter the land appearance considerably. Monitoring wells are constructed by drilling with a powered rig (hollow stem auger) to the desired depth (Figure 6.4b). The operator must be licensed

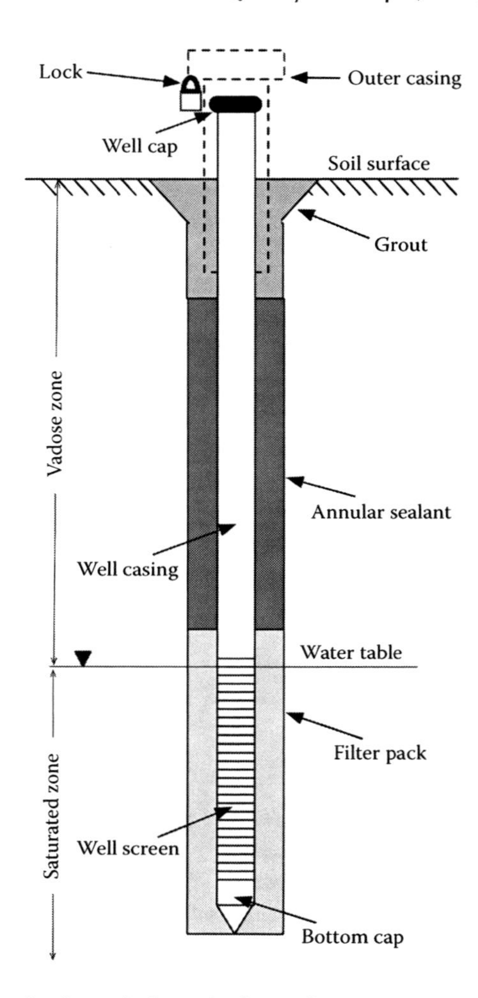

FIGURE 6.3 Schematic of a typical monitoring well.

according to local regulations. The use of a mud rotary drill should be avoided since it has the most potential for contaminating the well during drilling. The rig should be steam or pressure-washed between wells to avoid cross contamination. A person with hydrogeological field expertise must be present during the operation to document the materials encountered during the operation, record the water table depth, and collect soil samples. The filter pack material is selected based on the cumulative size distribution of the soil samples collected at the depth where the well screen will be placed. Typically, the filter pack material is selected to match the D70 (70 percentile of particle size cumulative distribution) of the soil sample. The well screen slot spacing is determined so that it retains between 85–100% of the filter pack material (Wilson, 1995). The well casing, screen, bottom, and top caps must be factory-clean and carried to the field in protective sleeves to avoid contamination of the well with foreign material during transport (Figure 6.4c). Bentonite chips are used as a common

FIGURE 6.4 Details of monitoring well construction: (a) Well materials including (top to bottom) solid and screened pipes, manhole, environmental plug, and bottom cap; (b) hollow stem auger drill rig; (c) well materials taken to the field, cleaned, and protected in plastic sleeves to avoid contamination; (d) uniform sand filter pack filling by vibrating the tube while pouring; (e) bentonite annular sealing with manhole set for grout finishing; and (f) finished well with neat cement grouting.

annular sealant, and cement as a grout (Figure 6.4e–f). The filter pack, bentonite, and cement must be filled from the bottom up using a tremie, or in the case of shallow wells with PVC pipes, by vibrating the well casing while slowly pouring the material to ensure a uniform packing without gaps (Figure 6.4d). After the wells are installed and the concrete is dry—typically 72 hours after pouring it—the monitoring well must be developed by overpumping. This procedure will clean the filter pack material from residues left during the well installation and ensure that there is a consistent connection between the well and the surrounding aquifer. A slug test must be performed to measure hydraulic conductivity if there are doubts about the condition of the well after installation. All monitoring wells must be surveyed and permanently labeled for easy identification. The installation procedure, well schematic with appropriate depths and materials, field notes, results from the samples taken during construction, surveying results, and identifiers must be logged and filed as part of the SOP and with regulatory agencies for review and approval as needed.

6.5 GROUNDWATER SAMPLING PRACTICE

6.5.1 Equipment and Supplies for Groundwater Sampling

Various sampling equipment and supplies (e.g., groundwater pumps, bailers, tubing, and sample containers) are needed to collect representative groundwater samples. Groundwater pumps are needed for both well purging and groundwater sampling. Sample containers are used to transport the sample from the field to laboratory with no loss of analyte. The type of sampler and sampling equipment selected depends on the type and location of well, depth to water from land surface, physical characteristics of the well, groundwater chemistry, and the target analytes. The different types of sampling equipment and supplies are discussed in subsequent sections.

6.5.1.1 Suction-Lift Pumps

Suction-lift pumps (e.g., peristaltic, centrifugal) are commonly used for groundwater sampling. They work by applying a vacuum to either the well casing or sample tubing that has been lowered into a well to the desired sampling depth. A variable speed peristaltic pump is typically used for shallow groundwater sampling, especially in wells that produce a small amount of water. Peristaltic pumps have the advantage over other pumps due to their few moving parts, easily replaceable tubing and heads, and portability. Centrifugal pumps can be used if the static water level in the well is less than 6 to 8 m below the land surface. Centrifugal pumps should be used for well development because they can pump a large quantity.

6.5.1.2 Submersible Pumps

Submersible pumps (positive pressure or other types of positive-displacement pumps) are becoming more common as costs decrease and portability increases. Submersible pumps are often preferred when the target analytes could be lost due to decrease in pressure caused by a vacuum (e.g., dissolved gasses, VOCs). Considerations when using submersible pumps are to:

- Place an antibacksiphon device above the pump head to prevent backflow from contaminating the groundwater.
- Avoid using submersible pumps in shallow wells (<1.5 m), as this could lower the water level below the pump head and dry out the motor.
- Do not use submersible pumps for well development, as this can ruin the pump or shorten its functional life.
- Do not use submersible pumps in wells deeper than 30.5 m.

6.5.1.3 Bailers

The use of bailers for groundwater sampling is not recommended unless other equipment is unavailable and unsuitable due to several reasons:

- May introduce atmospheric oxygen that precipitates metals, such as iron, or cause other chemical changes in the water sample, e.g., pH
- May agitate groundwater, which causes biases for volatile and semi-volatile organic analyses because of volatilization
- May result in fine particulates being resuspended in the aquifer due to the possible agitation
- May surge the well and loosen particulate matters in the annular space around the well screen
- May introduce dirt into the water column if it scrapes the sides of the casing wall

However, in practice, bailers sometimes are still used for a wide variety of analytes if certain requirements are met. Allowable use may include sampling for specific analytes where concentrations exceed regulatory action levels, when the purpose is to monitor effective treatment, and the regulatory program allows the use of bailers (see, for example, FDEP, 2008, Table FS2200-3). Also, bailers might be the only option available for sampling some deep groundwater systems. Bailers must be constructed of inert materials, free of analytes of interest. Stainless steel, Teflon, polyethylene (PE), and polypropylene (PP) bailers may be used to sample all analytes for the allowable cases. The use of disposable bailers is recommended when budget allows, especially in grossly contaminated sites. Lanyards used to lower and raise the bailer must be made of nonreactive, nonleachable material such as cotton twine, nylon, or stainless steel, or coated with Teflon, PE, or PP. Lanyards made of materials other than stainless steel, Teflon, or PE must be discarded after each monitoring well. When inserting and raising the bailer, it must be done slowly (maximum rate of 2 cm s^{-1} in the top of the column) to avoid disturbance in the water column and contamination with the well casing at different depths.

6.5.1.4 Containers

Containers used in groundwater sampling vary by analytes (Table 6.1). The sampling containers should be well washed, rinsed with distilled or deionized water, and dried before use. Recycling used bottles is not recommended. If sample bottles are reused, additional cleaning must be conducted that includes soaking in 3% reagent grade HCl at least 12 hrs.

6.5.1.5 Tubing and Other Materials

Tubing and materials for pump configurations are required and vary with analyte groups. Teflon, PE, or PP tubing is appropriate in most circumstances. Stainless steel or glass equipment is appropriate in most circumstances. For specific details, please refer to appropriate tubing selection and pump materials proposed by the U.S.

TABLE 6.1
Recommendations for Preservation, Holding Time, and Minimum Sampling Size of Water Samples

Water Parameters	Container[a]	Preservation	Holding Time	
			SM 21st ed[b]	USEPA Rule[c]
Alkalinity	P, FP, G	Cool, ≤6°C	24 hours	14 days
Ammonia	P, FP, G	Cool, ≤6°C H_2SO_4 to pH<2	7 days	28 days
Biochemical oxygen demand (BOD)	P, FP, G	Cool, ≤6°C	6 hours	48 hours
Boron	P, FP, or Quartz	HNO_3 to pH <2	28 days	6 months
Bromide	P, FP, G	None required	28 days	28 days
Chemical oxygen demand (COD)	P, FP, G	Cool, ≤6°C H_2SO_4 to pH <2	7 days	28 days
Chloride	P, FP, G	None required		28 days
Coliforms, total, fecal and *E. coli*	PA, G	Cool, ≤10°C 0.0008% $Na_2S_2O_3$		6 hours
Color	P, FP, G	Cool, ≤6°C	48 hours	48 hours
Fluoride	P	None required	28 days	28 days
Hardness	P, FP, G	HNO_3 or H_2SO_4 to pH <2	6 months	6 months
Hydrogen ion (pH)	P, FP, G	None required	15 minutes	15 minutes
Kjeldahl and organic N	P, FP, G	Cool, ≤6°C H_2SO_4 to pH<2	7 days	28 days
Metals, general	P, FP, G	HNO_3 to pH<2	6 months	6 months
Chromium VI	P, FP, G	Cool, ≤6°C pH = 9.3–9.7	24 hours	28 days
Mercury (CVAA)	P, FP, G	HNO_3 to pH<2	28 days	28 days
Mercury (CVAFS)	P, G, FP-lined cap	5 ml/*L* 12N HCl or 5 ml/*L* BrCl		90 days
Nitrate	P,FP, G	Cool, ≤6°C	48 hours	48 hours

Nitrite	P,FP, G	Cool, ≤6°C		48 hours
Nitrate–nitrite	P,FP, G	Cool, ≤6°C H_2SO_4 to pH <2	1–2 days	48 hours
Organic carbon	P,FP, G	Cool, ≤6°C HCl, H_2SO_4, or H_3PO_4 to pH <2		28 days
Orthophosphate	P,FP, G	Cool, ≤6°C	48 hours	Filter within 15 minutes & analyze within 48 hours
PCBs	G, FP-lined cap	Cool, ≤6°C	7 days	7 days before extraction; 1 year after extraction
Pesticides	G, FP-lined cap	Cool, ≤6°C pH 5–9	7 days	7 days until extraction; 40 days after extraction
Phosphorus, total	P,FP, G	Cool, ≤6°C H_2SO_4 to pH <2	28 days	28 days
Specific conductance	P,FP, G	Cool, ≤6°C	28 days	28 days
Sulfate	P,FP, G	Cool, ≤6°C	28 days	28 days
Sulfide	P,FP, G	Cool, ≤6°C Zinc acetate plus sodium hydroxide to pH >9	28 days	7 days
Surfactants	P,FP, G	Cool, ≤6°C		48 hours
Temperature		None required	15 minutes	15 minutes
Turbidity	P, FP, G	Cool, ≤6°C	24 hours	48 hours

[a] P—polyethylene; FP—fluoropolymer; G—glass; PA—any plastic that is made of a sterilizable material; LDPE—low density polyethylene.

[b] Based on USEPA (U.S. Environmental Protection Agency). 2007. Guidelines Establishing Test Procedures for the Analysis of Pollutants Under the Clean Water Act; National Primary Drinking Water Regulations; and National Secondary Drinking Water Regulations; Analysis and Sampling Procedures, Final Rule. Washington D.C.: Environmental Protection Agency.

[c] Based on Eaton, A.D., L.S. Clesceri, E.W. Rice, A.E. Greenberg, and M.A.H. Franson. 2005. *Standard Methods for the Examination of Water and Wastewater*. Washington, D.C.: American Public Health Association.

Geological Survey (USGS, 2003; see http://water.usgs.gov/owq/FieldManual/ as an example).

6.5.2 Purging

Purging is an important procedure for stagnant water removal and aquifer refill of the well (Figure 6.2) for obtaining representative groundwater samples. To properly purge a well, the amount of water in the well and a clear criterion for determining when the well has been refilled with fresh water from the aquifer must be known. A general summary for such procedures on groundwater sampling follows.

6.5.2.1 Initial Inspection

On site arrival, verify the identification of the monitoring well by looking at the markings, sign, or other designations when approaching the site; remove the well cover and all standing water around the top of the well before opening the well cap; inspect the exterior protective casing of the monitoring well for any damage and document it if it occurs; place a protective covering around the well head; and inspect the well lock to check if the cap fits tightly.

6.5.2.2 Water Level Measurements

Water level measurements should be taken to determine the total water volume of the well for purging purposes. Depth to water can be measured using an electronic probe or a weighted steel tape measure. Never lower the probe to the bottom of the well before purging and sampling since this could disturb the well water. The length of the water column can be determined by subtracting the depth of water from the total depth of the well. Measuring instruments should not be constructed of any material (e.g., lead) that could potentially contaminate a well and should be cleaned after use to prevent cross contamination between groundwater wells.

6.5.2.3 Calculating the Well Water Volume and Purging Equipment Volume

The total volume of water in the well can be calculated with the well diameter and the water depth:

$$V_w = 3.14 \times W_h \times (W_d/200)^2,$$

where V_w = total volume of water in the well in cubic meter, 3.14 = coefficient for area calculation, W_h = height of the water column in meter, and W_d = well diameter in centimeters.

A pumping rate can be calculated with a graduated container to measure the volume of water pumped out during a specified time period, and the total amount of time needed to purge each well volume can be estimated with total well water volume divided by the constant pumping rate. Alternatively, a totalizing flow meter can be connected to the pump tubing for obtaining pumped water volume. In this case, the reading on the flow meter prior to purging needs to be recorded. The USGS recommends removing three or more well volumes during the purging process (USGS, 2006).

TABLE 6.2
Water Stabilization Parameters Required for Complete Purging

Parameters	Criteria
Temperature	± 0.2°C (thermistor thermometer)
pH	± 0.1 standard units
Specific conductance	± 5.0% for ≤ 100 μS cm^{-1}
	± 3.0% for > 100 μS cm^{-1}
Dissolved oxygen	± 0.3 mg L^{-1}
Turbidity	± 10%, for turbidity < 100 TBY

Source: USGS, 2006. National Field Manual for the Collection of Water-Quality Data, Chapter A4. Collection of Water Samples. *Handbooks for Water-Resources Investigations*. V2. Reston, VA.

6.5.2.4 Well Purging

Well purging removes stagnant water from the well and conditions the sampling equipment with well water. Bailers are not recommended for the reasons presented in Section 6.5.1.2 and instead pumps must be used. The volume of water that is removed from the well during purging must be measured with a container or a flow-through cell. Well water recovery rate is the limiting factor to refill the well with fresh water from the aquifer for sampling. The recovery rate mainly depends on soil permeability and aquifer level. Wells with slow recovery rates (yields less than 100 mL/min) are not recommended for groundwater sampling (USGS, 2006). Measurement of water stabilization parameters can indicate whether or not the well water has recovered. Use a flow-through container or slightly insert certain measurement probes into the well water directly at the appropriate depth to measure the water stabilization parameters while purging. The rate of pumping should remain constant during purging. Purging is considered complete if five or more consecutive water measurements meet the criteria in Table 6.2. When purging a well, water is removed from the monitoring well and dispelled above ground. Care should be taken to ensure that the water purged does not flow back into the well. For most locations, the water may be pumped onto the ground surface a few meters from the groundwater monitoring well. However, field technicians should determine the appropriateness of this as some soils are very shallow and have high infiltration rates that could result in rapid infiltration and thus sample contamination. If groundwater is known or suspected to be contaminated, the purged water should be contained and disposed of properly.

6.5.3 Sampling Procedure

6.5.3.1 Sample Collection with Pump

After purging the well of stagnant water, a groundwater sample may be collected. The pumping rate should be the same for sample collection as it was for purging. Fluctuations in pumping rate affect sample quality (Gibs et al., 2000). If the sample is not collected within one hour of purging completion, the five water stabilization parameters (Table 6.2) need to be remeasured prior to sample collection. Additional purging is needed if the measured values differ by more than 10% of the initial stabilized measurements. To reduce the possibility of atmospheric contaminants, it is recommended that samples be collected in a sample chamber. A good example of how to construct a sample chamber can be found in the USGS National Field Manual (USGS, 2003). When collecting the sample particular attention must be given to the following:

1. Collect the sample directly into the sampling container and do not use intermediate containers to reduce possible contamination. In addition, to avoid contamination, handle the sampling equipment as little as possible, minimize the equipment that is exposed to the sample, minimize aeration of samples collected for volatile organic carbon (VOC), and reduce the sampling pump flow rates to ≤ 100 ml per minute when collecting VOC samples. To minimize air in the sample collection container, bottom-fill the vial and then slowly withdraw the sample tubing; overfill the vial, leaving a convex meniscus, and place the cap on so that no air bubbles are visible. Slowly turn the sample vial upside down to inspect for bubble formation. If there are bubbles present, discard vial and repeat with new one. Do not pour out the sample and try to resample with the same vial, as this could bias the results.
2. For inorganic constituent sampling, collect samples from the effluent tubing connected to the pump. Generally the same pump is used to collect samples from multiple wells. All materials that come in contact with the groundwater sample (usually this is the tubing and pump head) must be changed or decontaminated between sampling sites.

6.5.3.2 Sample Collection with Bailers

Bailers may be used instead of a pump for sampling groundwater. However, as previously stated, specific approval from regulatory or governing authority is required to use bailers.

1. The bailer should be handled as little as possible to reduce possible contamination. When using a bailer, wear sampling gloves, remove the bailer from its protective wrapping just before use, attach a lanyard of appropriate materials to move and position the bailer, and do not allow the bailer or lanyard to touch the ground or other surface that may contaminate the bailer.
2. The bailer should be rinsed before a sample is collected. If both a pump and a bailer are to be used to collect samples, collect pump sample first and rinse the exterior and interior of the bailer with sample water from the pump before removing the pump. Discard the excess water from the bailer appropriately.
3. Raise and lower the bailer gently with minimal contact with the well casing to minimize resuspension of particulate matter in the well and the water column, which can change the sample turbidity.
4. The depth to lower the bailer should not exceed the depth of the water column to guarantee that it does not touch the well bottom.
5. The bailer should be raised and lowered carefully and slowly to approximately reach the same depth each time.

6. A device to control the flow from the bottom of the bailer should be used, and be sure to take it off prior to lowering it in the water column. Once a bailer is filled with water, discard the first few inches of water before collecting the groundwater sample. To fill a sample container, release water from the bailer and allow the sample to slowly flow down the inside of the container.
7. Once samples have been collected, remeasure the DO, pH, temperature, turbidity, and specific conductivity.

6.5.3.3 Quality Control (QC) of Groundwater Sampling

For each sampling event, QC samples (e.g., equipment blanks, field blanks) must be collected to ensure or verify the sampling results throughout the entire procedure of water sampling, handling, and pretreatment for chemical analyses. QC sampling requirements are generally outlined by the funding agency for which the work is being performed and/or the regulatory authority. All involved in the sampling process should be familiar with the SOP that outlines QC requirements for a particular project. SOPs are discussed in detail in Chapter 4 and with surface water examples provided in Chapter 14.

6.5.3.4 Order in Sample Collection

Often multiple constituents are to be analyzed requiring the collecting of multiple groundwater samples from one site. The order for collecting the samples should be outlined in the project SOP. Field blank samples can be collected right before the actual sample is collected. The equipment blank samples can be collected after all samples have been collected and all equipment and tubing have been washed and rinsed with distilled or deionized water (see Chapter 4 for more information on sample blanks). Based on USGS recommendations, actual samples most sensitive to handling should be collected first, followed by those less sensitive to handling (USGS, 2004) as in the following order:

1. Organic compounds—raw (whole water or unfiltered) samples first, followed by filtered samples
2. Volatile organic compounds (VOCs)
3. Pesticides, herbicides, polychlorinated biphenyls (PCBs), and other agricultural and industrial organic compounds.
4. Organic carbon (TOC)
5. Inorganic constituents, nutrients, radiochemicals, isotopes: filtered samples first, followed by raw samples
   a. Trace and major element cations
   b. Separate-treatment constituents (e.g., mercury, arsenic, selenium)
   c. Nutrients, major anions, and alkalinity
   d. Radiochemicals and isotopes. Samples for additional field parameter measurement (independent of purging data)

6.5.3.5 Additional Considerations for Groundwater Sampling

The field team should always keep in mind that once delivered to the ground surface, the groundwater samples are exposed to atmospheric conditions, which may alter the physical and chemical characteristics significantly (e.g., temperature, pressure, redox status). In turn, these changes may result in chemical and biological reactions, for instance, oxidation of organic and inorganic substances, such as nitrate, sulfide, and dissolved metals (Stumm and Morgan, 1996). Thus, the contact time of the samples to air should be minimized. A few sampling guidelines related to groundwater sample collection follow:

1. Caps should be kept on sample containers until the moment they are ready to be filled, and recap these containers immediately upon being filled.
2. Samples should be collected directly from the discharge tubing. Funnels or other intermediate vessels are not allowed to prevent introducing potential errors and/or bias due to possible cross-contamination.
3. Avoid contact of the samples and containers with ground surface, sampling equipment, and human skin (appropriate gloves should be worn).
4. Special protocols must be followed for collecting VOC samples. Forty ml vials with zero-headspace associated with flow rate control at approximately 200 to 250 ml per min are recommended for sample collection. Once the VOC samples have been collected, store them upside down to detect formation of bubbles during handling.
5. To collect a zero-headspace sample, the groundwater is collected directly from the pump discharge tubing or the grab sampling device. The container should be held in an angle to let the water gently flow down the inside container along the wall. Once the container is filling, slowly straighten the container to vertical to fill it up until positive meniscus forms on the top of the vial. Avoid overfill if the container has been preacidified for preservation because overfill will wash out chemical preservatives contained in the vial. Cap the vial hand-tight immediately without disturbing the meniscus.

REFERENCES

Bedient, P.B., H.A. Rifai, and C.J. Newell. 1999. *Ground Water Contamination Transport and Remediation*. 2nd ed. Prentice Hall PTR: Upper Saddle River, NJ.

Brassington, R. 1998. *Field Hydrogeology*. John Wiley & Sons, Chichester (England).

Canter, L.W. 1996. *Nitrates in Groundwater*. CRC Press: Boca Raton, FL.

Eaton, A.D., L.S. Clesceri, E.W. Rice, A.E. Greenberg, and M.A.H. Franson. 2005. *Standard Methods for the Examination of Water and Wastewater*. Washington, D.C.: American Public Health Association.

Eckhardt, D.A.V., and P.E. Stackelberg. 1995. Relation of ground-water quality to land use on Long Island, New York. *Ground Water* 33(6):1019–1033.

FDEP (Florida Department of Environmental Protection). 2008. FS 2200 Groundwater Sampling. http://www.dep.state.fl.us/labs/qa/sops.htm.

Gibs, J., Z. Szabo, T. Ivahnenko, and F.D. Wilde. 2000. Change in field turbidity and trace-element concentrations during well purging. *Ground Water*, v. 38, no. 4, pp. 577–488.

Helena, B., R. Pardo, M. Vega, E. Barrado, J.M. Fernandez, and L. Fernandez. 2000. Temporal evolution of groundwater composition in an alluvial aquifer (Pisuerga river, Spain) by principal component analysis. *Water Res.* 34:807–816.

Kolpin, D.W. 1997. Agricultural chemicals in groundwater of the Midwestern United States: Relations to land use. *J. Environ. Quality* 26:1025–1037.

Nielsen, D.M., and G.L. Nielsen. 2007. *The Essential Handbook of Ground-Water Sampling*. Boca Raton: CRC Press.

Office of Technology Assessment (OTA). 1984. Protecting the Nation's Groundwater from Contamination: Volume II. Washington, DC: US Congress, OTA-O-276.

Patrick, R., E. Ford, and J. Quarles. 1987. *Groundwater Contamination in the United States*. 2nd ed. University of Pennsylvania Press: Philadelphia, PA.

Pionke, H.B., and J.B. Urban. 1985. Effect of agricultural land use on groundwater quality in a small Pennsylvania watershed. *Ground Water* 23:68–80.

Power, J.F., and J.S. Schepers. 1989. Nitrate contamination of groundwater in North America. *Agriculture, Ecosystems and Environment* 26:165–187.

Puls, R.W., and M.J. Barcelona. 1996. Low flow (minimal-drawdown) groundwater sampling procedures. EPA/540/5-95/504, Washington, D.C.: Groundwater Issue. USEPA.

Reilly, T.E., K.F. Dennehy, W.M. Alley, and W.L. Cunningham. 2008. Ground-water availability in the United States. Circular No. 1323 p 70, US Geological Survey.

Ryker, S.J., and J.L. Jones.1995. Nitrate Concentrations in Ground Water of the Central Columbia Plateau USGS Open-File Report 95-445. http://wa.water.usgs.gov/pubs/ofr/ofr95-445/.

Spalding, R.F., and M.E. Exner. 1993. Occurrence of nitrate in groundwater—a review. *J. Environ. Quality* 22:392–402.

Strebel, O., W.H.M. Duynisveld, and J. Bottcher. 1989. Nitrate pollution of groundwater in western Europe. *Agric. Ecosyst. Environ*. 26(3–4):189–214.

Stumm, W., and J.J. Morgan. 1996. *Aquatic Chemistry: Chemical Equilibria and Rates in Natural Waters*. New York: John Wiley & Sons.

Thornton, D., S. Ita, and K. Larsen. 1997. Broader use of innovative groundwater access technologies. In *Proceedings of the Hazwaste World Superfund XVIII Conference*, Vol. 2. 639–646. Washington, D.C.: E.J. Krause Co.

USEPA (U.S. Environmental Protection Agency). 2007. *Guidelines establishing test procedures for the analysis of pollutants under the Clean Water Act; national primary drinking water regulations; and national secondary drinking water regulations; analysis and sampling procedures, final rule*. Washington D.C.: Environmental Protection Agency.

USGS. 2003. National Field Manual for the Collection of Water-Quality Data, Chapter A2. Selection of Equipment for Water Sampling. Handbooks for Water-Resources Investigations. V2. Reston, VA.

USGS. 2004. National Field Manual for the Collection of Water-Quality Data, Chapter A5. Processing of Water Samples. Handbooks for Water-Resources Investigations. V2.1. Reston, VA.

USGS. 2006. National Field Manual for the Collection of Water-Quality Data, Chapter A4. Collection of Water Samples. Handbooks for Water-Resources Investigations. V2. Reston, VA.

Wilson, N. 1995. *Soil Water and Ground Water Sampling*. Boca Raton: Taylor & Francis.

Zhang, W.L., Z.X. Tian, N. Zhang, and X.Q. Li. 1996. Nitrate pollution of groundwater in northern China. *Agric. Ecosyst. Environ*. 59(3):223–231.

7 Sampling Pore Water from Soil and Sediment

Yuncong Li, Kati W. Migliaccio, Meifang Zhou, and Nicholas Kiggundu

CONTENTS

7.1 INTRODUCTION

Because pore water fills the spaces between soil or sediment particles, sampling this water is a useful tool for water quality monitoring, management, and research. Dissolved contaminants move through pores to groundwater and sometimes surface waters. Soils/sediments have the ability to alter water chemistry through chemical (adsorption, precipitation, etc.), physical (filtration, retention, etc.), and biological (degradation, etc.) processes. Some contaminants may never reach groundwater because of their interdiction by these processes, while others that do not interact are often found in groundwater. Analysis of pore water provides information on contaminant fate and can provide early warning for contaminates that may potentially reach groundwater and contaminate the aquifer. Pore water sampling can be of use for almost all common pollutants as shown by Wilson and Artiola (2004) who listed 15 possible pollution sources divided into three classes: industrial, municipal, and agricultural. Industrial pollution sources include surface impoundments, landfills, land treatments sites, underground storage tanks, mine drainage, and waste piles, while municipal sources comprise sanitary landfills, septic systems, oxidation ponds, artificial recharge facilities, wetlands, and urban runoff drainage wells. Pore water sampling can be used to monitor pollution from agricultural sources such as fertilizer bioavailability and leachability in and below the root zone, pesticide mixing sites, and in-field retention ponds. Sampling pore water, often referred to as interstitial water, is also widely used for sediments in wetlands, lakes, and oceans.

Unlike in ground- and surface water monitoring, federal and state governments provide no or few, if any, standard operating procedures for pore water sampling. Consequently, a variety of methods have been used and reported. This chapter will present methods commonly used for soil and sediment pore water sampling.

7.2 DEFINING SAMPLING PROJECT GOALS

The utility of pore water sampling for predicting water quality is largely dependent on the project objectives and the nature of contaminants and soils/sediments. To develop a study plan with clear objectives, key questions must be asked to design the project. Why is pore water sampling needed and what will be accomplished? Who will use the monitoring results and which variables should be measured? What is the budget for the project and who is responsible for sampling? Typical sampling goals for pore water include (1) monitoring the distribution of contaminants throughout the vadose zone without disturbing soil/sediment, (2) serving as an early warning for waste disposal facilities or underground storages, (3) determining bioavailability of nutrients and toxicity of contaminants in the plant root zone, (4) evaluating effects of sediment on surface water above and groundwater below, (5) evaluating the effectiveness of pollution prevention practices such as Best Management Practices (BMPs), and (6) investigating processes of accumulation, decomposition, and preservation of organic matter, cycle of minerals, and growth of microbial communities in soil and rocks. However, pore water sampling, while important, cannot replace standard groundwater sampling. Moreover, well-chosen, scientifically valid protocols must be observed, since no pore water quality standards have been established by federal or state governments.

7.3 SITE SELECTION AND SAMPLE PREPARATION

Sites for pore water sampling should be selected based on project objectives and field conditions. Useful information containing factors that need to be considered for the placement and location of soil/sediment pore water samplers is provided by Wilson

(1995) and Burton and Pitt (2002). In selecting and preparing a pore water sampling site, the following criteria should be considered: (1) sampling sites should include the areas that represent the greatest and least impacts from the pollutant source; (2) a certain minimum of sampling sites and locations may be required to allow for statistical analysis (these should be selected to represent the spatial variations present); (3) thickness of soil/sediment and depth to bedrock should be observed for determining sampling depth; (4) accurate description of soil/sediment properties such as texture, hydraulic conductivity, hardpan, etc.; (5) sampling equipment should be designed to minimize the influence of soil or sediment heterogeneity; (6) plant root interferences should be avoided due to their release and uptake of inorganic and organic nutrients; and (7) oxidation of sediment pore water should be avoided and changes of temperature and pressure should be minimized to avoid a shift in ion exchange, adsorption/desorption, and precipitation/dissolution equilibrium.

7.4 SOIL PORE WATER SAMPLING

This section focuses on in-situ soil pore water sampling that allows repeated measurements at the same exact site. Soil pore water sampling is typically conducted in an agricultural field, at a waste disposal facility, or in other landscape settings where solutes in pore water are of interest. Two primary methods, suction and gravity water collection lysimeters, are used for in situ collection of soil pore water.

7.4.1 Suction Lysimeters

Suction (or tension) lysimeters generally consist of a collection tube with a porous cup attached at one end. Although the porous cup may be composed of different materials, ceramic is the most common. The top end is sealed air-tight by a rubber stopper through which pass two tubes, one for applying suction and the other that extends to the bottom for extracting solution. The lysimeter is inserted into the soil so that the ceramic tip is located adjacent to the area of interest for soil pore water sampling (Figure 7.1), and the collection tube and stopper extends above the soil surface. Suction is applied through a narrow plastic tube inserted through the top cap (or stopper) of the lysimeter. Pore water samples are collected using another narrow plastic tube once suction has extracted water solution through the porous tip into the collection tube. Because of its ease of use and low cost, this type of lysimeter has been historically used to collect soil water samples from both saturated and unsaturated soils and from varying soil depths (Wagner, 1962; Van der Ploeg and Beese, 1977; Litaor, 1988; Swistock et al., 1990). The primary physical limitation of this method is that samples are collected at designated points in time rather than continuously so that results only reflect a series of snapshots in time and not the full range of characteristics over the entire period of interest. Suction lysimeters are also limited by their small area of influence immediately surrounding the ceramic tip (Talsma et al., 1979). In addition, the nature of the soil structure, particularly where there are many macropores or other heterogeneities, may prevent representative soil water sampling during saturated conditions due to bypass flow (Shaffer et al., 1979; Grossmann and Udluft, 1991).

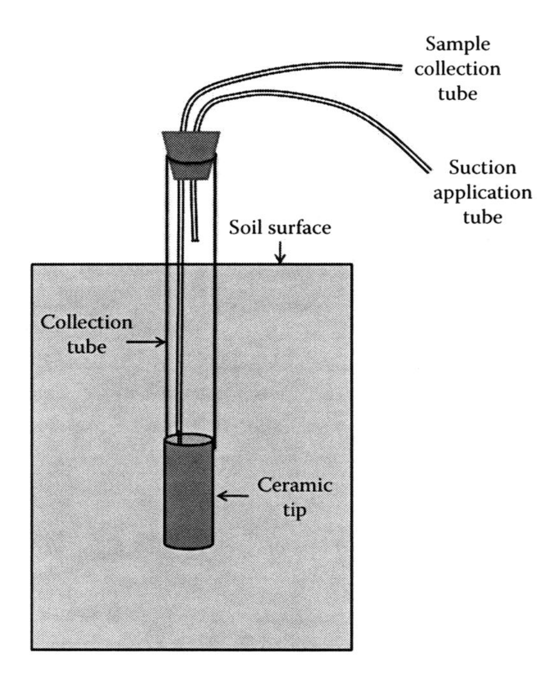

FIGURE 7.1 Schematic of a suction lysimeter.

The chemical makeup (kaolin, talc, alumina, ball clay, and other fieldspathic minerals that are often assumed to be inert) of the ceramic tip can alter the composition of the soil water collected (Van der Ploeg and Beese, 1977; Litaor, 1988; Wilson and Artiola, 2004; Soilmoisture, 2007). While sorption of solute ion(s) is a major problem with ceramic tension lysimeters (Hansen and Harris, 1975; Nagpal, 1982; Grossmann and Udluft, 1991), other chemical reactions such as solute ion adsorption and precipitation can also influence solution composition and pH, sorption capacity of the cup, suction applied, and sampling rate (Hansen and Harris, 1975; Nagpal, 1982; Grossmann and Udluft, 1991). Pore water collected in this way may vary in constituents, depending on extraction time used for sample collection (Van der Ploeg and Beese, 1977; Swistock et al., 1990).

Despite a plethora of reports in the literature regarding the limitations of ceramic cup suction lysimeters, Beier and Hansen (1992), who compared ceramic and polytetrafluoroethene (PTFE) cup lysimeters for sampling leachate that contained sodium (Na), potassium (K), calcium (Ca), aluminum (Al), ammonia (NH_4), and nonpurgeable organic carbon (NPOC), found that neither lysimeter altered the samples nor retained solutes on the cup. Levin and Jackson (1977), who compared micro hollow fiber and ceramic cup lysimeters for sampling leachate containing Ca, magnesium (Mg), and orthophosphate (PO_4–P), found that neither extractor altered leachate chemical composition. Similarly, Haines et al. (1982) observed no differences in PO_4–P in a forest ecosystem measured using ceramic and zero tension lysimeters. Other studies have shown significant phosphorus (P) adsorption on ceramic cups (Hansen and Harris, 1975; Severson and Grigal, 1976; Zimmermann et al., 1978; Nagpal, 1982; Bottcher et al., 1984; Andersen, 1994). With increasing P concentration, Hansen and Harris (1975) showed that up to 110 mg of P could be sorbed by a single ceramic cup in a laboratory experiment, an observation supported by Litaor (1988), who stated that soil water samplers are not suitable for use in studies involving P due to P adsorption on the ceramic cups, but no quantitative values were given. This controversy may stem from the differences in chemical composition/source of the ceramic materials used in the manufacture of the various lysimeters and their interaction with the solution sampled (Hughes and Reynolds, 1990). Continued use of ceramic tension lysimeters is possibly due to economic or physical limitations of other methods for collecting soil pore water (e.g., Brye et al., 2002; Pregitzer et al., 2004; Bajracharya and Homagain, 2006; Lentz, 2006). Although the above discussion has been focused on ceramic lysimeters, other materials (stainless steel and PTFE) have been used as the porous extractive interface (tip) of suction lysimeters, but their low bubbling pressures, 20 to 60 kPa and 7 to 20 kPa, respectively, are a limitation (Wilson and Artiola, 2004) requiring near-saturated conditions for sampling pore water.

7.4.2 Gravity Water Collection Lysimeters

Soil pore water can also be sampled using gravity collection lysimeters or free drainage samplers that rely primarily on natural forces to collect pore water from macropore flow, as, for example, when water supplied (by irrigation or rainfall) exceeds field capacity (Zhu et al., 2002). Because properly maintained gravity water collectors sample accumulated water and therefore leachate for a particular location, a measurement of water volume leached and constituent loads transported to the depth of the lysimeter can be obtained and are often used in research and monitoring applications (e.g., Jemison, 1994; Shipitalo and Edwards, 1994; Brown et al., 1999). As the volume of water collected by gravity lysimeters is influenced by water tension differences between the soil directly above and surrounding the lysimeter (Jemison and Fox, 1992), correction factors may be needed to obtain a true value for real leachate volume (Zhu et al., 2002). Nevertheless, they are still used in research and monitoring programs to assess leachate. Depending on the application, one of three main variations (pan, bucket, or wicking lysimeters) is used.

7.4.2.1 Pan Lysimeter

The most common type of gravity lysimeter is the pan lysimeter (Jordan, 1968; Parizek and Lane, 1970; Essington, 2003). Use of a pan lysimeter to collect leachate requires a fairly complex installation process (Figure 7.2) involving the following steps: (1) excavation of a pit adjacent to the area of interest, (2) lateral excavation of a portion of the soil next to the pit where the pan will be placed so that the soil above the pan remains undisturbed, (3) placement of the pan into the laterally excavated area and top filling with nonreactive porous media so that the fill contacts the minimally disturbed soil above, (4) backfilling around the pan, (5) angling and placement of the pan to ensure that water solution (or leachate) is collected by a plastic tube and transported to a collection device located in the pit, and (6) attaching an air vent line and a sampling line to the collection device to allow for aboveground collection of leachate (after backfilling the pit; Essington, 2003).

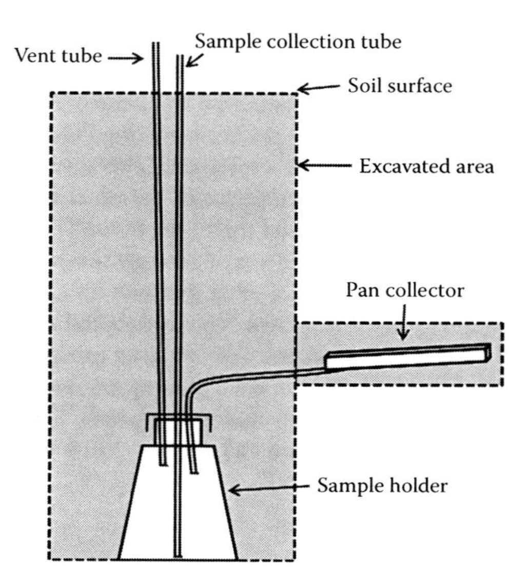

FIGURE 7.2 Schematic of a pan lysimeter.

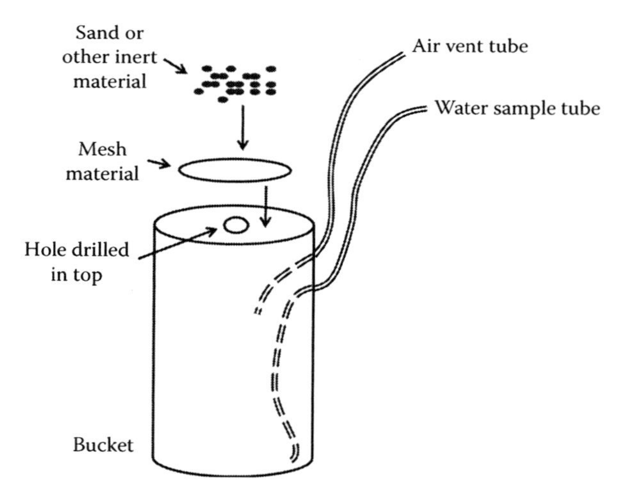

FIGURE 7.3 Schematic of a bucket lysimeter.

7.4.2.2 Bucket Lysimeter

A less-commonly used, but simple, lysimeter method is the bucket lysimeter (Figures 7.3 and 7.4; Migliaccio et al., 2009). Bucket lysimeters are typically buried underneath the soil zone of interest and thus require soil disturbance. Bucket lysimeters are ideal for locations in which soils would be disturbed as part of a management practice or where pan lysimeters are not practical. Bucket lysimeters consist of a collection container topped with a water-permeable surface made of a fine metal mesh or other material overlain with sand that collects gravitational water as it percolates

FIGURE 7.4 Lysimeter with sand placed in collecting plate and flexible tubing attached for sample collection.

through the permeable top. This water is then sampled by means of a tube using a peristaltic pump or similar mechanism. Exact placement of bucket lysimeters should be made after consideration of the impact of soil disturbance on results (considering sampling objectives).

7.4.2.3 Wicking Lysimeter

Wick lysimeters allow for the collection of leachate under nonsaturated conditions (Hornby et al., 1986). The basic principle is the same as for pan and bucket lysimeters except for the incorporation of a self-priming wick that exerts suction on the soil directly above the lysimeter. The advantage of this device is that it overcomes the water suction difference between the pan or bucket and adjacent soil to minimize tension-related bypass flow (Zhu et al., 2002).

7.4.3 Example 1—Soil Pore Water Sampling Using Bucket Lysimeters

Project objective: Evaluating effects of irrigation and fertilizer best management practices (BMPs) in reducing nutrient leaching.

Key question: Will optimized irrigation and fertilizer rates reduce fertilizer leaching below the root zone?

Monitoring: Collecting leachates below the corn root zone and measuring water volume and N and P concentrations in leachate.

Project design: A complete block design included soil suctions for irrigation rates (5 and 15 kPa) and P fertilizer (granular triple super phosphate) rates (0, 50 and 100 kg P ha^{-1}). Each treatment had 4 replicates of 2-row corn in a rectangular plot of 9 m^2 (1.5 m wide × 6 m long). A zero-tension bucket lysimeter installed between plant rows in the middle of each plot before planting was used to collect the leachate below the corn root zone. Wang et al. (2005) published the detailed project report.

Sampler installation: To install a lysimeter, a 50-cm across hole was dug to a depth of 55 cm from the soil surface. A lysimeter was placed so that the top edge of the storage bucket was at a depth of 20 cm from soil surface. The excavated soils were backfilled to the collection pan and sampling hoses were buried with aluminum cans. The buried lysimeters were located using a metal detector after planting. Figures 7.5 through 7.7 show

FIGURE 7.5 Auguring a cylindrical hole into the soil or limestone bedrock to depth so that the top of the lysimeter is below crop roots.

FIGURE 7.6 Installing a bucket lysimeter.

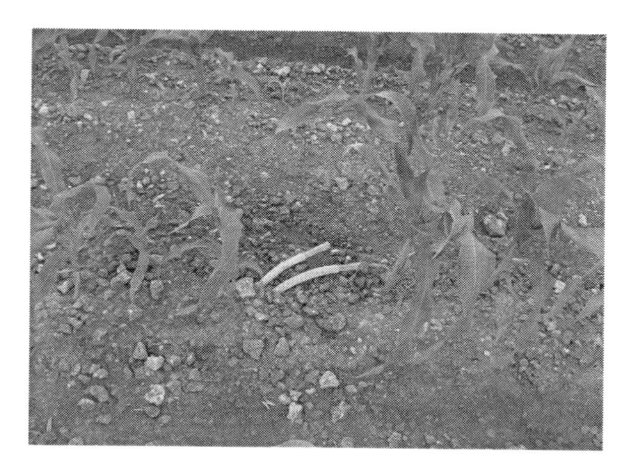

FIGURE 7.7 After installing the lysimeter, the two flexible tubes are left protruding from the ground. Water collected in the lysimeter is removed using the two tubes: one tube serves as an air vent while the other is connected to a peristaltic pump. It is important to secure (but not to an airtight seal) the exposed tube ends to ensure that dirt, debris, or insects do not enter the tubing. If the air vent tube is airtight, no leachate will be collected in the lysimeter irrespective of the rainfall/irrigation water applied.

the installation of the bucket lysimeter. The specifications, construction, operation, and maintenance of lysimeters are described in Section 7.4.2.2 and by Migliaccio et al. (2009).

Sampling: Leachate in lysimeters was collected using a small peristaltic pump at 15, 30, 45, 60, 75, 90, 105, 120, and 135 days after planting. Leachate volumes were measured at sampling and a 250 mL subsample was stored for chemical analysis. Leachate samples were analyzed for pH, NH_4-N, nitrate (NO_3-N), PO_4-P, total Kjeldahl nitrogen (TKN), and total P. The chemical methods for sample extraction/digestion and analysis are described in Chapter 10.

7.5 SEDIMENT PORE WATER SAMPLING

Sediment pore water extraction methods fall into two groups: ex situ (laboratory) and in situ (field). The former requires collection of a sediment core from the field and transport to a laboratory with subsequent separation of the pore water from the sediment by pressurization (squeezing) or centrifugation; for the latter, the pore water is collected in the field using a sampler often referred to as a pore water equilibrator or dialyzer. The most common in situ extraction methods are dialysis (peeper) and micropiezometer (sipper) methods, whereas centrifugation is the most reliable for sampling large volumes of pore water at sites (Mason et al., 1998; Bloom et al., 1999). However, both the centrifugation and squeeze methods are not suitable for wetlands with coarse sediments and/or dense root systems because in the former, pore water readily drains out of the coarse sediment (Berg and McGlathery, 2001) and in the latter, nutrients leak from damaged roots inside the core (Howes et al., 1985; Henrichs and Farrington, 1987; Hines et al., 1994). In addition, these methods are very difficult to implement, requiring extensive training and an on-site or near-site mobile laboratory with a glove box for processing pore water (such as slicing cores and expressing or filtering of the centrifugate) in an inert atmosphere (N_2 or Ar) to prevent potential oxidation. To decrease variability and optimize the volume of pore water for analysis, multiple cores must be collected and processed due to the high degree of heterogeneity in pore water concentrations (Mason et al., 1998) and/or low pore water yields in the squeeze method. Temperature (Fanning and Pilson, 1971) and pressure (Murthy and Ferrel, 1972) artifacts are problems in the squeezing procedures. Pore water sulfide concentrations from the centrifugation (Dacey and Wakeham, 1985) and squeeze methods (Hines et al., 1989) are lower than those from the sipper method due to volatilization and/or oxidation of H_2S during the centrifuging/squeezing process. Due to the limitations of laboratory methods, field methods (sipper and peeper) have been developed to collect pore water in situ to minimize the sampling artifacts created during collection and processing of sediment cores for pore water.

7.5.1 Micropiezometer (Sipper)

Although this sipper method is much easier than the centrifugal and squeeze methods to implement, it lacks the depth resolution routinely achieved with centrifugation required for consolidated sediments with a well-defined water/sediment interface (Bufflap and Allen, 1995). Furthermore, only a few mL of pore water is extracted from the active sediment layers without surface water breakthrough, requiring the use of micro-analytical techniques even for single constituent analyses. When the resolution of concentration micro-gradients is not the objective of a study, such as biogeochemical processes in pore water, a modified sipper method can collect sufficient water for all the analytes to be measured using standard methods. This modified sipper can integrate pore water composition over 4 ± 2 cm of top sediment wherein biogeochemical processes are at a maximum.

7.5.1.1 Original Sipper

The sipper originally designed by USGS (Wentz et al., 2003) for Hg research projects is very similar to the sippers used for nutrients, trace metals, and Hg studies by other researchers (Dacey and Wakeham, 1985; Berg and McGlathery, 2001; Branfireun, 2004; Figure 7.8). Pore water is relatively easy to collect via a slotted Teflon probe (sipper) deployed at the desired depth. The main disadvantage of the original sipper is that it cannot collect sufficient pore water for multianalytes analyses without surface water breakthrough.

7.5.1.2 Modified Sipper Method

The modified sipper was designed and tested by a mercury (Hg) research team from the South Florida Water Management District and Tetra Tech (Fink et al., 2007; Struve et al., 2007). Four new features that were added to the modified sipper (Figure 7.9) facilitate a substantial increase in the pore water sample volume for Hg methlylation collected from the active sediment layer without surface water breakthrough. The

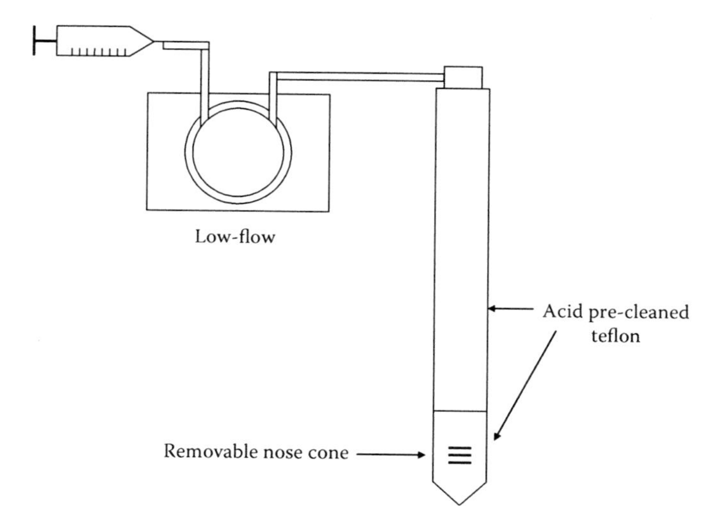

FIGURE 7.8 Original sipper (not to scale).

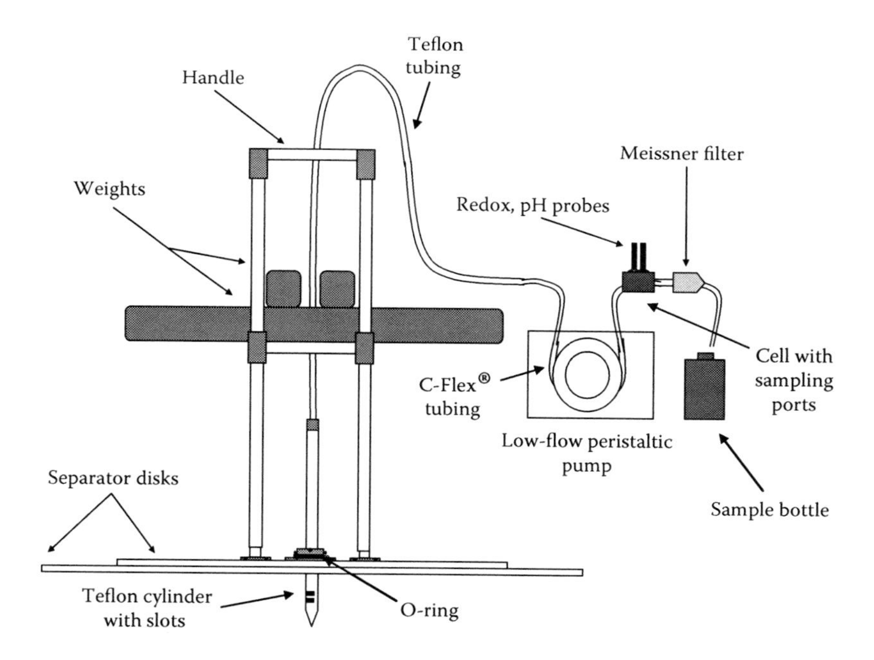

FIGURE 7.9 Modified sipper (not to scale).

first new feature is the addition of a 75 cm dia. × 2.5 cm thick molded disk composed of high-density polyethylene "starboard." The sipper probe is inserted through a 2.5 cm mounting hole in the center of the disk, then it is sealed with an o-ring and fixed at the desired depth with three sets of nylon screws. The second is a PVC handle to permit vertical insertion of the probe into the subsurface sediment layer, even in relatively deep water conditions in a wetland. The third is a set of five equally-distributed 12-kg weights laid across the handle cross-bars to ensure that a uniform pressure is exerted on the sediment to seal off the water/sediment interface, expel surface water below the disk, and prevent inadvertent collection of surface water, while compressing the loose flocculated surface layer and attaining a depth equivalent to a constant sediment bulk modulus across sampling sites. The high weight density minimizes the effect of buoyancy on uniform pressure application at sites in deep versus shallow water. The fourth new feature is a low-volume flow-through cell with sampling ports interspersed between the sample collection tubing on the other side of the peristaltic pump and the sample collection bottle. Microprobes are inserted into the cell for continuous measurement of redox potential and pH during sampling. This provides for a continuous verification of the absence of surface water breakthrough and/or oxidation.

The sipper probe is connected to a Masterflex peristaltic pump with Teflon tubing. Standard C-Flex tubing is passed through the pump head and joined to the Teflon tubing by inserting the smaller diameter tubing into the larger for a pressure seal. A high-surface area 0.45 μm Meissner capsule filter certified for ultra-trace metals analysis is connected to the C-Flex and Teflon tubing with an acid-precleaned Teflon connector. The flow-through sampling cell with sampling ports for the redox and pH probes is connected to the Teflon tubing with a Teflon connector and a Teflon tube leading from the last Teflon connector to the sample bottle.

To initiate sampling, the sipper system is purged with roughly 0.03 L pore water to remove pore water that could have been mixed with surface water during the insertion of the sipper probe into the sediment and to flush the sipper system. Roughly 0.5 L of filtered pore water is collected, respectively, at each site for total Hg (THg), methyl Hg (MeHg), sulfide (S^{2-}), ferrous (Fe^{2+}), ferric (Fe^{3+}), total Fe, sulfate (SO_4^{2-}), chloride (Cl^-), total manganese (Mn), dissolved organic carbon (DOC), Ca^{2+}, and Mg^{2+} analyses, in that order, to minimize the influence of surface water intrusion at the beginning and end of the sampling event. Specifically, separate bottles are collected for THg and MeHg (~200 mL); S^{2-} (15 mL); Fe^{2+}, Fe^{3+} (15 mL), and SO_4^{2-}, Cl^-, TFe, TMn, DOC, Ca^{2+}, and Mg^{2+} (~200 mL). The modified sipper was designed and tested for an Hg research/monitoring project. Pore water collected by this method is also suitable for nutrient research/monitoring if the depth profile is not the goal but large sample volumes are needed for determination of all required analytes using standard methods.

7.5.2 Peeper

Dialysis pore water samplers (peepers) are well-established tools for in situ sediment pore water sampling in wetlands (Teasdale et al., 1995). Pore water sampling devices based on diffusive equilibration were first developed by Hesslein (1976). Concentration profiles across the surface water/sediment interface can be obtained using a peeper, which provide valuable information on biogeochemical processes taking place at the surface water/sediment interface.

7.5.2.1 Peeper Design and Material Selection

A typical peeper is shown in Figure 7.10. When designing a peeper, the following parameters should be considered: (1) sample volume (V) (typically 5–20 mL) should be large enough to allow for analysis of all required compounds; (2) surface area of the cell window (A, cm^2) should be large enough to maximize diffusion of molecules into the cell and minimize the influence of sediment heterogeneity; and (3) sample vertical interval (the distance between each cell, typically 1–3 cm) should be small enough to provide sufficient depth profile resolution. The equilibration time required for water inside the cells to reach equilibrium with the surrounding pore water is controlled by the design factor ($F = V/A$; Brandl and Hanselmann, 1991) and also by species diffusivity, solid-phase adsorption, temperature, and porosity (Carignan, 1984). The equilibration time varies from 3 to 20 days (Carignan, 1984) and 2 weeks is probably a safe compromise for tropical and subtropical regions.

Polymethymethacrylate (acrylic, perspex, and plexiglass) is the most common material used for peeper construction (Teasdale et al., 1995). Oxygen should be removed from the sampler by a degassing prior to insertion in the sediment to avoid oxidation of some ions, such as Fe(II) and Mn(II) (Carignan et al., 1994). When pore water contains very high sulfide and very low Fe and Mn levels, such as that in the Everglade peat sediments, the effect of oxygen in peeper construction materials on pore water will be minimum. Membrane selection is also important with use of biologically inert materials such as polysulfone, PVC, and Teflon being preferable to biodegradable cellulose membranes (Carignan et al., 1985). The pore size of the membranes is typically 0.45 and 0.2 μm.

7.5.2.2 Peeper Preparation, Deployment, and Retrieval

New peepers should be cleaned in low P content detergent to remove any oil or grease, soaked together with membranes in dilute acid, and finally washed with deionized or distilled (DI) water before use. Place peepers and covers in a DI water bath and fill

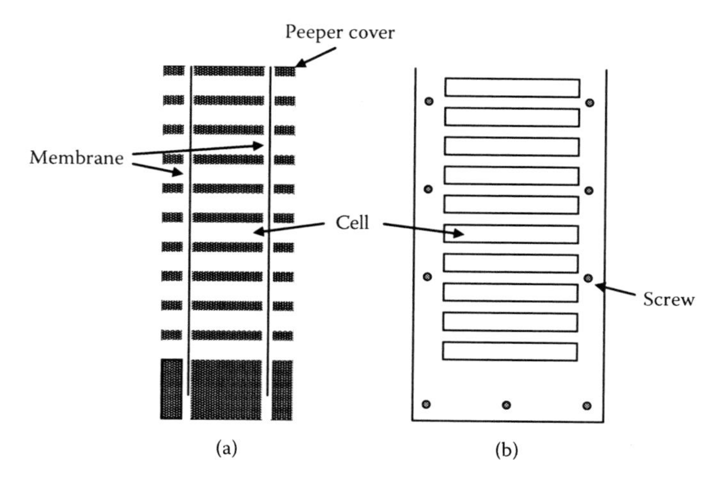

FIGURE 7.10 Typical peeper: (a) side view, (b) top view (not to scale).

the chambers with DI water, taking care to remove all air bubbles from the peeper cells. Submerge and line up the membrane between the peeper and its cover in the DI water bath. Use plastic (such as nylon) screws to assemble and make the peeper airtight. If the peeper is to be deployed in anoxic sediments with low reductant concentrations in an oligotrophic water body, a degassing step such as soaking the peeper (before and after assembly) in an N_2 bubble bath should be implemented. The assembled peeper must be stored in deoxygenated DI water during transport from the laboratory to the field.

Before deploying peepers in the field, a sampling site that is representative of the target study area must be selected. In most wetlands, peepers can be hand-deployed. If the water body is very deep, a scuba diver may be needed. Deployment is generally at a depth of 10 cm above or below the surface water/sediment interface where there is the greatest interest in studying biogeochemical processes or at other points of interest. In both cases, care must be taken to minimize disturbance to the sediments. In order to secure and aid retrieval, peepers can be attached by a rope to an anchor and/or marked by a float. After being deployed for 14 or 20 days in warm or cold regions, respectively, peepers are removed from the sediment and cleaned by agitating in water or using a water pistol to remove sediment particles on the membrane surface. Pore water in all cells from the anoxic to the oxic end of the peeper must be transferred into sample bottles/containers as soon as possible (within 5–20 min) to prevent oxidation. This is achieved by piercing the membrane with either a plastic micropipette tip or a needle and withdrawing pore water by either a pipette or a syringe. For trace metal studies, a plastic micropipette tip and pipette should be used to prevent contamination from the metal syringe needle.

Since the volume of a typical peeper cell is small (~10 mL), the subsample volume for each pair of analytes (PO_4/ NO_2, NO_x/NH_4) may be just sufficient for one injection into the flow injection analysis (FIA) system. A typical subsampling protocol for the Everglades wetland pore water nutrient study is as follows: (1) 4~5 mL pore water subsample is placed into a 14 mL centrifuge tube for PO_4/ NO_2 determination; (2) 4~5 mL pore water subsample is acidified with a drop of dilute H_2SO_4 to $1.9 < pH < 1.3$ in a 14 mL centrifuge tube for NOx/NH_4 analyses; and (3) ~0.8 mL pore water subsample is placed into a 1 mL vial for CL^-/SO_4^{2+} measurement using ion chromatography (IC).

7.5.3 Example 2—Sediment Pore Water Sampling Using a Modified Sipper

Project objective: Evaluate the modified sipper method in large volumes of pore water for an Hg project and use pore water data to find the sulfur species concentration breakpoint for predicting the tendency to form MeHg.

Key question: Will the modified sipper collect sufficient pore water to analyze all required constituents using standard methods and achieve comparable results for conservative and ultra-trace constituents as the centrifugation method?

Monitoring: Measuring pore water Eh/pH and redox-sensitive and ultra-trace constituents (Hg) in pore water (Figure 7.11).

Project design: Monthly pore water samples were collected from 10 sampling sites in the three treatment cells for 6 months. Each of the three treatment cells had three sediment sampling sites except cell 1, which had 4. One site was used to take triplicate pore water samples using the modified sipper method to verify the field reproducibility of the method. Pore water was also collected from three sites (one site in each treatment cell) once for comparison of the modified sipper and centrifuge methods.

Sampler deployment: After transporting the modified sipper to the site using an airboat (Figures 7.12 and 7.13), the probe was inserted into the sediment. Five weights were laid across the handle cross-bars. Figure 7.14 shows the deployment of a modified sipper. The specifications and operation of the modified sipper are described in Section 7.5.1.2.

Sampling: About 500 mL of pore water samples were collected from each site using the modified sipper connected to a Masterflex peristaltic pump. The "Dirty Hands and Clean Hands" technique (USEPA method 1669) was used throughout for pore water collection. The pore water was analyzed for THg, MeHg, S^{2-}, Fe^{2+}, Fe^{3+}, SO_4^{2-}, Cl^-, TMn, DOC, Ca^{2+}, and Mg^{2+}.

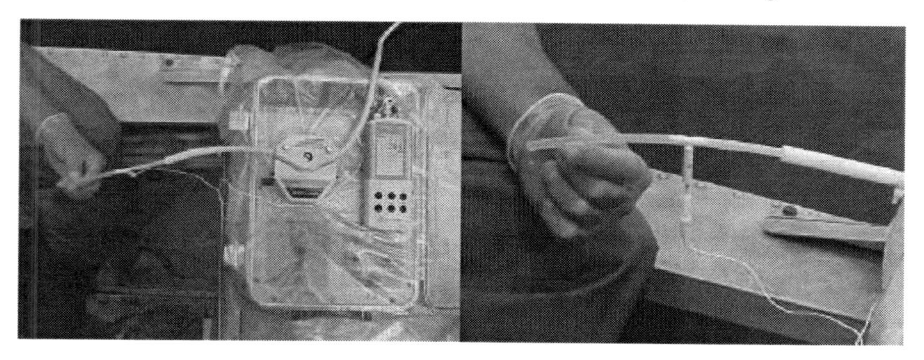

FIGURE 7.11 Monitoring pore water Eh/pH continually using in-line flow-through probes during pore sample collection.

FIGURE 7.12 Transporting modified sipper to a site using an airboat.

FIGURE 7.13 Moving the modified sipper from airboat to a sampling site.

FIGURE 7.14 Deploying the modified sipper at the sampling site with a set of five, equally distributed 12 kg weights laid across the handle cross-bars.

7.6 SUMMARY

Pore water sampling can provide useful information for understanding soil/sediment pore water chemistry. The sampling program should be designed for the specific objectives of the project. Selection of sampling sites, instruments, and methods should fit specific purposes. There is no universal standard operating procedure for pore water sampling. Pore water samplers discussed in this chapter are the most commonly used and are compared in Table 7.1. Two case examples provide details

TABLE 7.1
Comparison of Common in situ Pore Water Samplers

Sampler	Strength	Weakness	Sample Volume	Availability	Estimated Cost (US$)
Suction or tension lysimeter	Widely used for saturated or unsaturated soils and sediments; easy to assemble and install; can repeat sampling	No leachate volume measurement; small sampling area; affected by preferential flow; performs poorly with fine textured soil or sediment; possible alteration of water chemistry by ceramic cups especially low solute concentrations	Various 0.5–1000 ml	Commercially available with ceramic, stainless steel, and PTFE tips	$100–500
Pan lysimeter (free drainage sampler)	Most common lysimeter for soil water; zero tension; easy to assemble; measures volume of leachate	Only used for unsaturated soil; affected by preferential flow; hard to install without disturbing soil	Large amount	Not commercially available, but easy to assemble	$50–100
Bucket lysimeter	Zero tension; easy to assemble; measures volume of leachate	Only use for unsaturated soil; affected by preferential flow; disturbs soil during installation	Large amount	Not commercially available, but easy to assemble	$50–100
Wicking lysimeter	Eliminates effect of preferential flow; zero tension; measures volume of leachate	Only used for unsaturated soil; disturbs soil during installation	Large amount	Not commercially available, but easy to assemble	$50–100
Micropiezometer (Sipper)	Easy to use	No depth resolution; small sample volume	1–30 ml	Commercially available and also fabricated by researchers	$50–100
Modified sipper	Large sample volume; hands-free sampling; consistent insertion angle; reproducible sampling depth	No depth resolution	500 ml	Not commercially available, but easy to assemble	$100–500

on the construction of samplers, site selection, installation of samplers, and sampling. These examples may be easily modified to fit into (various) monitoring programs.

REFERENCES

Andersen, R.B. 1994. Dissolved inorganic and organic phosphorus in soil water from an acid forest soil collected by ceramic and PTFE soil water samplers. *Bull. Environ. Contam. Toxicol.* 53:361–367.

Bajracharya, R.M., and A. Homagain. 2006. Fabrication and testing of a low-cost ceramic-cup soil solution sampler. *Agric. Water Manag.* 84:207–211.

Beier, C., and K. Hansen. 1992. Evaluation of porous cup soil-water samplers under controlled field conditions—comparison of ceramic and PTFE cups. *J. Soil Sci.* 43:261–271.

Berg, P., and K.J. McGlathery. 2001. A high-resolution pore water sampler for sandy sediments. *Limnol. Oceanogr.* 46:203–210.

Bloom, N.S., G.A. Gill, S. Cappellino, C. Dobbs, L. Mcshea, C. Driscoll, R. Mason, and J. Rudd. 1999. Speciation and cycling of mercury in Lavaca Bay, Texas, sediments. *Environ. Sci. Technol.* 33:7–13.

Bottcher, A.B., L.W. Miller, and K.L. Campbell. 1984. Phosphorus adsorption in various soil-water extraction cup materials—effect of acid wash. *Soil Sci.* 137:239–244.

Brandl, H., and K.W. Hanselmann. 1991. Evaluation and application of dialysis pore water samplers for microbiological studies at sediment-water interfaces. *Aquat. Sci.* 53:55–73.

Branfireun, B.A. 2004. Does microtopography influence subsurface pore water chemistry? Implications for the study of methylmercury in peatlands. *Wetlands* 24:207–211.

Brown, V.A., J.J. McDonnell, D.A. Burns, and C. Kendall. 1999. The role of event water, a rapid shallow flow component, and catchment size in summer stormflow. *J. Hydrol.* 217:171–190.

Brye, K.R., T.W. Andraski, W.M. Jarrell, L.G. Bundy, and J.M. Norman. 2002. Phosphorus leaching under a restored tallgrass prairie and corn agroecosystems. *J. Environ. Qual.* 31:769–781.

Bufflap, S.E., and H.E. Allen. 1995. Sediment pore water collection methods for trace metal analysis: A review. *Water Res.* 29:165–177.

Burton, G.A., and R.E. Pitt. 2002. *Stormwater Effects Handbook.* Boca Raton, FL: Taylor & Francis.

Carignan, R. 1984. Interstitial water sampling by dialysis: Methodological notes. *Limnol. Oceanogr.* 29:667–670.

Carignan, R., F. Rapin, and A. Tessier. 1985. Sediment porewater sampling for metal analysis—A comparison of techniques. *Geochim. Cosmochim. Acta.* 49:2493–2497.

Carignan, R., S. St-Pierre, and R. Gächter. 1994. Use of diffusion samplers in oligotrophic lake sediments: Effects of free oxygen in sampler material. *Limnol. Oceanogr.* 39:468–474.

Dacey, J.W., and S.G. Wakeham. 1985. Effect of sampling technique on measurements of pore water constituents in salt marsh sediments. *Limnol. Oceanogr.* 30:221–227.

Essington, M.E. 2003. *Soil and Water Chemistry: An Integrative Approach.* Boca Raton, FL: CRC Press.

Fanning, K.A., and M.E.Q. Pilson. 1971. Interstitial silica and pH in marine sediments: Some effects of sampling procedures. *Science* 173:1228–1231.

Fink, L.E., D.M. Struve, M. Zhou, P. Zuloaga, and R.M. Keyser. 2007. An in situ micropiezometer (Sipper) for large-volume pore water sample collection from poorly consolidated hydrosoil or sediment. South Florida Water Management District, West Palm Beach, FL.

Grossmann, J., and P. Udluft. 1991. The extraction of soil-water by the suction-cup method—a review. *J. Soil Sci.* 42:83–93.

Haines, B.L., J.B. Waide, and R.L. Todd. 1982. Soil solution nutrient concentrations sampled with tension and zero-tension-lysimeters—Report of discrepancies. *Soil Sci. Soc. Am. J.* 46:658–661.

Hansen, E.A., and A.R. Harris. 1975. Validity of soil-water samples collected with porous ceramic cups. *Soil Sci. Soc. Am. J.* 39:528–536.

Hansen, J.W., and B. Lomstein. 1999. Leakage of ammonium, urea, and dissolved organic nitrogen and carbon from eelgrass *Zostera marina* roots and rhizomes during sediment handling. *Aquat. Microb. Ecol.* 16:303–307.

Henrichs, S.M., and J.W. Farrington. 1987. Early diagenesis of amino acids and organic matter in two coastal marine sediments. *Geochim. Cosmochim. Acta.* 51:1–15.

Hesslein, R.H. 1976. An in situ sampler for close interval pore water studies. *Limnol. Oceanogr.* 21:912–914.

Hines, M.E., S.L. Knollmeyer, and J.B. Tugei. 1989. Sulfate reduction and other sedimentary biogeochemistry in a northern New England salt marsh. *Limnol. Oceanogr.* 34:578–590.

Hines, M.E., G.T. Banta, A.E. Giblin, J.E. Hobbie, and J.B. Tugel. 1994. Acetate concentrations and oxidation in salt-marsh sediments. *Limnol. Oceanogr.* 39:140–148.

Hornby, W.J., J.D. Zabcik, and F. Crawley. 1986. Factors which affect soil-pore liquid: A comparison of currently available samplers with two new designs. *Groundwater Monit. Rev.* 6:61–66.

Howes, B.L., J.W.H. Dacey, and S.G. Wakeham. 1985. Effects of sampling technique on measurements of pore water constituents in salt marsh sediments. *Limnol. Oceanogr.* 30:221–227.

Hughes, S., and B. Reynolds. 1990. Evaluation of porous ceramic cups for monitoring soil-water aluminum in acid soils—Comment. *J. Soil Sci.* 41:325–328.

Jemison, J.M., Jr. 1994. Nitrate leaching from nitrogen-fertilized and manured corn measured with zero-tension pan lysimeters. *J. Environ. Qual.* 23:337–343.

Jemison, J.M., Jr., and R.H. Fox. 1992. Estimation of zero-tension pan lysimeter collection efficiency. *Soil Sci.* 154:85–94.

Jordan, C.F. 1968. A simple, tension-free lysimeter. *Soil Sci.* 105:81–86.

Lentz, R.D. 2006. Solute response to changing nutrient loads in soil and walled ceramic cup samplers under continuous extraction. *J. Environ. Qual.* 35:1863–1872.

Levin, M.J., and D.R. Jackson. 1977. Comparison of in situ extractors for sampling soil-water. *Soil Sci. Soc. Am. J.* 41:535–536.

Litaor, M.I. 1988. Review of soil solution samplers. *Water Resour. Res.* 24:727–733.

Mason, R., N. Bloom, S. Cappellino, G. Gill, J. Benoit, and C. Dobbs. 1998. Investigation of pore water sampling methods for mercury and methylmercury. *Environ. Sci. Technol.* 32:4031–4040.

Migliaccio, K.W., Y.C. Li, H. Trafford, and E.A. Evans. 2009 (updated), 2006. A simple lysimeter for soil water sampling in South Florida. ABE361, Agricultural and Biological Engineering Department, FL Cooperative Extension Service, IFAS, UF. URL: http://edis.ifas.ufl.edu/AE387.

Murthy, A.S.P., and R.E. Ferrell, Jr. 1972. Comparative chemical composition of sediment interstitial waters. *Clays Clay Miner* 20:317–321.

Nagpal, N.K. 1982. Comparison among and evaluation of ceramic porous cup soil-water samplers for nutrient transport studies. *Can. J. Soil Sci.* 62:685–694.

Parizek, R.R., and B.E. Lane. 1970. Soil-water sampling using pan and deep pressure-vacuum lysimeters. *J. Hydrol.* 11:1–21.

Pregitzer, K.S., D.R. Zak, A.J. Burton, J.A. Ashby, and N.W. MacDonald. 2004. Chronic nitrate additions dramatically increase the export of carbon and nitrogen from northern hardwood ecosystems. *Biogeochem.* 68:179–197.

Severson, C.R., and D.F. Grigal. 1976. Soil solution concentrations: Effect of extraction time using porous ceramic cups under constant tension. *Water Resour. Bull.* 12:1161–1170.

Shaffer, K.A., D.D. Fritton, and D.E. Baker. 1979. Drainage water sampling in a wet, dual-pore soil system. *J. Environ. Qual.* 8:241–246.

Shipitalo, M.J., and W.M. Edwards. 1994. Comparison of water movement and quality in earthworm burrows and pan lyisimeters *J. Environ. Qual.* 23:1345–1351.

Soilmoisture. 2007. *Operation Instruction Manual for 1900 Soil Water Sampler*. Santa Barbara, CA: Soil Moisture Equipment Corp.

Struve, D.M., L.E. Fink, M. Zhou, P. Zuloaga, and R.M. Keyser. 2007. The correlations between methylmercury in pore water and surface/pore water chemical properties using a sulfur species concentration breakpoint. South Florida Water Management District, West Palm Beach, FL.

Swistock, B.R., J.J. Yamona, D.R. Dewalle, and W.E. Sharpe. 1990. Comparison of soil-water chemistry and sample-size requirements for pan vs tension lysimeters. *Water Air Soil Pollut.* 50:387–396.

Talsma, T., P.M. Hallam, and R.S. Mansell. 1979. Evaluation of porous cup soil-water extractors: Physical factors. *Aust. J. Soil Res.* 17:414–422.

Teasdale, P.R., G.E. Batley, S.C. Apte, and I.T. Webster. 1995. Pore water sampling with sediment peepers. *Trends Anal. Chem.* 14:250–256.

Van der Ploeg, R.R., and F. Beese. 1977. Model calculations for the extraction of soil water by ceramic cups and plates. *Soil Sci. Soc. Am. J.* 41:466–470.

Wagner, G.H. 1962. Use of porous ceramic cups to sample soil water within the profile. *Soil Sci.* 94: 379–386.

Wang, X., Y.C. Li, R. Muñoz-Carpena, P. Nkedi-Kizza and T. Olczyk. 2005. Effect of zeolitic soil amendment on phosphorus leaching in a sweet corn field. *Soil Crop Sci. Soc. Florida Proc.* 64:55–59.

Wentz, D., M. Brigham, M.M. DiPasquale, W. Orem, D. Krabbenhoft, G. Aiken, and M. Corum. 2003. *Porewater Collection Protocols for the NAWQA Mercury Study*. Middleton, WI: USGS.

Wilson, L.G., and J.F Artiola. 2004. *Environmental Monitoring and Characterization*. New York: Elsevier.

Wilson, N. 1995. *Soil Water and Ground Water Sampling*. Boca Raton, FL: Taylor & Francis.

Zhu, Y., R.H. Fox, and J.D. Toth. 2002. Leachate collection efficiency of zero-tension pan and passive capillary fiberglass wick lysimeters. *Soil Sci. Soc. Am. J.* 66:37–43.

Zimmermann, C.F., M.T. Price, and J.R. Montgomery. 1978. A comparison of ceramic and Teflon in situ samplers for nutrient pore water determinations. *Estuar. Coast. Mar. Sci.* 7:93–97.

8 Field Measurements

David Struve and Meifang Zhou

CONTENTS

8.1 INTRODUCTION

Modern water quality monitoring programs often include some field measurements that are measured at the time and location of water sample collection. Field measurement, as opposed to laboratory measurement, may be used if the analyte is unstable once removed from the water body or simply because such measurements are more cost effective or convenient to determine in the field. Typical field-measured water quality parameters include pH, specific conductivity, dissolved oxygen (DO), redox potential, and temperature, which may be assayed using individual or multiparameter portable instruments. Depending on the goals of a specific water quality monitoring program, other measurements (for which field test kits are available) may also be performed. Test kits covering a wide range of water quality parameters are commercially available. However, these are generally less sensitive with less rigorous quality control than those commonly performed in an analytical laboratory.

With the advent of powerful microcomputers, novel software, data loggers, wireless communication networks, and new robust sensor designs, devices for autonomous real-time or near real-time field measurements have also been developed. These systems can be placed near or into a given water body and are programmed to conduct water quality tests automatically on a prescribed schedule or when triggered remotely. After connection to a wireless network, real-time measurements are instantly relayed to a central location for immediate operational decisions in response to water quality changes.

8.1.1 Definition of Terms

Much terminology has developed and evolved to describe various field measurements common for use in water quality evaluation. Possibly the most common, in situ monitoring, has been used to describe a wide variety of measurements assessed directly in a water body. Although the term *in situ monitoring* is useful, it does not distinguish among the types of direct measurement systems and the types of calibration, quality control, system control, and measurement frequencies employed.

Quality control is the set of procedures for ensuring that measurements meet the minimum accuracy requirements for a monitoring network. Typically, these procedures require measurements of water samples of known concentrations to determine if the field technique is sufficiently accurate. With certain types of field measurement instruments, standard samples of known concentration are analyzed before and after taking field measurements. These checks are known as *pre- and postcalibration checks.*

Accuracy is the degree of closeness of a measured quantity to its actual (true) value. Accuracy is closely related to *precision*, also called *reproducibility* or *repeatability.* Precision is the degree to which further measurements produce the same or similar results. *Measurement uncertainty* describes the range and distribution that encompasses the true value of a given parameter. Measurement uncertainty is discussed in more detail in Chapter 12.

Sondes are one of the many devices used for field measurement of physical conditions. These devices may consist of a single sensor or an array of sensors. Often the sonde includes a *data logger*, which is an electronic component that allows for storage and retrieval of measurements. A more complex system, an *autonomous real-time field analyzer*, typically includes sensors and/or detectors, a data logger, a sample delivery system, and containers of reagents and quality control standards that allow the system to collect measurements automatically on a fixed schedule or when triggered remotely. In addition to taking measurements, these systems are self-calibrating, which allows for a high degree of accuracy and precision to be maintained over longer time periods.

A *field test kit* is a complete set of reagents and a measurement system that is portable for assessing selected field water quality constituents. An range of these kits (which vary widely in accuracy and precision) are commercially available.

8.1.2 Importance of Field Analytical Techniques and Analyzers

Simple field measurements such as pH, specific conductance, temperature, and dissolved oxygen that are routinely performed for many monitoring networks provide general information on water quality. When an autonomous real-time field analyzer is used, field measurements may be more important than those in the laboratory because they can be obtained more quickly, at a greater frequency, and at significantly lower cost than conventional sampling and laboratory analyses. In addition, these systems can provide data to a central location where real-time operational decisions or regulatory actions can be made. Test kits for field measurements are generally less accurate than laboratory assays because the measurements are less sensitive and have higher levels of uncertainty due to limited quality control, uncontrolled test conditions, and inferior techniques for calibration and sample analysis. Regardless of the technique employed to produce field measurements, failure to perform calibration and quality control reduces the certainty, accuracy, and precision in the measurement value relative to those made in a properly qualified analytical laboratory. Users of water quality data must be aware of the procedures used and the associated measurement uncertainty. If conducted properly, field measurements can be as useful as laboratory assays, but without proper quality control they are inferior.

8.2 TYPES OF FIELD ANALYSIS

Field measurements of water quality can be divided into three categories: (1) manual techniques; (2) automated techniques that do require additional chemical manipulations; and (3) automated techniques that do not require additional chemical manipulations. Manual field techniques require a sample to be removed from the water body and assayed for the analyte of interest. The procedures may involve an appropriate sensor to directly measure the analyte or wet chemistry where reagents are added prior to spectrophotometric, titrimetric, or colorimetric measurement. Automated techniques without additional chemical manipulations consist of measurements made directly in the water body by sensors targeted to the analytes of interest, or where an automatically withdrawn sample is analyzed spectrophotometrically or with a specialized sensor. Automated techniques with additional chemical manipulations include a host of measurements on an automatically withdrawn sample by application of reagents, digestion, or combustion in preparation for analysis of the analyte(s) of interest (Hunt and Wilson, 1986; Johnson et al., 2007).

8.2.1 Manual Techniques

Many test kits and instruments are available for field testing water samples for a large number of analytes. Manufacturers such as Hach, LaMotte, and Chemetrics distribute test kit products. The quality and availability of many field test kits have been summarized (Burton and Pitt, 2002) according to parameters such as method, manufacturer and kit name, cost, analysis time, precision, recovery, and potential problems.

Depending on the quality level desired, kits are available for many possible analyte concentrations and levels of sophistication as illustrated by the example in Table 8.1

TABLE 8.1
Field Test Kits for Orthophosphate Analysis

Method	Concentration Range (mg/L)	EDL[a] (mg/L)	Manufacturer and Kit Name
Test Strip	0–50	—	HACH Test Strip
Color Wheel	0–1		
	0–5		
	0–50	—	HACH Color Disc
Color Chart	0, 1, 2, 4	—	LaMotte TesTabs Color Chart
Comparator			
-ascorbic acid	0, 0.2, 0.4, 0.6, 0.8, 1, 1.5, 2	0.2	LaMotte Octet Comparator with Axial Reader
Comparator	0–1	0.05	Metetrics CheMet
-stannous chloride	1–10		
Comparator			
-stannous chloride	0.05, 0.1, 0.2, 0.3, 0.4, 0.6, 0.8, 1	0.05	LaMotte Octet Comparator with Axial Reader
Colorimeter			
-stannous chloride	0.1–2.64	—	Metetrics Vacu-vialsials
Colorimeter			
-ascorbic acid	0.02–3	0.02	HACH Pocket Colorimeter II

[a] EDL: estimated direction limit.

for orthophosphate. As can be seen, concentrations range from 0.02 to 50 mg/L with the kits offering simple color charts, comparators, or small-scale spectrophotometric techniques for measurement.

To select an appropriate manual technique for field measurement, the expected range of concentrations of the analyte of interest and the quality objectives must be known. For example, a test kit might be suitable for orthophosphate concentration range of 1 to 10 mg/L in a fish pond. In another example, for the determination of Arsenic (As) in drinking water, the World Health Organization (WHO) guideline for As concentration in drinking water is 10 μg/L; the sensitivity of most As field test kits was not sufficient, and performance of those kits had generally been unsatisfactory (Feldmann, 2008). However, two new test kits capable of detecting As concentration in water near 10 μg/L were evaluated and determined to be suitable for As surveillance and remediation (Steinmaus et al., 2006). The keys to successful manual measurements are selecting the proper kit, following the procedures for the test kit as closely as possible, and minimizing environmental factors that might increase measurement uncertainty.

8.2.2 Automated Techniques without Additional Chemical Manipulations

Automated techniques that do not require additional chemical manipulations are commonly used in many modern water quality monitoring networks. Often referred to as sensors or probes, they include DO, pH, temperature, conductivity, turbidity, and fluorescence probes, as well as ion selective electrodes (ISE) and UV nitrate sensors. Sondes equipped with multiple sensors are available from a number of manufacturers (e.g., Hydrolab, YSI, and Greenspan). When equipped with data loggers, data values collected are stored in the instrument until retrieved by a field technician. Like all analytical equipment, they must be calibrated on a regular basis including precalibration (initial calibration and initial calibration verification) and postcalibration (continuing calibration verification) and checked with quality control samples to document their performance. A typical sonde from YSI (see Figure 8.1) is equipped with sensors for the determination of pH, temperature, DO, and specific conductance. Figure 8.2 shows a sampling technician with a Sonde deployed directly in the water body. The long cable facilitates deployment from an embankment, bridge, dock, piling, etc. Typically, a few minutes are required after deployment to allow the sensors to equilibrate to the surrounding environment for accurate and precise measurements. Once equilibrated, a measurement cycle can be initiated and collected data can be stored in the data logger or viewed from a visual display.

For example, an optical nitrate sensor that measures the UV absorption spectrum from 217 to 240 nm is used to directly assay nitrate in seawater where the lower levels of organic materials and sediment particles (compared to those in fresh water) pose little interference. Johnson et al. (2006) used a UV nitrate sensor for over 2 years of monitoring bay sea water and reported that 65% of the collected data were useful. Although the nitrate concentrations were accurate to ± 2 μM (standard error of estimate for regression versus bottle samples measured in laboratory), the average relative difference can be expected to be > 20% since the nitrate concentrations were very low (below the practical quantization limit [PQL] of the sensor) in some sea waters.

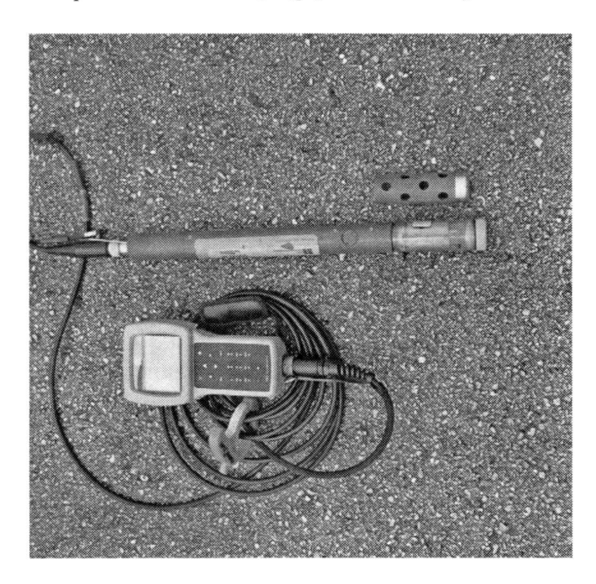

FIGURE 8.1 A typical sonde from YSI.

FIGURE 8.2 Sonde field deployment.

Using multi-wavelength UV adsorption spectra, some corrections for the interference from organic materials and sediments can be made. Nevertheless, using the UV nitrate sensor for long-term nitrate monitoring in fresh waters with relatively high levels of turbidity and organic matter is very challenging. Filtering samples are necessary for the accurate measurements. Pellerin et al. (2009) successfully measured nitrate in a river every 30 min for 5 days with an optical UV sensor after water was pumped through a 10 μm and then 0.2 μm membrane filter to remove the particulate materials.

Sondes and other sensors can be deployed for long-term monitoring in a flow-through mode. Glasgow et al. (2004) used a peristaltic pump to pass estuarine water through a flow-through system equipped with a multi-parameter sonde (YSI 6600) with pH, DO, temperature, conductivity, chlorophyll fluorescence, and turbidity probes. Because the sensors were not permanently submersed, the flow-through system reduced biofouling and increased deployment cycle time.

Instrument measurement stability and interfering substances are the major challenges in using ISEs for long-term monitoring. Muller et al. (2003) used a flow-through system to deliver both water samples and standard solutions to the Ion-selective electrode (ISE) probe flow cells. Lack of stability due to electrode fouling was eliminated by periodic calibration; the Nicolsky–Eisenmann equation was used to correct for interferences from potassium on ammonia, and chloride and bicarbonate on nitrate. Ammonia, nitrate, and pH were measured every 12 min, and a grab sample was collected every other day for bicarbonate, potassium, and chloride chemical analysis. The ISE flow cell was cleaned approximately every month, and new electrodes were installed because with longer deployments, the calibration slopes suddenly decreased, most probably due to biofouling.

8.2.3 Automated Techniques with Additional Chemical Manipulations

Automated techniques with additional chemical manipulations are the most sophisticated routine water quality systems used to measure analytes in the field. These systems can be portable but more typically are secured for longer periods of time at a single location to collect measurements on a regular schedule or when triggered remotely. These systems are generally autonomous in that all standards and/or quality control solutions required to operate independently and self-calibrate regularly can be equipped with wireless communications to transfer data remotely.

Colorimetric method is commonly used in field measurement with either a batch or flow injection analyzer (FIA). A flow-through sample introduction system is required to deliver the sample to the analyzer. Since the tubing size and flow cell path of a field FIA are typically very small, sediments are generally removed from the sample introduction system to prevent malfunction. Total P (TP) and/or N (TN) are typically measured using a batch analyzer that has relatively large size tubing and/or flow cell path. The sample line of the TP/N analyzer should be periodically checked and, when appropriate, replaced to prevent clogging. A system manufactured by Greenspan (Warwick, Australia) designed specifically for the determination of reactive and total P is shown in Figures 8.3 and 8.4. To minimize environmental factors that affect accuracy, this temperature-controlled system is completely

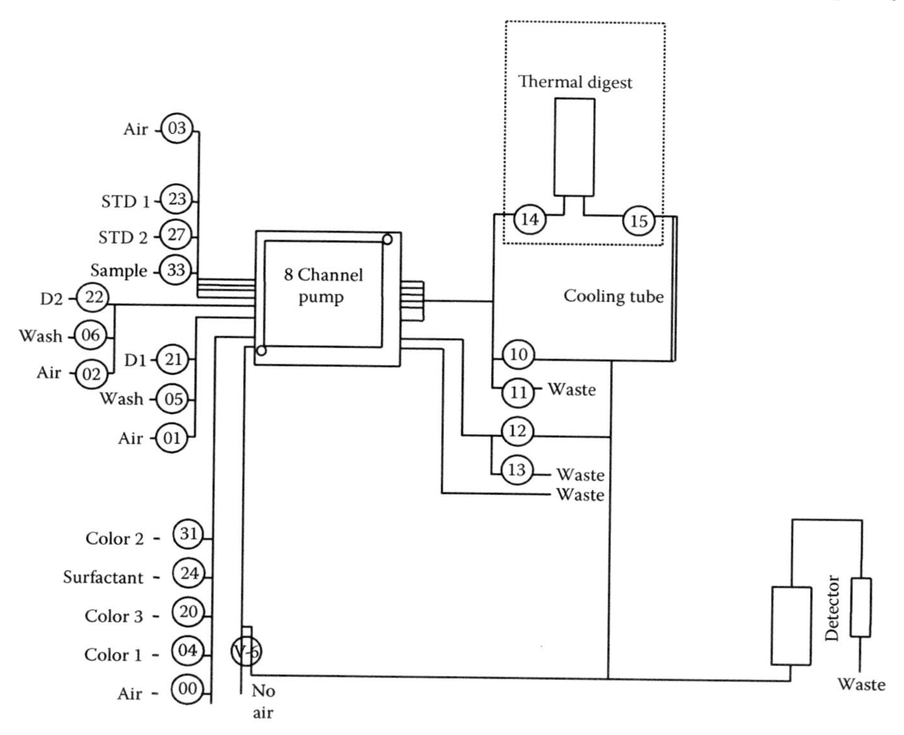

FIGURE 8.3 Schematic diagram of Greenspan remote P analyzer.

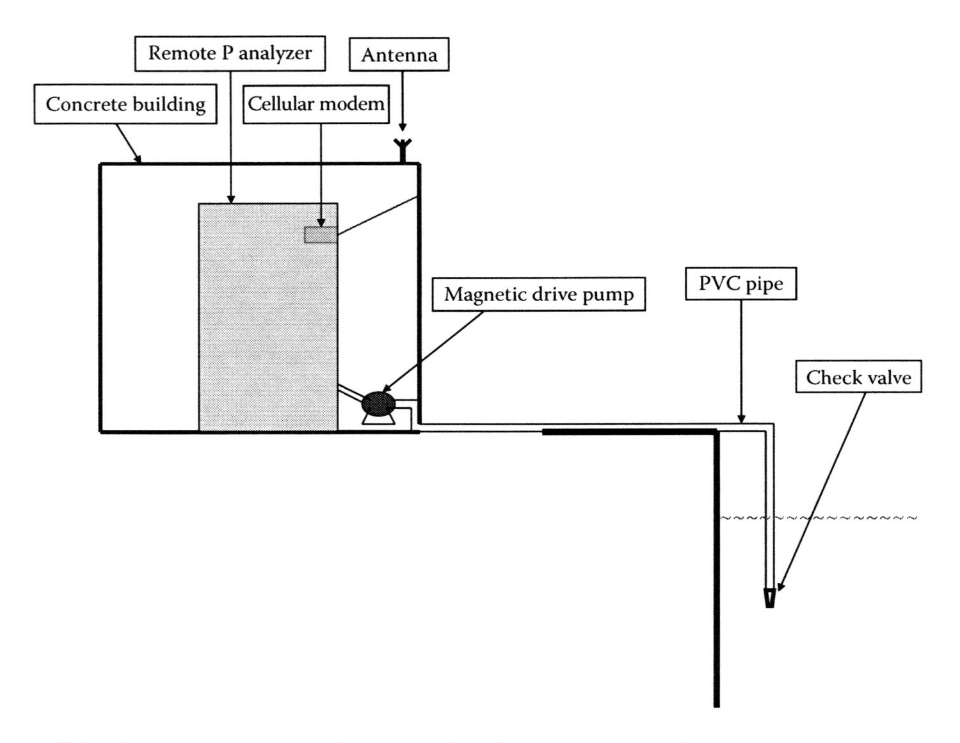

FIGURE 8.4 Components and configuration of a typical remote P analyzer.

autonomous and regularly self-calibrates, operating for periods of up to 3 months without any human intervention. Internal sensors to detect leaks from liquid transport systems and monitor internal temperatures are also incorporated. These sensors are monitored autonomously by the system and in the event of a leak the system shuts down automatically and sends an alarm signal to the central monitoring location. Several Greenspan remote TP analyzers have been deployed in various locations in South Florida; one of these is at the outlet of storm water treatment area 1 west (STA1 west) where 2 years of TP and total reactive P (TRP) data have been collected (Figure 8.5).

In the laboratory, FIA generally uses a peristaltic pump to deliver the standards, sample, and reagents, whereas in the field, a battery-powered peristaltic pump or several solenoid pumps are used. Fewer reagents and standards are consumed in FIA utilizing solenoid pumps since it is a stop- (or pulsed-) flow FIA. Chapin et al. (2004) used a field nitrate FIA that had seven solenoid pumps for propelling the sample, standards, and reagents. Prior to analysis, coarse particles were removed by a 10 μm filter but fine particles were likely retained by the packed Cd column. In-line filters often decrease sample flow rate as deployment time/sediment loading increase, and changes in sample flow rate can negatively affect instrument performance. In the study, the nitrate calibration curve was linear up to 100 μM. One standard (below or near 100 μM) and one blank were analyzed every 20 h for the calibration and check. The temperature effect on the color development of the pink color azo dye

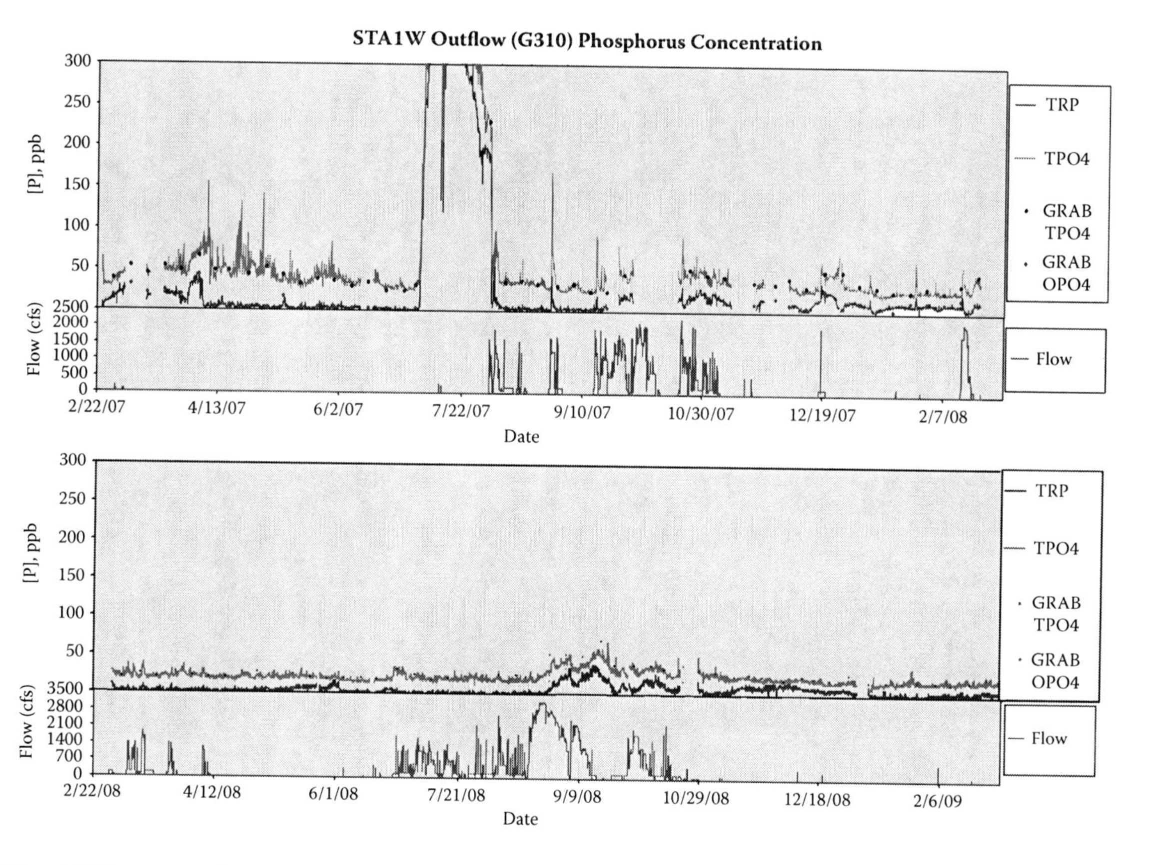

FIGURE 8.5 TP and TRP concentrations in STA G310 from a remote P analyzer.

was minimized by using a long reaction time (55 minutes) before the absorbance reading. Holm et al. (2008) adapted the YSI 9600 nitrate analyzer, which also uses seven micro-solenoid pumps to deliver the sample, standard, and reagents to autonomously analyze Cu by chemiluminescence in situ using a 0.45 μm pre-filter and a 0.45 μm syringe filter for inline filtration. Because flow rates vary among solenoid pumps and flow rates change during deployment due to changing flow characteristics of in-line filters with time, a correction factor was needed. This correction factor was estimated by running one standard with all three pumps (standard, blank, and sample) before and after each deployment. A linear interpolation was used to estimate the time dependence of the correction factor.

Plant et al. (2009) developed an in situ ammonium analyzer for estuarine and sea water. The analyzer is based on gas diffusion conductivity detection and uses micro-solenoid pumps to propel sample/reagents in a pulsed-flow mode through polytetrafluoroethylene (PTFE) tubing (ID 0.8 mm). Although the temperature effect on conductivity was corrected by temperature compensation, complete elimination of the effect was not possible unless the temperature was nearly constant or controlled. This is because the efficiency of gas diffusion through the cell usually increases as temperature increases. Gardolinski et al. (2002) used two 12 V battery-powered peristaltic pumps to deliver sample/standard and reagents in a nitrate FIA system using a 0.45 μm membrane filter during a 2-day shipboard deployment, a quartz wool packed prefilter, and a 0.45 μm syringe filter in 5 hours of submersible deployment from the side of the vessel. The calibration curve was completed daily using one standard after every three samples.

Lyddy-Meaney et al. (2002) used a modified FIA to save reagent usage for reactive P analysis. A 12 V DC high-capacity peristaltic pump was used to continuously deliver a water sample through a 0.2 μm tangential-flow filter at a flow rate of 260 mL/min into an FIA system at 1.5 mL/min, and color reagents (10 μL) were injected into the sample stream using computer-controlled miniature solenoid valves. Because the residence time of sample and reagent in the reaction zone was very short, $SnCl_2$ instead of ascorbic acid was used as the reducing agent. This method is not suitable for estuarine water because widely varying chloride concentrations will cause P results to be inconsistent.

8.3 APPLICATION OF FIELD ANALYSIS TECHNIQUES

The various field techniques described previously can be used for monitoring water quality in a variety of ways. Manual techniques are commonly used for synoptic surveys where the purpose is to quickly assess a water body or watershed. Automated techniques, such as water quality sondes, are being utilized for a variety of traditional monitoring networks. For short-term deployments, sondes may be used to make measurements at a single location from several hours up to a week or more so that a variety of environmental and/or hydrologic conditions can be monitored. Sondes and/or field FIAs can be deployed shipboard to determine spatial or profile (distance or depth) distribution of water quality parameters in a river, lake, or sea. Automated real-time or near real-time analyzers may be used for water quality monitoring where they are deployed at fixed remote locations for periods of a year or more to assess long-term trends and/or the effects of operational decisions or management practices.

8.3.1 Synoptic Surveys and Traditional Monitoring Networks

Commonly, field measurements are used for a synoptic survey of a water body or watershed using field test kit measurements and/or sondes to develop rough estimates of water quality and to assess current conditions. For such an initial assessment, maximum accuracy and precision are not generally needed, but if areas of concern are identified, further follow-up sampling or development of a more sophisticated traditional monitoring network can be justified. This approach may also help in the identification of point sources of pollution that warrant regulatory action and/or more sophisticated additional monitoring.

Traditional monitoring networks typically include field measurements to assess overall water quality. In this case, field measurements are commonly made for water quality parameters that are not amenable to laboratory measurement due to sample instability or cost. Field measurements routinely performed for traditional monitoring networks include pH, conductivity, temperature, DO, and redox potential. These measurements are usually made with a sonde equipped with the requisite sensors and a data logger to store the data until field sampling is complete. All sensors are calibrated before the field trip and the performance of the sonde is verified using pre- and postcalibration checks. In the event of the sonde failing the post-calibration check, the data may be considered unusable or must be qualified to indicate that a problem occurred. These field measurements are also extremely useful for the analytical laboratory responsible for testing samples, as measurements of conductivity, for example, can be useful in estimating the relative concentrations of various ions prior to analysis.

8.3.2 Short-Term Deployments

Short-term deployments of a sonde and/or a field analyzer at a fixed location to collect field measurements are useful where monitoring water quality at a high frequency may be necessary to account for weather or hydrologic events, tide changes, or discharge. Typically, the sonde is anchored to a fixed site below the water surface and programmed to conduct analyses continuously or at a fixed frequency. The data may be stored in the sonde's internal data logger or transmitted directly to a central location for monitoring.

The field measurement of spatial or profile (depth or distance) distribution of water quality parameters of a water body is often accomplished by a short-term deployment. Hodge et al. (2005) used several off-the-shelf instruments (e.g., YSI multi-parameter probes [YSI6600], flow injection analyzer, a GPS, and computer with LabView program) to develop a portable water quality monitoring system that could be mounted on a vessel to measure pH, temperature, specific conductance, DO, turbidity, chlorophyll-a fluorescence, nitrate, and orthophosphate during vessel transits. This system provided a spatially referenced "snapshot" of water quality conditions but was not suitable for long-term deployment. If filtration is required in the sample introduction system, most field FIA systems (see Section 8.2.3) may be only suitable for short-term deployment unless an alternative sediment removal tool such as settling or tangential filtration is incorporated into the instrument configuration.

8.3.3 Long-Term Fixed Deployments

Long-term in situ deployment provides the greatest challenges, particularly with regard to sample filtration and delivery, biofouling, calibration, temperature control, reagent and standard stability, reliability, ease of maintenance, quality assurance, data storage, and remote communication. Long-term deployment is becoming more popular as more robust equipment becomes available, as it provides a continuous data series during all types of hydrologic and weather events and avoids the need for frequent personnel visits to gather samples for traditional laboratory testing.

Secure installation of equipment is extremely important for this type of deployment to prevent damage from vandals, wildlife, insects, and biota so that accurate and precise measurements can be collected and recorded over time. Adequate shelter to protect instrumentation from wind and rain together with backup power may also be required to ensure a continuous data set.

For long-term deployment of in situ analyzers, it is critical to have internal self-calibration and an integrated performance monitor so that analyzer problems (such as drift due to biofouling or leaks in the liquid transport systems) can be detected. Long-term stability of standards and reagents should be evaluated in the laboratory before field deployment. The standards should be analyzed to recalibrate the instrument periodically during long-term deployment.

8.4 GENERAL ASPECTS OF FIELD ANALYSIS

As with all water quality monitoring, the cost of resources requires justification for the techniques that will be used, the frequency of analysis, and the quality objectives of the monitoring scheme. Before measurements of water quality are undertaken, monitoring objectives should be clearly identified. Consideration of the quality objectives will lead to the proper choice of techniques, frequency of measurement, procedures to evaluate the performance of the monitoring, and designs for the sampling systems.

8.4.1 Objectives

The quality objectives that are most important for determining the proper field techniques to be used for a given water quality monitoring scheme should include a determination of the required accuracy, precision, and allowable measurement uncertainty, permitting a choice of field technique and quality control procedures. When these quality objectives are so stringent that field testing is not a viable alternative, traditional sample collection and analysis by a qualified laboratory may be required.

Other monitoring objectives may be important in selection of a proper field analysis technique. Requirements for high-frequency monitoring at remote locations under variable environmental and/or hydrologic conditions may require that equipment for field measurements be deployed at a fixed location for an extended period. In these circumstances, cost of equipment, power sources, potential for biofouling, ability to provide accurate and precise readings for an extended period, and possibility of damage from the elements or vandals must also be considered.

8.4.2 Justification for Measurement Schemes

The justification for a given field measurement scheme may be from a regulatory requirement, but in many cases, it is created as part of the monitoring plan design. Modeling techniques to estimate the required measurement frequency and quality objectives to accurately characterize a hydrologic or weather event are often employed to assist in planning. Synoptic surveys to evaluate general water quality conditions prior to the design of a monitoring plan may also be useful. For example, high-frequency monitoring techniques with long-term deployments of field analyzers may not be necessary if modeling or initial synoptic surveys show little or no variation in water quality over time. Additionally, surrogate tests, such as frequent measurements of specific conductivity, may be sufficient to characterize water quality, with only a full suite of laboratory tests being conducted on a less frequent basis.

8.4.3 Choice of Field Analysis Technique

Once the plan for a monitoring design is developed along with the required quality objectives, an appropriate field technique can be chosen. Table 8.2 shows a series of common field techniques along with their advantages and disadvantages.

8.4.4 Evaluation of System Performance

Before a new technology is chosen for field measurement, the instrument must be evaluated both in the laboratory and in the field. The evaluation criteria (which may differ depending on the instrument) typically include method detection limit (MDL) determination, quality control (QC) recovery, digestion efficiency if applicable, standard solution and standard response (instrument drift) stability, and comparisons between field and standard laboratory methods. Some of the evaluation results for the Greenspan P remote analyzer are shown in Tables 8.3 and 8.4 and Figures 8.6 and 8.7.

The USEPA Advanced Monitoring Systems Center (http://www.epa.gov/etv/vt-ams.html) carries out the in situ technology verification program for surface/ground/drinking water. While most multiparameter sensors/sondes evaluated met the data quality objectives of the respective field measurements, the only two nutrient analyzers evaluated needed improvement before they could be approved for in situ surface water monitoring. The Alliance for Coastal Technologies (ACT, http://www.act-us.info/) is responsible for coastal/sea water in situ analyzer verification. While several sensors for measuring parameters such as turbidity and DO have been evaluated, no truly reliable in situ nutrient analyzers are available for coastal/sea water monitoring.

Once field techniques have been chosen, a system to evaluate their performance should be developed. Where simple manual or automated techniques are employed, standard statistical process control, where the results of quality control samples are

TABLE 8.2
The Field Analytical Methods

Method	Advantages	Disadvantages	Application(s)	Manufacturer(s) or Brands
Test Kits	Fast, cheap, no formal training needed	Low sensitivity; lack of quality control	Yes/no type survey, Well water As	Hach, Lamotte, Chemetrics
Sonde	Stabile, easy to operate, readily available from manufacturers	Difficult to do the self-calibration during deployment. Monthly or weekly visit required for long-term deployment	Short- to long-term	YSI, Hydrolab, Greenspan
ISE	No reagent usage, low power usage	Interference from other ions. Stability problem. Monthly or weekly visit required for long-term deployment	Short- to long-term	Greenspan
UV nitrate analyzer	Stabile, no reagent usage, low power usage	Interference from organics and sediments. Sensitivity may not be adequate for low nitrate sea water. Monthly or weekly visit required for long-term deployment	Long-term in sea water	ISUS
FIA-solenoid pumps	Low usage of reagents and standards, flexibility in color development time	The flow rate difference in different pumps; two on board standards; lack of temperature control; problem with in-line filtration	Short-term, shipboard or submerged	YSI
FIA-peristaltic pump	Easy to modify from lab instrument	Pump tubing easy to wear out and in-line filter easy to over load due to continuous flow and flow rate may decrease over time. Color development problem due to short reaction residual time and lack of temperature control. Two on-board standards	Short-term, shipboard or submerged	—
Batch analyzer-syringe pump	Low usage of reagents and standards, flexibility in color development time, temperature control in color development	One on-board standard, problem with in-line filtration, temperature control only for color development; the flow configuration problem for Cd column; extensive wash required for the syringe pump to prevent the acid from color reagent to degrade the Cd column	Short-term, submerged	Microlab
Batch analyzer-P analyzer (Greenspan)	Temperature control, fault detect, less problem with sediments in water sample, up to 3 months without field visit. Three on board standards	Need sufficient power to operate and a good water sample delivery system	Long-term on the bank/shore	Greenspan

Note: All sensors/analyzers are subject to the biofouling problem especially during submerged deployment in coastal and estuarine waters.

TABLE 8.3
Standard (STD), QC Recoveries, Method Blank (MB), and Method Detection Limit (MDL)

Sample Name	OPO4 Recovery (%)	TP Recovery (%)	n	MDL (μg/L)	MB (μg/L)
10 μg/L	98 ± 4	115 ± 5	32	1.2	—
25 μg/L	98 ± 2	105 ± 3	8	—	—
100 μg/L	99 ± 2	101 ± 2	51	—	—
150 μg/L	102 ± 2	102 ± 2	8	—	
200 μg/L	103 ± 1	103 ± 1	7	—	—
MB (OPO4)	—	—	23	—	0 ± 0
MB (TP)	—	—	23	3.7	1.7 ± 0.8
Old STD[a] 100 μg/L	98				
Old STD 200 μg/L	100				

[a] The old standard was measured for P concentration after 3 months of deployment using the laboratory method.

TABLE 8.4
TP Recoveries of Inorganic and Organic Phosphates

Phosphate	Recovery (%)	n
Organic Phosphate		
Phytic acid	101 ± 0.6	9
Adenosine-5-triphosphate	91 ± 4	6
β-glycerol phosphate	102 ± 0.7	8
Inorganic Phosphate		
KH_2PO_4	101 ± 1.6	51
Inorganic Polyphosphates		
Tripolyphosphate	92 ± 1.2	26
Trimetaphosphate	91 ± 1.2	8
Average recovery		96

charted and checked against acceptable control ranges, provides a sufficient performance evaluation. This type of analysis, along with a robust program for preventative maintenance of the field testing equipment, ensures that measurements consistently meet the quality objectives for the monitoring program.

For more sophisticated monitoring schemes using autonomous analyzers deployed in the field for long periods, the evaluation of system performance requires a more rigorous approach. Along with statistical process controls mentioned above, a preventative maintenance program that includes site visits to inspect for biofouling

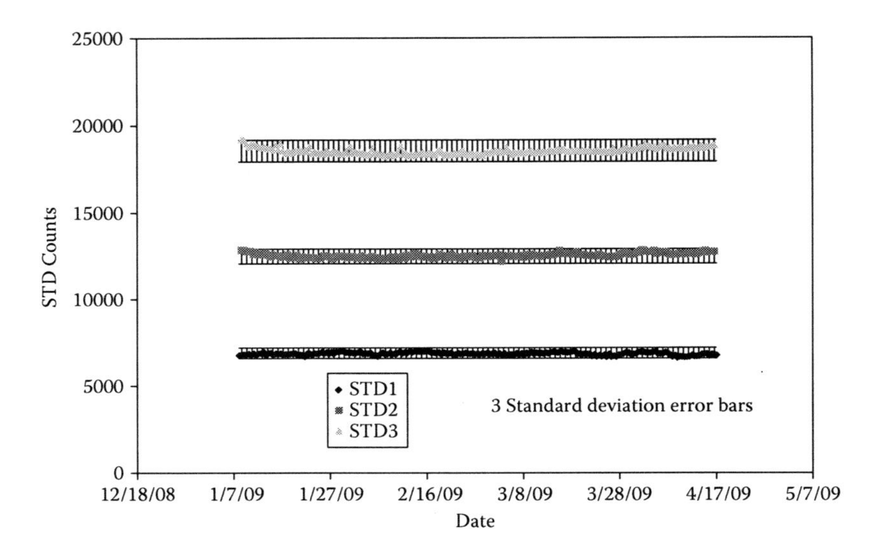

FIGURE 8.6 The standard response with 3 standard deviation error bars during 3 months' deployment.

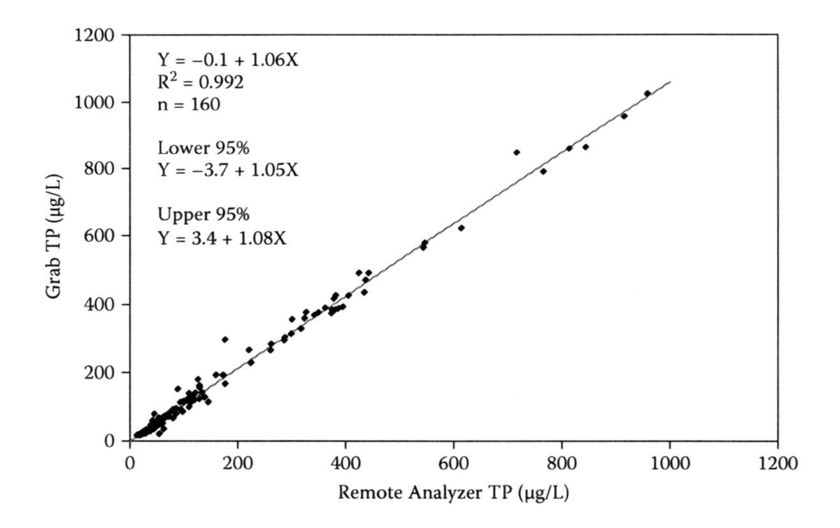

FIGURE 8.7 The TP results from the grab sample and Greenspan remote analysis.

around sample intakes, vandalism, and proper functioning of supporting equipment are likely to be necessary.

8.4.5 Sample Introduction

Two types of in situ sample introduction systems are commonly used: (1) a non-flow-through system where the sensor with a protective guard and/or shutter is directly immersed into water and (2) a flow-through system comprising a pump, pipes, sample inlet (such as coarse-screened check/foot valve), and an optional filter. The selection of a particular system depends on the type of in situ analyzer. If the sensor can be submerged without the need for periodic calibration, the non-flow-through system may be suitable, but if the analyzer needs shelter or periodic calibration, the sample must be delivered from the measuring point to the analyzer located in a shelter.

The design and care of sample introduction systems for autonomous field analyzers deserves special attention because environmental and hydrologic conditions at the measuring site can have serious impacts. Consequently, to obtain high-quality time series measurements, a robust design is essential. For example, if the sample introduction system is clogged by vegetation and/or other biological material (biofouling) or tubing in an analytical instrument is obstructed by sediment, all subsequent measurements may not represent the actual water quality and render any data interpretation invalid. For measurements at remote locations, significant effort and/or time is needed to visit the site to repair equipment. Proper care in the design of the sample introduction system can help to avoid this problem.

Minimizing biofouling in the water delivery system and/or on the surface of the sensor is a design criterion for sample introduction systems especially for the non-flow type. In highly productive coastal and estuarine waters, sensor surfaces may be overgrown within a few days to weeks by microbial and algal films or barnacles that degrade performance by creating microenvironments that alter chemical concentrations, block optical paths and flow to sensing surfaces (Johnson et al., 2007), or obstruct the flow of water through the delivery system. Several antifoulant compounds such as tributyl tin (TBT), polymers, and bromine have been used but with limited success in the non-flow-through system. Wrapping the guard with fine copper mesh (hydrolab or YSI) or using a copper shutter can prevent premature fouling of the sensors (Chavez et al., 2000). The disadvantage of this practice is that while the copper surface prevents biofouling, it reacts with sulfur, sulfide, and ammonia (Davis, 2001) increasing the absorbance in the UV region (Johnson and Coletti, 2002) and can adsorb/precipitate phosphate on the copper oxide coating that forms. For a flow-through system, the pump must have a sufficiently high flow velocity to not only minimize fouling in the sample delivery system and/or on the sensor surface but also to ensure sufficient volume to rinse the sample delivery system before a subsample is introduced into the analyzer.

When using an FIA for in situ measurements, sediment particles in the water must be removed before introduction into the reaction manifold. Because the internal diameter of tubing in the FIA manifold is usually very small (0.8 or 0.5 mm), blockage caused by sediment particles could make the results unreliable. Consequently, a high capacity pump and filter are often incorporated into the instrument configurations. A tangential-flow filtration membrane has been shown to increase filter life and minimize cake formation (Benson et al., 1996; Hanrahan et al., 2001). The resulting high cross-flow velocity of the liquid not only improves the scouring effect but also inhibits the buildup of particulate material and decreases the residence time of nutrients/metals in the filter, thus reducing the extent of any reactions. The life of a tangential filter at high flow rate was observed to be about 1 week (Benson et al., 1996). Since the sample flow rate in the FIA system is much lower than that in the tangential, a flow spilt should be installed between the tangential filter and FIA (Lyddy-Meaney et al., 2002). Other low capacity in-line filters such as a 0.45 μm syringe filter can be used to remove sediment particles (Gardolinski et al. 2002; Holm et al., 2008). The deployment time for FIA with a low capacity filter is usually very short and filter clogging may occur after only a few hours of operation. Accumulated particles on the filter may also affect data reliability because the particles not only decrease sample flow rate but can become active sites for adsorption/desorption and dissolution/precipitation with the compound of interest at low flow rates. In addition to not being suitable for long-term deployment, the filtration system may also not be reliable during a storm event since the high sediment load in runoff water may quickly cause blockage.

In the case of total P or N, large diameter tubing and a flow cell are used to prevent clogging so that no filtration is required. A "blank" or "baseline" analysis composed of the sample without addition of color reagent can be made to correct for the effect of sediment on the absorbance reading in the colormetric determination.

Regardless of the methods used to handle particulates, during each field visit, the sample introduction system should be checked and any sediment or biofouling should be removed. The same protocol applies after a storm/flooding event since large amounts of sediments from runoff may enter the sample delivery system. Additionally, even when preventative measures are in place, the analyzer intake line should be replaced periodically, as constant contact with the sample will likely cause at least some degradation of the tubing and/or accumulation of inorganic/organic/biological materials inside of tubing that over time may negatively impact the measurement system. A typical sample flow-through introduction system (see Figure 8.8) is equipped with a coarse filter/screen to prevent biological materials from entering the system, while the sample collection point is below the surface to eliminate the interference of floating vegetation.

8.4.6 Temperature Control

Uncontrolled temperature of the measurement system is also a limitation in many field instruments. Although most sensors/detectors can be temperature-compensated within certain temperature/concentration ranges, they may not function under extreme temperatures. Also, temperature significantly affects color development in many nutrient and trace metal analyses and ammonia diffusion in FIA conductivity detectors. In laboratory FIA methods, temperature is controlled above the ambient temperature to speed color development; however, in most field instruments there is little control as the power required to heat the system is not available. Consequently, development of maximum color intensity and nitrate reduction to nitrite in Cd columns takes longer at low temperatures (midnight) compared to higher temperatures (midday). To overcome this problem, Chapin et al. (2004) used a long reaction time (55 min) to minimize temperature effects on the pink azo-dye color development in field nitrate analysis. On the other hand, the long reaction time limits the number of samples that can be analyzed daily, increases potential interferences, and may impact color stability. Stability of some reagents and standards can also be affected by highly variable ambient temperatures. If

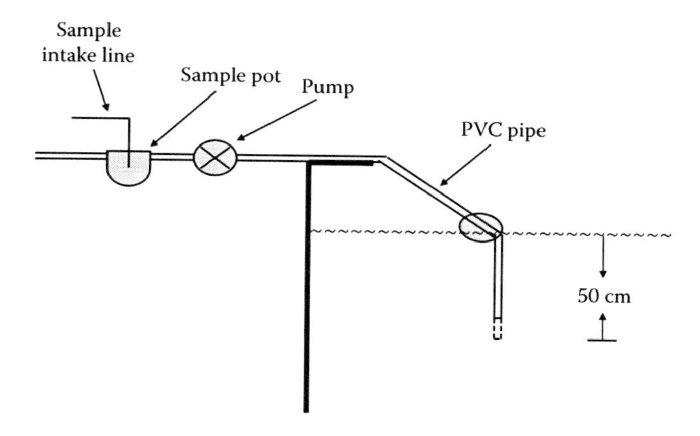

FIGURE 8.8 A typical sample flow-through introduction system.

power is not a limiting factor at a monitoring site, field instruments and reagents should be housed in a temperature-controlled cabinet or at a minimum a well-ventilated shelter to minimize heat accumulated during hot summers and extreme temperature variation.

8.4.7 Calibration

Because instrument calibration and data quality are directly linked, periodic self-calibration helps to ensure accurate and precise measurements, especially during long-term deployment. For batch and FIA field analysis of nutrients and trace metals, a minimum of three standards that cover the entire range of expected values are typically used. But most other field instruments use either two points (one blank and one high standard) or one point (high standard only) to calibrate the instrument because of the limited space available for storing standards and/or because limited solenoid pumps or valves are available for transferring standards/reagents to the measurement system. In any case, the use of only one high concentration standard is likely to reduce accuracy at low levels since variations in instrument sensitivity are common in field instrumentation. Common causes for loss of sensitivity include (a) bio or chemical fouling, (b) variations (usually increasing) in color development time at low concentration, (c) temperature effects, and (d) reagent degradation that increases baseline absorbance. Consequently, results at low concentrations for virtually all field instrumentation have large uncertainty unless low level standards are analyzed on a regular basis to account for the factors described above.

8.4.8 Various Design, Installation, and Operational Considerations

Before deployment of field water quality monitoring systems, a careful assessment of the environment and the design features of the system must be completed. Depending on the environment, these systems may be subjected to extremes in temperature and precipitation, bird and insect nesting, biofouling, or corrosive conditions.

Consequently, they must be designed to protect sensitive electronics and sensors from these potentially deleterious conditions.

The way in which the measurement system is installed at the monitoring location can also be an important factor in obtaining representative water quality data. When expensive monitoring equipment is used, security to prevent potential damage from vandals and recreational activities such as water sports, hunting, or fishing is essential. Installations in open water may attract fish and fishing lines. Consequently, they should be as smooth as possible without exposed wires or other protuberances with which fishing tackle can easily become entangled. Where hurricanes or tornadoes may occur, heavy prefabricated concrete shelters together with backup power supplies for measurement equipment may be necessary to withstand these extreme conditions.

The operational schedule for the analyzer must be chosen to accommodate the frequency of water quality measurements required to characterize all relevant hydrologic and weather conditions. Failure to have a sufficient number of measurements during a period of high flow or during a short-term spike in concentration may result in a failure to properly measure the concentration variations during an event. In deploying where events have not been previously measured, a trial operational period with a relatively high frequency of measurement may be employed initially until the concentration profile over time is well-understood (after which the measurement frequency can be reduced).

8.4.9 Data Handling, Quality Control, and Presentation of Results

Because field measurement monitoring networks can produce vast amounts of data, provisions for data handling and validation are critical. Acceptance criteria for field parameters measured by sondes, which are listed in Table 8.5, are similar to those for the excellent and good ratings in the USGS accuracy rating, which is based on combined fouling and calibration drift (Wagner et al., 2006). Although quality control criteria and procedures are well developed for laboratory nutrient and metal analyses, different procedures should be established for field methods.

Currently, The NELAC Institute (TNI) is developing standards and pending standards for field measurements which can be found at: http://www.nelac-institute.org/standards.php?pab=1_1#pab1_2. These standards for field measurements are more flexible and method-specific than those for laboratory measurements. Typically, only on-board calibration and not quality control standards are used for field measurement instruments due to the limited space, pumps, pump channels, or valves for storing or delivering standards. The on-board standards in a field instrument usually serve a dual purpose, namely, as calibration and quality control checks. The standards should be analyzed in the laboratory before and after deployment to ensure that the concentrations of standards are correct and stable (Table 8.3). If the standards are known to be stable under field conditions, the stability and performance of the instrument can be checked by plotting standard responses versus time for each batch of standards/reagents together using statistical process control (Figure 8.6). Any result outside the confidence limits has questionable value and results generated after unacceptable calibration results should be rejected.

TABLE 8.5
Acceptance Criteria and Accuracy Rating of Field Parameters Measured by Sonde

Field Parameter	Acceptance Criteria[a]	Ratings of Accuracy[b]			
		Excellent	Good	Fair	Poor
Temperature	± 0.2°C	≤ ± 0.2°C	> ± 0.2–0.5°C	> ± 0.5–0.8°C	> ± 0.8°C
Specific conductance	± 5%	≤ ± 3%	> ± 3–10%	> ± 10–15%	> ± 15%
Dissolved oxygen	± 0.3 mg/L	≤ ± 0.3 mg/L or ≤ ± 5%, whichever is greater	> ± 0.3–0.5 mg/L or > ± 5–10%, whichever is greater	> ± 0.5–0.8 mg/L or > ± 10–15%, whichever is greater	> ± 0.8 mg/L or > ± 15%, whichever is greater
pH	± 0.2 units	≤ ± 0.2 units	> ± 0.2–0.5 units	> ± 0.5–0.8 units	> ± 0.8 units
Turbidity	0.1–10 NTU: ±10% 11–40 NTU: ±8% 41–100 NTU: ±6.55 >100 NTU: ±5%	≤ ± 0.5 turbidity units or ± 5%, whichever is greater	> ± 0.5–1.0 turbidity units or > ± 5–10%, whichever is greater	> ± 1.0–1.5 turbidity units or > ± 10–15%, whichever is greater	> ± 1.5 turbidity units or > ± 15%, whichever is greater

[a] SFWMD and FDEP.
[b] USGS.

Because field measurement systems, especially during long-term deployments, generate quantities of data in excess of that which can easily be presented and interpreted in tabular form, statistical processing after the quality control step allows hourly, daily, weekly, monthly, yearly, or per storm records of minimum, maximum, and either mean or median to be presented. The results can also be presented for a given event or time frame and plotted together with other water quality or hydrological/meteorological data such water flow rates, tides, or precipitation amounts to observe the relationships between the hydrologic/meteorological conditions and the concentration profiles for the analyte(s) of interest.

When high-quality field measurements are made at high frequency and presented along with hydrologic and meteorological information, this is perhaps the most compelling data that can be presented in support of the effort necessary to make the measurements. And for many this type of data is the "holy grail" of water quality monitoring, as it provides information that can be used without the assumptions required when data is collected at less frequent intervals or less than optimal hydrologic conditions. As more autonomous, robust, and cost-effective field sensors become available for a wide variety of water quality parameters, the quest for water quality data during all ambient conditions may someday be realized, supplanting the need for conventional sample collection and laboratory analysis.

REFERENCES

Benson, R.L., Y.B. Truong, I.D. McKelvie, B.T. Hart, G.W. Bryant, and W.P. Hilkmann. 1996. Monitoring of dissolved reactive phosphorus in wastewaters by flow injection analysis. Part 2. On-line monitoring system. *Water Res.* 30: 1965–1971.

Burton, G.A., and R.E. Pitt. 2002. *Stormwater Effects Handbook.* Boca Raton, FL: Lewis Publishers.

Chapin, T.P., J.M. Caffrey, H.W. Jannasch, L.J. Coletti, J.C. Haskins, and K.S. Johnson. 2004. Nitrate sources and sinks in Elkhorn Slough, California: Results from long-term continuous in situ nitrate analyzers. *Estuaries* 27: 882–894.

Chavez, F.P., D. Wright, R. Herlien, M. Kelley, F. Shane, and P.G. Strutton. 2000. A device for protecting moored spectroadiometers from biofouling. *J. Atmos. Ocean. Technol.* 17: 215–219.

Davis, J.R. 2001. *Copper and Copper Alloys.* ASM International Handbook Committee. Cleveland, OH: ASM International.

Feldmann, J. 2008. Onsite testing for arsenic: Field test kits. *Rev. Environ. Contam. Toxicol.* 197: 61–75.

Gardolinski, P.C.F.C., A.R.J. David, and P.J. Worsfold. 2002. Miniature flow injection analyser for laboratory, shipboard and in situ monitoring of nitrate in estuarine and coastal waters. *Talanta.* 58: 1015–1027.

Glasgow, H. G., J.M. Burkholder, R.E. Reed, A.J. Lewitus, and J.E. Kleinman. 2004. Real-time remote monitoring of water quality: A review of current applications, and advancements in sensor, telemetry, and computing technologies. *J. Exp. Mar. Biol. Ecol.* 300:409–448.

Hanrahan, G., M. Gledhill, P.J. Fletcher, and P.J. Worsfold. 2001. High temporal resolution field monitoring of phosphate in the River Frome using flow injection with diode array detection. *Anal. Chim. Acta.* 440: 55–62.

Hodge, J., B. Longstaff, A. Steven, P. Thornton, P. Ellis, and I. McKelvie. 2005. Rapid underway profiling of water quality in Queensland estuaries. *Mar. Pollut. Bull.* 51: 113–118.

Holm, C.E., Z. Chase, H.W. Jannasch, and K.S. Johnson. 2008. Development and initial deployments of an autonomous in situ instrument for long-term monitoring of copper (II) in the marine environment. *Limnol. Oceanogr. Meth.* 6: 336–346.

Hunt, D.T.E., and A.L. Wilson. 1986. *The Chemical Analysis of Water: General Principles and Techniques*. 2nd ed. The Royal Society of Chemistry, London.

Johnson, K.S., and L.J. Coletti. 2002. In situ ultraviolet spectrophotometry for high resolution and long-term monitoring of nitrate, bromide and bisulfide in the ocean. *Deep-Sea Res. I.* 49: 1291–1305.

Johnson, K.S., L.J. Coletti, and F.P. Chavez. 2006. Diel nitrate cycles observed with in situ sensors predict monthly and annual new production. *Deep-Sea Res. I.* 53: 561–573.

Johnson, K.S., J.A. Needoba, S.C. Riser, and W.J. Showers. 2007. Chemical sensor networks for the aquatic environment. *Chem. Rev.* 107: 623–640.

Lyddy-Meaney, A.J., P.S. Ellis, P.J. Worsfold, E.C.V. Butler, and I.D. McKelvie. 2002. A compact flow injection analysis system for surface mapping of phosphate in marine waters. *Talanta.* 58: 1043–1053.

Muller, B., M. Reinhardt, and R. Gachter. 2003. High temporal resolution monitoring of inorganic nitrogen load in drainage waters. *J. Environ. Monit.* 5: 808–812.

Pellerin, B.A., B.D. Downing, C. Kendall, R.A. Dahlgren, T.E.C. Kraus, J. Saraceno, R.G.M. Spencer, and B.A. Bergamaschi. 2009. Assessing the sources and magnitude of diurnal nitrate variability in the San Joaquin River (California) with an in situ optical nitrate sensor and dual nitrate isotopes. *Freshwater Biol.* 54: 376–387.

Plant, J.N., K.S. Johnson, J.A. Needoba, and L.J. Coletti. 2009. NH_4-digiscan: An in situ and laboratory ammonium analyzer for estuarine, coastal, and shelf waters. *Limnol. Oceanogr. Meth.* 7: 144–156.

Steinmaus, C.M., C.M. George, D.A. Kalman, and A.H. Smith. 2006. Evaluation of two new arsenic field test kits capable of detecting arsenic water concentrations close to 10µg /L. *Environ. Sci. Technol.* 40**:** 3362–3366.

Wagner, R.J., R.W. Boulger, C.J. Oblinger, and B.A. Smith. 2006. Guidelines and standard procedures for continuous water-quality monitors-Station operation, record computation, and data reporting. U.S. Geological Survey Techniques and Methods 1–D3. http://pubs.water.usgs.gov/tm1d3 (accessed December 3, 2009).

9 Laboratory Qualifications for Water Quality Monitoring

Yuncong Li, Meifang Zhou, and Jianqiang Zhao

CONTENTS

9.1 INTRODUCTION

Water quality monitoring program success depends on the quality of sampling and laboratory analysis. An ideal water quality monitoring plan should include an integrated team with water sampling staff and laboratory personnel. However, most monitoring programs have separate teams for water sampling and sample analysis, and some programs dispatch their samples to an analytical laboratory. Usually, the water quality sampling team is not familiar with laboratory analytical techniques and vice versa. Accurate laboratory analysis will not overcome mistakes in the sample collection process. If laboratory personnel have the knowledge of where and how samples are collected, atypical samples could be rejected when received or before analysis. If field personnel know how field sample preservation affects laboratory analysis, the frequency of sample rejection in the laboratory would be lower. (Examples of sample conditions that can cause rejection in the field or laboratory are listed in Table 9.1.) Therefore, the sampling and laboratory analytical teams need to understand both sampling procedures and laboratory analyses. Because water quality sampling has been described in detail in Chapters 4, 5, 6, 7, and 8, the sole purpose of this chapter is to provide information on the requirements for laboratories that analyze samples for water quality monitoring programs. The chapter covers laboratory accreditation, establishment of an accredited laboratory, and selection of an analytical laboratory if an in-house laboratory is not available.

TABLE 9.1
Rejection Criteria for Water Quality Samples in Field and Laboratory

Observation(s)	Field and Laboratory Actions
1. Presence of insect(s)	Field personnel must discard and recollect. Laboratory should reject at login. Data are flagged only if contaminant was present in sample bottle and the sample was analyzed.
2. Presence of vegetation particulates	Field personnel may discard and recollect the sample if it is deemed not representative of the sampling site. Laboratory may reject at login if there is an excessive amount of vegetation particulates in the sample that could interfere with analysis. If analyzed, laboratory will enter comments onto the data.
3. Presence of unusually large amount of sediment	The presence of sediment or particulates in a sample may be representative of some areas or typical for some projects (e.g., construction areas, routine areas after a storm or pumping event, or unsettled areas). Field personnel must determine whether a sample is representative and must document the observed condition in the field notes, which need to be referred to by laboratory staff unless the field personnel and/or their supervisors are contacted to verify the sample. If it is deemed that the amount of sediment present in the sample would interfere with analysis, the sample may not be processed. Any unusual condition of received samples must be noted in the login comments. If analyzed, comments should be inserted in the data sheet.
4. Presence of potential contaminant was observed but values < MDL	Data qualification is not required. Presence of potential contaminant must be noted in the comments for the sample.
5. H_2SO_4 preserved nutrient sample pH <1.3 or >2	Field personnel must discard and recollect. Laboratory staff should reject at login. Check the pH of the sample using pH strip (0–2.5).
6. Nutrient sample preserved with HNO_3	Field personnel must discard and recollect. Laboratory staff should reject at login. Check the sample with nitrate test strip.

9.2 WHAT IS AN ACCREDITED WATER QUALITY LABORATORY?

Accredited water quality laboratories are those that are accredited by an organization responsible for ensuring the attainment of acceptable standards for laboratory analysis. Laboratory accreditation and certification, which are indicators of a quality laboratory, are sought for a variety of reasons: (1) to validate or implement a quality system based on national or international standards and (2) to improve the ability to consistently produce valid results. Many environmental laboratories are accredited based on standards of the International Organization for Standardization (ISO) or national or regional adaptations of the ISO standards. Differences in these standards and organizations follow.

9.2.1 International Organization for Standardization Standards

The International Organization for Standardization (ISO) was founded in 1974 and is a network of 162 national standardization institutes. Most of its member institutes are governmental or government-mandated agencies while some are private sector entities. The American National Standards Institute (ANSI) represents the United States at the ISO. The ISO standards including management and technical requirements are developed by technical committees comprised of national delegations of experts from the industrial, technical, and business sectors. ISO/IEC 17025 (general requirements for the competence of testing and calibration laboratories) is the main standard used for laboratory accreditation (ISO/IEC 17025:2005 (E)) (ISO, 2005). It was initially issued by the ISO in 1999 and was updated in 2005. Management requirements deal with the operation and effectiveness of the quality management system within the laboratory, while technical requirements cover the competence of staff, methodology, and test/calibration of equipment. A laboratory has to establish and document its quality management system based on this standard before it can be considered for accreditation. The ISO does not provide accreditation service but rather an accreditation body accredits a laboratory based on compliance with the ISO standards. A copy of the ISO standards can be purchased from the ANSI at http://webstore.ansi.org.

9.2.2 National Environmental Laboratory Accreditation Conference Standards

Many water quality laboratories in the United States are accredited by the National Environmental Laboratory Accreditation Program (NELAP), which is administered by the NELAC Institute (TNI), a nonprofit organization. The current requirements for laboratories to be accredited under NELAP (http://www.nelac-institute.org) was created based on ISO/IEC 17025 and released in 2003, and is often referred to as the "2003 NELAC standard" (Parr, 2008). TNI has published a new set of standards that will replace the 2003 NELAC standard in 2011. The NELAP authorizes state governmental agencies as accreditation bodies to which laboratories seeking accreditation must apply. For example, environmental laboratories in Florida must apply for NELAP accreditation through the Florida Department of Health, which is the state accreditation body. While some states may opt to only accredit laboratories for chemistry and microbiology under the drinking water program, others may elect to operate a comprehensive program, which includes many types of analyses for hazardous waste, waste water, drinking water, air, and soil. If the accreditation body in a particular state does not offer accreditation for testing in conformance with a particular field of accreditation, laboratories may obtain primary accreditation for that particular field of accreditation from any other NELAP accreditation body. Figure 9.1 shows the steps required for laboratory accreditation in the United States.

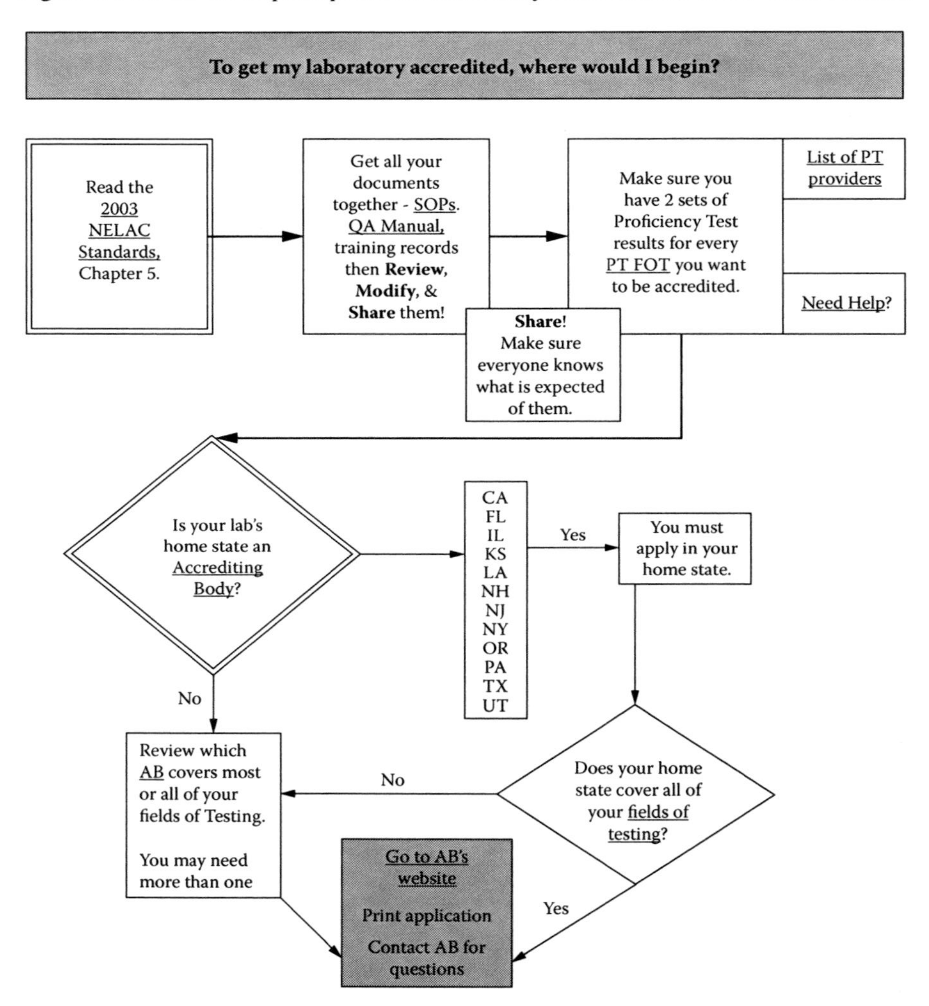

FIGURE 9.1 Steps for laboratory accreditation in the United States (adapted from http://www.nelac-institute.org). AB = accreditation body (the organization granting your accreditation); FOT = fields of testing. Each combination of matrix (nonpotable water, drinking water, solid and chemical materials), method/technology, and analyte is considered an FOT. PT FOT—FOTs that have PT requirements for accreditation; SOP—standard operating procedures; QA Manual—document that contains or addresses the lab-specific collection of policies, procedures, and quality related documents. (Adopted from the NELAC Institute Web site, www.nelac-institute.org.)

After the laboratory is accredited and receives certification, the Environmental Testing Laboratory Certificate is displayed at all times in a prominent place in the laboratory. Accredited laboratories have to pass at least two proficiency tests from recognized proficiency test (PT) sample providers per year. An accreditation agency will inspect an accredited laboratory at least once every 2 years. All laboratory certifications expire on July 1 of each year and can be renewed annually.

9.2.3 Department of Defense Environmental Laboratory Accreditation Program

Effective October 1, 2009, laboratories who are seeking to perform testing in support of the Department of Defense (DOD)'s environmental restoration programs need to be accredited in accordance with the USDOD Environmental Laboratory Accreditation Program (ELAP; USDOD, 2009a). This program requires laboratories to comply with the USDOD Quality System Manual for Environmental Laboratories (DOD QSM). The USDOD QSM manual is based on the NELAC quality systems standard (Chapter 5), which provides guidelines for implementing the international standard, ISO/IEC 17025. While the standard is based on NELAC requirements, laboratories providing services to the USDOD are not required to be accredited by NELAP (USDOD, 2009b). However, if the laboratory is not NELAC accredited to perform work for the USDOD in support of the environmental restoration program, accreditation under the USDOD ELAP must be obtained because NELAP accreditation does not satisfy specific requirements of the USDOD ELAP. More information for USDOD ELAP can be found at http://www.navylabs.navy.mil.

9.3 HOW TO ESTABLISH AN ACCREDITED WATER QUALITY LABORATORY?

Some water quality monitoring programs may have their own analytical laboratories or have funds to construct permanent, temporary, or mobile laboratories. These water quality laboratories can be formed for limited analyses of water samples with minimum funds. For example, an environmental manager for a vegetable packing house in Belize may want to monitor the quality of water discharged from the facility. Because there is no analytical laboratory available in the region, water samples have to be shipped out of country for analysis, causing sample holding time problems and increased shipping costs. A simple, temporary lab could be established for monitoring sensitive water quality parameters such as pH and bacteria. Water pH should be measured immediately and water samples for total coliforms and *Escherichia coli* should be analyzed as soon as possible after collection or at least within 30 hours for drinking water samples and 6 hours for source water samples. Many other constituents also require holding time to be less than 72 hours (see Chapter 6, Table 6.1). Many analyses such as pH, specific conductance, and turbidity can be easily measured in a temporary laboratory. Instructions for establishing a water quality laboratory follow.

9.3.1 Laboratory Personnel

A water quality laboratory should have qualified personnel to carry out required analyses. A minimum of three types of laboratory staff may be needed for a water quality analytical laboratory:

1. The laboratory technical director acts as the day-to-day supervisor of laboratory operations, including the monitoring of quality control and quality assurance (QA/QC) and data validation in the laboratory. The person should have a bachelor degree or at least an associate degree in chemical, environmental, and/or biological sciences with at least 2 years of experience in water quality analysis.
2. The QA officer is accountable for data quality and conducts internal audits. In a small laboratory with limited personnel, the QA officer can have dual responsibilities as deputy technical director or laboratory technician. These duties should not affect his/her performance on data quality control.
3. Laboratory technicians who prepare and analyze samples and use instruments should have education, training, and experience related to laboratory analysis. In addition, laboratory assistant(s) may be used to clean glassware, prepare samples, enter data, or other duties working under the supervision of a laboratory technician. The laboratory should have a policy for personnel training especially for method updates and data integrity.

9.3.2 Laboratory Spaces

The facility does not need to be large and sophisticated, but it should have adequate space for the following:

1. Sample receipt and preparation.
2. Sample storage.
3. Chemical and waste storage.
4. Office or data handling and storage.
5. A fume hood with a good exhaust system with positive air pressure that should run continuously. Typically this space should include an acid and corrosives vented storage cabinet located under the fume hood.
6. General chemistry area for sample digestion and filtration with laboratory countertops of chemical-resistant material (e.g., epoxy resin, solid phenolic, high-pressure plastic laminate, stainless steel, natural stone) and vinyl or concrete.
7. Instrumentation housing, which may need to be temperature-, humidity-, and dust-controlled. Special exhaust systems may be required for some instruments. Analytical balances should be set up on a stable table or countertop not subjected to flowing air.
8. Area for storing gases that have to be secured.
9. Washing area with sink, drain, drying racks, and deionized or distilled water.
10. A separate lunchroom if possible. The laboratory should never be used as a kitchen or dining area.

9.3.3 Sophisticated Laboratory Instruments, Equipment, and Supplies

9.3.3.1 Sophisticated Laboratory Instruments

Laboratory instruments depend on the requirements of water quality monitoring programs. While some parameters only need a simple and inexpensive meter, others, especially for chemical analyses, may require expensive and sophisticated instruments. Instrument requirements for analysis of selected water quality parameters based on USEPA-approved methods are listed in Table 9.2. The most common sophisticated instruments for laboratory chemical analysis include UV–VIS spectrophotometer, autoanalyzer (AA: continuous flow autoanalyzer [CFA] or flow-injection autoanalyzer [FIA]), discrete analyzer, atomic absorption spectrophotometer (AAS), inductively coupled plasma atomic emission spectrometer (ICP-AES), total carbon analyzer (TOC analyzer), ion chromatography (IC), gas chromatography (GC), and high performance liquid chromatography (HPLC).

9.3.3.1.1 UV–VIS Spectrophotometer

Previously known as a colorimeter, the UV–VIS spectrophotometer has had a long history of application in chemical analysis. The main components of the instrument include a light source (a beam of monochromatic light), a sample cell, and a detector used to measure transmittance or absorbance by comparing the intensity of light before and after passing through a sample. It is easy to use and inexpensive. Thus, a simple spectrophotometer may be the most practical instrument for many small laboratories or water quality monitoring programs. Many standard methods have been developed for UV–VIS spectrophotometers. Continuous flow, flow injection, and discrete analyzers all evolved based on the principle of the UV–VIS spectrometry plus automation.

9.3.3.1.2 Autoanalyzer

An automated spectrophotometer uses a special flow technique CFA or FIA where a segmented-flow analyzer with a multichannel peristaltic pump mixes samples with chemical reagents in a continuously moving stream. The CFA was first introduced as the technicon autoanalyzer in the late 1950s, followed by flow injection analyzers (FIA) in 1970s. The CFA systems separate samples with air bubbles and are most commonly used. The FIA analyzer is based on similar principles to CFA, but the sample is injected into the carrier stream without the introduction of air bubbles. Edwards et al. (2004) provide a detailed discussion of CFA and FIA. CFAs and FIAs are now available from Alliance Instruments (www.alliance-instruments.com), Astoria-Pacific International (www.astoria-pacific.com), FIAlab Instruments (www.flowinjection.com), Lachat Instruments (www.lachatinstruments.com), OI Analytical (www.oico.com), SEAL Analytical (www.seal-analytical.com), and Skalar (www.skalar.com).

9.3.3.1.3 Discrete Analyzer

Discrete analyzers are automated spectrophotometers that use a robot arm to transfer water samples and mix them with reagents in a cuvette before measuring the density of color. Discrete analyzers are easy to use, can run a large number of samples per day with fewer reagents, and generate less waste compared to UV-VIS. However,

TABLE 9.2
Selected USEPA-Approved Standard Methods and Instruments Used for Water Quality Analysis[a]

	USEPA Methods[b]		SM 21[st]ed[c]	
Water Parameters	**Number**	**Instruments[d]**	**Number**	**Instruments**
1. Alkalinity	310.2	Autoanalyzer	2320 B	pH meter
2. Aluminum	200.5 Rev 4.2	ICP-AES	3113 B	GFAAS
	200.7 Rev 4.4	ICP-AES	3120 B	ICP-AES
	200.8 Rev 5.4	ICP-MS		
	200.9 Rev 2.2	GFAAS		
3. Ammonia	350.1	Autoanalyzer		
4. Arsenic	200.5 Rev 4.2 1	ICP-AES	3113 B	GFAAS
	206.5	AAS	3114 B	AAS
	200.8 Rev 5.4 1	ICP-MS		
	200.9 Rev 2.2 1	GFAAS		
5. Calcium	200.5 Rev 4.2	ICP-AES	3113 B	GFAAS
	200.7 Rev 4.4		3120 B	ICP-AES
			3500-Ca B	—
6. Chemical oxygen demand (COD)	410.3	—		
	410.4	Autoanalyzer		
7. Chloride	300.0 Rev 2.1	IC	4110 B	IC
	300.1 Rev 1.0	IC	4500-Cl-B	Microammeter
			4500-Cl-D	Microammeter
8. Coliforms, total	1604	Incubator	9221 B	Incubator
			9221 D	
			9222 B	
			9222 C	
			9223	
9. Copper	200.5 Rev 4.2	ICP-AES	3111 B	AAS
	200.7 Rev 4.4	ICP-AES	3113 B	GFAAS
	200.8 Rev 5.4	ICP-MS	3120 B	ICP-AES
	200.9 Rev 2.2	GFAAS		
10. *Escherichia coli*				
11. Fluoride	300.0 Rev 2.1	IC	4110 B	IC
	300.1 Rev 1.0	IC	4500-F-B	Distillation
			4500-F-C	pH meter
			4500-F-D	UV-VIS
			4500-F-E	Autoanalyzer
12. Hardness	130.1	Autoanalyzer		
13. Kjeldahl & organic N	351.1	Autoanalyzer		
	351.2	Autoanalyzer		
14. Iron	200.5 Rev 4.2	ICP-AES	3111 B	AAS
	200.7 Rev 4.4	ICP-AES	3113 B	GFAAS
	200.9 Rev 2.2	GFAAS	3120 B	ICP-AES

TABLE 9.2 (continued)
Selected USEPA-Approved Standard Methods and Instruments Used for Water Quality Analysis[a]

	USEPA Methods[b]		SM 21[st]ed[c]	
Water Parameters	**Number**	**Instruments[d]**	**Number**	**Instruments**
15. Lead	200.5 Rev 4.2	ICP-AES	3113 B	GFAAS
	200.8 Rev 5.4	ICP-MS		
	200.9 Rev 2.2	GFAAS		
16. Magnesium	200.5 Rev 4.2	ICP-AES	3120 B	ICP-AES
	200.7 Rev 4.4	ICP-AES	3111 B	AAS
17. Manganese	200.5 Rev 4.2	ICP-AES	3111 B	AAS
	200.7 Rev 4.4	ICP-AES	3113 B	GFAAS
	200.8 Rev 5.4	ICP-MS	3120 B	ICP-AES
	200.9 Rev 2.2	GFAAS		
18. Mercury	200.8 Rev 5.4	ICP-MS	3112 B	AAS
	245.1 Rev 3.0	AAS		
	245.2	Autoanalyzer		
	245.7	CVAFS		
19. Nitrate	300.0 Rev 2.1	IC	4110 B	IC
	300.1 Rev 1.0	IC	4500-NO_3-D	pH meter
	325.2 Rev 2.0	Autoanalyzer	4500-NO_3-E	UV-VIS
	352.1	UV-VIS	4500-NO_3-F	Autoanalyzer
20. Orthophosphate	300.0 Rev 2.1	IC	4110 B	IC
	300.1 Rev 1.0	IC	4500-P-E	UV-VIS
	365.1 Rev 2.0	Autoanalyzer	4500-P-F	Autoanalyzer
21. Pesticide—Atrazine	505 Rev 2.1	GC-ECD		
	507 Rev 2.1	GC-NPD		
	508.1 Rev 2.0	GC-ECD		
	525.2 Rev 2.0	GC-MS		
	551.1 Rev 1.0	GC-ECD		
22. pH	150.1	pH meter	4500-H-B	pH meter
	150.2			
23. Phosphorus, total	365.3	UV-VIS		
	365.4	Autoanalyzer		
24. Potassium				
25. Sodium	200.5 Rev 4.2	ICP-AES	3111 B	AAS
	200.7 Rev 4.4	ICP-AES		
26. Specific conductance			2510 B	Conductivity meter
27. Sulfate	300.0 Rev 2.1	IC	4110 B	IC
	300.1 Rev 1.0	IC	4500-SO_4-C	Oven
	375.2 Rev 2.0	Autoanalyzer	4500-SO_4-D	Oven
			4500-SO_4-E	UV-VIS
			4500-SO_4-F	Autoanalyzer

(continued)

TABLE 9.2 (continued)
Selected USEPA-Approved Standard Methods and Instruments Used for Water Quality Analysis[a]

	USEPA Methods[b]		SM 21st ed[c]	
Water Parameters	**Number**	**Instruments[d]**	**Number**	**Instruments**
28. Total dissolved solids	120.1	Conductivity meter	2540 C	Oven
29. Total organic carbon (TOC)			5310 B	TOC analyzer
			5310 C	—
			5310 D	—
30. Turbidity	180.1 Rev 2.0	Turbidimeter	2130 B	Turbidimeter

[a] Based on Guidelines Establishing Test Procedures for the Analysis of Pollutants Under the Clean Water Act; National Primary Drinking Water Regulations; and National Secondary Drinking Water Regulations; Analysis and Sampling Procedures; Final Rule (USEPA 2007)

[b] USEPA methods can be found at the Web site of Test Method Collection, http://www.epa.gov/osa/fem/methcollectns.htm.

[c] Based on the Standard Methods for the Examination of Water & Wastewater, 21st Edition (Eaton et al., 2005).

[d] AAS—atomic absorption spectrophotometer; AES—atomic emission spectrometer; CVAFS—cold vapor atomic fluorescence spectrometer; GC—gas chromatography; ECD—electron capture detector; NPD—nitrogen-phosphorus detector; MS—mass spectrometry; GFAAS—graphite furnace atomic absorption spectrometry; IC—ion chromatography; ICP—inductively coupled plasma; and UV-VIS—UV-VIS spectrophotometer.

they are much more expensive than a regular spectrophotometer. Currently, only a few standard methods have been developed for discrete analyzers and method detect limits (MDLs) for most elements are still higher than these for flow analyzers. Several instrument manufacturers market discrete analyzers: Astoria-Pacific International, OI Analytical, SEAL Analytical, Skalar Analytical, Systea Scientific (www.easychem.com), and Westco Scientific Instruments (www.westcoscientific.com).

9.3.3.1.4 Atomic Absorption Spectrophotometer (AAS)

Since its introduction in 1955, AAS has become one of the most common methods for elemental analysis. In a flame atomic absorption spectrometer (FAAS), the water sample is aspirated, aerosolized, and mixed with combustible gases (e.g., acetylene, air, nitrous oxide), then vaporized and atomized in a flame at temperatures of 2100 to 2800°C. The atoms in the sample are transformed into free, unexcited ground state atoms, which absorb light at specific wavelengths. A light beam from a lamp whose cathode is made of the element of interest is passed through the flame. The amount of light absorbed is proportional to the concentration of the element in the sample. If an AAS is equipped with a graphite furnace (electrically heated atomizer) instead of a standard burner head, the instrument is called a *graphite furnace atomic absorption spectrophotometer* (GFAAS), which has better sensitivity and detection limits and requires much smaller sample volumes than FAAS for most elements. However, it is more expensive and not as easy to operate as FAAS. Many scientific instrument companies market AAS.

9.3.3.1.5 Inductively Coupled Plasma Atomic Emission Spectrometer (ICP-AES)

First introduced in 1973, the ICP-AES, which consists of an inductively coupled plasma source and an optical spectrometer, has become increasingly popular for metal analysis. An argon plasma source is used to produce excited atoms and ions that emit light at specific wavelengths depending on the element present. This light is detected by an optical spectrometer and compared to standards to calculate the concentrations of the elements of interest. The ICP-AES has time-saving multi-element capability and large linear calibration ranges with less interference, but is more expensive to purchase and maintain than FAAS or GFAAS. Csuros and Csuros (2002) provide detailed descriptions and comparisons of AAS and ICP-AES.

9.3.3.1.6 Ion Chromatography (IC)

Developed in the 1970s, IC uses ion-exchange resins to separate ionic components in water samples based on their interaction with the resin. Sample solutions pass through a pressurized chromatographic column where ions are absorbed by resins. While a mobile phase or eluent (e.g., sodium hydroxide or potassium hydroxide) passes through the column, the absorbed ions separate from the resin and pass through a detector (conductivity detector, electrochemical detector, or UV director). The retention time of different species determines the ionic concentrations in the sample. The IC has been effectively used to determine chloride, fluoride, nitrite, nitrate, sulfate, and phosphate, but detection limits for phosphate are not as satisfactory as those for other anions. It is still not widely used for analyses of cations. Tabatabal et al. (2004) provide a detailed discussion of the IC for environmental analyses. Main manufacturers of ion chromatographic systems are Dionex (www.dionex.com), Lachat (www.lachatinstruments.com), Metrohm (www.brinkmann.com), and Waters (www.waters.com).

9.3.3.1.7 Total Carbon Analyzer (TOC Analyzer)

Since its development in the 1950s, high-temperature catalytic oxidation (HTCO) is one of the two commonly used methods for determination of total or dissolved organic carbon in water samples (Wurl and Sin, 2009). HTCO has become the most popular method since the late 1980s because of improvements in the methodology. Organic carbon is usually measured as the nonpurgable organic form, which is determined by acidifying samples to a pH less than 2.0 and sparging with CO_2-free air or oxygen to remove inorganic carbon. The sample is then injected into a combustion tube, which has been packed with oxidation catalyst and heated to 680°C. Several (3–5) pure platinum wire gauzes (~0.5 cm size cubes) are usually placed at the top of the column bed instead of quartz wool for seawater TOC analysis to prevent the formation of a salt cake on the top of the column bed, which causes degradation of the peak shape. The organic carbon in water, present in various forms and oxidation states, is oxidized to carbon dioxide (CO_2). The CO_2 is then measured directly by a nondispersive infrared detector. The peak area count is proportional to the TOC concentration of the sample. The HTCO method has been described in detail by Wurl and Sin (2009) and several commercial TOC analyzers based on the HTCO principle have been evaluated by Dafner and Wangersky (2002).

9.3.3.1.8 Gas Chromatography (GC)

First appearing on the market in 1955, the GC technique has advanced and is now widely used in the separation field. Generally, organic compounds with relatively high volatility can be analyzed by GC. When the sample is injected into the head of the GC column, a selected mobile phase (carrier gas) elutes the sample through the stationary phase where the separation of analytes occurs. At this point, the separated analytes are detected with either (a) flame ionization detector (FID), (b) a nitrogen-phosphorus detector (NPD), or (c) an electron-capture detector (ECD). With the FID method, the sample effluent from the capillary column is mixed with hydrogen and air and ignited. Organic compounds burning in the flame produce ions and electrons, which conduct electricity through the flame measured by an electrometer. This detector is most useful for analysis of many organic compounds in water samples. With the NPD, a highly specific thermionic detector for organically bound nitrogen and phosphorus electrically heats a glass bead containing an alkali metal until electrons are emitted. These electrons are captured by stable intermediates to form hydrogen plasma in which the column effluent is ionized. A polarizing field directs these resulting ions to a collector anode creating a current. Many pesticides can be analyzed with NPD; they are atrazine, captan, chlorpyrifos, DDVP, diazinon, eradicane, malathion, naled, parathion, and pirimiphos-methyl. With the ECD method, an electron from the emitter causes ionization of the carrier gas and the production of a burst of electrons. ECD is important for the detection and determination of chlorinated insecticides, such as endosulfan, dicofol, and tetradifon. PerkinElmer (www.perkinelmer.com), Thermo Scientific (www.thermo.com), Agilent Technologies (www.agilent.com), and Restek (www.Restek.com) are vendors for GC instruments.

9.3.3.1.9 High Performance Liquid Chromatography (HPLC)

Though HPLC was developed about two decades later than GC, it was rapidly improved by the development of column packing materials and on-line detectors. In the late 1970s, new methods including reverse phase liquid chromatography allowed for improved separation between very similar compounds. Unlike in GC where the mobile phase does not react with the analytes, the mobile phase used in HPLC reacts with analyte molecules and with the stationary phase of the analytical column. Therefore, the separation or elution in the column is a process where the mobile phase, the stationary phase, and the compounds of interest are all interactive. The ability of the solute to interact selectively with both the stationary and mobile phases in HPLC provides opportunities to achieve the desired separation. The most commonly used detectors in HPLC include UV adsorption and fluorescence detectors, which can detect nanogram (10^{-9} g) and picogram (10^{-12} g) quantities of a wide variety of materials, respectively. Most of the detectors employed in HPLC are nondestructive so that sample components can be collected easily as they pass through the detector, making it possible to use online double detectors for multiresidue analysis. Herbicides that can be detected with a UV detector include bromacil, cyanazine, diuron, imidacloprid, metalaxyl, norflurazon, and simazine. Pesticides of the carbamate class are commonly derivatized with a postcolumn module and analyzed with a fluorescence detector.

9.3.3.1.10 Mass Spectrometry (MS)

In the middle of the 20th century, MS was developed for measuring a variety of materials and compounds ranging from the identification of hydrocarbon mixtures in the petroleum industry and speciation of atoms and molecules on surfaces to studies of environmental contaminants (Millard, 1978). The fundamental principle of MS is based on the mass-to-charge ratio of the mobile ions formed in an electric spark. The instrument consists of an inlet system, an ion source, a mass analyzer, and a detector. The most commonly used inlet systems are gas chromatography and liquid chromatography. Various types of mass spectrometers are available from Waters, PerkinElmer, Thermo Scientific, Agilent Technologies, and Varian (www.varianinc.com).

9.3.3.2 Basic Laboratory Equipment and Supplies

The follow list provides a summary of basic laboratory supplies needed for water quality analyses.

1. Bench-top pH and conductivity meters with electrodes. Some combination meters can be used for both measurements and many hand-held meters often do not work well.
2. Analytical balance with precision of 0.0001 g for preparing standards, weighing small samples, calibrating pipettes, and other activities. A set of weighing papers, dishes, and spatula may be needed.
3. Top loading balance with precision of 0.01 g for preparing reagents and weighing large samples.
4. Digestion block and/or an autoclave digestion system for digesting samples.
5. Ovens for drying samples and chemicals.
6. Hotplates and stirrers for sample preparation.
7. Refrigerators and freezers for storing samples and standards.
8. Power backup systems for major instruments, refrigerators, and freezers.
9. Incubator for biological analyses.
10. Autoclave for biological and chemical analyses.
11. Muffle furnace for sample digestion.
12. Automatic titrator for titration that can be replaced with regular burettes if funding is an issue.
13. Pipettes for dispensing reagents and solutions. They can be grade A glass pipettes, hand-held or automatic pipettes (which should be calibrated every three months) with different size tips.
14. Glassware: beakers, bottles for DI water, and reagent and standard storage (plastic and glass), funnels, graduated cylinders, digestion tubes, grade A volumetric flasks, petri dishes, and desiccators.
15. Water purification system for reagent grade water, which has low electrical conductivity (American Society for Testing and Materials standard is <0.056 μS/cm for type I and the ISO Standard is <0.1 μS/cm for grade 1). Deionized water can be obtained from an ion exchange column system,

which is relatively inexpensive and removes dissolved inorganic ions effectively, but is ineffective for organic compounds, particles, or bacteria. A distillation system can remove a broad range of contaminants and produce sterilized distilled water but many contaminants are carried to some extent into the condensate; it requires careful maintenance to ensure purity and consumes large amounts of tap water for cooling and electrical energy for heating. Distilled and deionized (DD) water or double D water can be obtained by sequentially treating water by distillation and deionization.

16. A laboratory should have access to analytical reagent grade chemicals and high purity standards. ACS reagent chemicals meet the specifications of the American Chemical Society (ACS). Analytical reagents can be used in a wide variety of analytical techniques for quality control, research, and development. Trace metal grade HNO_3 should be used for trace metal analysis. Most purity standards that can be purchased from vendors are traceable to the national standard/certified reference materials.
17. Glassware should be cleaned according to the following procedures: (1) properly dispose of any waste contents into a waste container or down the sink, according to the waste disposal rules (see waste management/equipment); (2) remove all labels and rinse the glassware once with tap water; fill a plastic bucket with deionized or distilled water and add one capful of a laboratory grade "Liqui-Nox" detergent (skip this step for the glassware in trace metal analysis); (3) soak the glassware for at least 1 h; rinse with deionized or distilled water 3 to 4 times or until the soapy residue is no longer present; (4) soak the glassware overnight in 5% HCl for at least 4 h (soak the glassware in 25% HNO_3 for trace metal analysis); (5) rinse the glassware with deionized or distilled water 5 times; and (6) store clean and dry glassware in a designated place.
18. Waste management/equipment: Most samples (or digested samples), such as total phosphorus and total Kjeldahl nitrogen (TKN), use $CuSO_4$ and not mercury oxide as the catalyst, hence reagents and standards pose little threat to the environment when disposed of down the sink with tap water for dilution and rinsing. A limestone bath may be required to neutralize acid in samples, reagents, and standards before disposal into the sewage system. Laboratories sometimes generate hazardous wastes that must be disposed of according to federal, state, and local regulations governing waste management. Expired trace metals standard stock should be kept in the original bottle for pick up by a waste management company. Organic solvents used in sample extraction (such as acetone for chlorophyll extraction), HPLC mobile phases, and reagents (phenol in ammonia analysis) should be stored in a waste drum for later disposal; a login book should be kept to track the volume and estimated concentrations of wastes.
19. Laminar flow cabinets and/or fume hoods, depending on the nature of the laboratory research: Laminar flow cabinets are needed in order to provide a sterile environment for working with samples. Fume hoods are needed

in order to work safely with potentially hazardous fumes. Both types of equipment should be inspected and re-certified annually.

20. Safety equipment: Hand-held chemical emergency fire extinguisher (ABC type), eyewash and deluge shower and chemical spill kit with easy access, gloves, safety goggles, and first aid supplies are needed. Material Safety Data Sheets (MSDS) should also be made readily available for all chemicals used or analyzed in the laboratory. Burton and Pitt (2002) provide a detailed discussion of laboratory safety issues.

9.3.4 Developing a Laboratory Quality Manual and Obtaining a Laboratory Accreditation

The NELAP requires that all accredited laboratories have a quality manual to set up detailed procedures for QA/QC protocols in laboratory analyses, to provide evidence of management's commitment to quality, to show compliance with accreditation requirements, and to communicate information about the quality management system to staff members, customers, and assessors or inspectors. An example of the table of contents of a quality manual is provided in Appendix 9.1. Briefly, the quality manual includes a description of the laboratory's organizational structure, a quality policy statement, the policies for each of the quality system essentials, references supporting processes, procedures, forms, and records. It should be accessible to all staff and reflect laboratory practices. After a quality manual is completed and before application of laboratory accreditation is started, the laboratory has to pass two single-blind, single-concentration proficiency testing (PT) studies. The PT samples should be ordered from the Proficiency Testing Provider Accreditor (PTPA) approved PT providers and analyzed following the quality manual. The procedure for applying for laboratory accreditation is presented in Figure 9.1 and detailed information for NELAP laboratory accreditation can be found at http://www.nelac-institute.org. For laboratories that have never been accredited, the process usually takes 12 to 18 months.

9.4 HOW TO SELECT AN ANALYTICAL LABORATORY

If a water quality program does not have its own laboratory, samples can be sent to government, university, or private laboratories for analysis. The laboratory selected must meet the requirements of water quality monitoring programs. Recommendations have been presented by Wilson (1995) and Nielsen and Nielsen (2007). In general, the following criteria should be considered when selecting a laboratory:

1. The laboratory should be accredited by NELAP or other accreditation body that follows ISO/IEC 17025 standards. Some laboratories are only certified for either drinking or nonpotable water, while others are certified for both. The list of NELAP-accredited laboratories can be found at http://www.nelac-institute.org/accred-labs.php.

2. The laboratory must use USEPA-approved or other standard methods for water analyses. The USEPA methods can be found at http://www.epa.gov/osa/fem/methcollectns.htm. Since the standard method usually needs to be adjusted for best performance and fit a specific instrument, the laboratory should develop a standard operating procedure (SOP) based on USEPA-approved or other standard methods. Such SOPs can be found within the 2003 NELAC Standard, page 208 (http://www.nelac-institute.org/docs/2003nelacstandard.pdf). Chapter 10 provides a more detailed discussion of analytical methods.
3. If a certified laboratory is not available, it is still necessary to enquire whether a potential laboratory can meet requirements of NELAC or ISO/IEC 17025.
4. To assure reliable results, the laboratory should have a quality manual that should also cover many other issues including qualifications of the laboratory staff.
5. Sample transport issues should be considered and discussed with the laboratory. The laboratory may or may not receive and analyze samples after working hours or on the weekend. It is particularly important for analyses that must occur within the holding time.
6. The laboratory should be visited and inspected to become familiarized with the facility and to meet the laboratory technicians.
7. The laboratory should be willing to work with water sampling personnel to discuss any issues related to sample analysis and reporting, as well as the necessary sample preservation and protocols to be followed prior to submission to the lab.

9.5 SUMMARY

Laboratory analyses are critical to water quality programs as many parameters cannot be accurately or easily assessed in the field. Laboratories are evaluated and may or may not receive special accreditation; accredited laboratories should be used when at all possible to ensure the credibility of the analyses results. It is also possible for water quality monitoring programs to establish their own mobile, temporary, or permanent laboratories. A detailed summary of current equipment commonly used in analyses is provided as well as a summary of basic laboratory supplies for those interested in establishing a laboratory.

ACKNOWLEDGMENTS

Authors wish to thank Jerry Parr from the NELAC Institute for reviewing the chapter and providing information on laboratory accreditation and also thank Sam Allen and Yin Chen for reviewing the chapter.

REFERENCES

Burton, G.A., and R.E. Pitt. 2002. *Stormwater Effects Handbook: A Toolbox for Watershed Managers, Scientists, and Engineers*. Boca Raton: Lewis Publishers.

Csuros, M., and C. Csuros. 2002. *Environmental Sampling and Analysis for Metals*. Boca Raton: Taylor & Francis.

Dafner, E.V., and P. J. Wangersky. 2002. A brief overview of modern directions in marine DOC studies: Part I.—Methodological aspects. *J. Environ. Monit.* 4: 48–54.

Eaton, A.D., L.S. Clesceri, E.W. Rice, A.E. Greenberg, and M.A.H. Franson. 2005. Standard methods for the examination of water and wastewater. Washington, D.C.: American Public Health Association.

Edwards, A.C., M.S. Credder, K.A. Smith, and A. Scott. 2004. Continuous-flow, flow-injection, and discrete analysis. In *Soil and Environmental Analysis*, ed. K.A. Smith and M.S. Cresser, 137–187. New York: Marcel Dekker.

ISO (International Organization of Standardization). 2005. ISO/IEC 17025:2005 (E), Second Edition. 2005-05-15. General requirements for the competence of testing and calibration laboratories. Geneva: International Organization of Standards.

Millard, B.J. 1978. *Quantitative Mass Spectrometry*. London: Heyden Publishing.

Nielsen, D.M. and G.L. Nielsen. 2007. *The Essential Handbook of Ground-Water Sampling*. Boca Raton: Taylor & Francis.

Parr, J.L. 2008. History and future of laboratory accreditation. The NELAC Institute, Presented at the Water Environment Federation's Annual Technical Exhibition and Conference (WEFTEC), October 20, 2008. http://www.nelac-institute.org/docs/pubs/NELAC%20history2008.pdf. (accessed December 30, 2009).

Tabatabal, M.A., N.T. Basta, and S.V. Karmarkar. 2004. Ion chromatography. In *Soil and Environmental Analysis*, ed. K.A. Smith and M.S. Cresser, 189–234. New York: Marcel Dekker.

USDOD (U.S. Department of Defense). 2009a. Fact Sheet: DoD Environmental Laboratory Accreditation Program. http://www.navylabs.navy.mil/Archive/DoDV3.pdf. (accessed December 30, 2009).

USDOD (U.S. Department of Defense). 2009b. Department of Defense Quality System Manuel for Environmental Laboratories. DoD Environmental Data Quality Workgroup, Version 4.1. http://www.navylabs.navy.mil/QSM%20Version%204.1.pdf. (accessed December 30, 2009).

USEPA (Environmental Protection Agency). 2007. Guidelines establishing test procedures for the analysis of pollutants under the Clean Water Act; national primary drinking water regulations; and national secondary drinking water regulations; analysis and sampling procedures, final rule. Washington D.C.: Environmental Protection Agency.

Wilson, N. 1995. *Soil Water and Ground Water Sampling*. Boca Raton: Taylor & Francis.

Wurl, O., and T.M. Sin. 2009. Analysis of dissolved and particulate Organic Carbon with the HTCO Technique. In *Practical Guidelines for the Analysis of Seawater*, ed. O. Wurl 34–48. Boca Raton: CRC Press.

APPENDIX 9.1 AN EXAMPLE OF THE TABLE OF CONTENTS FOR A QUALITY MANUAL

(Adapted from the NELAC Institute Web site, www.nelac-institute.org.)

TABLE OF CONTENTS

SECTION 3—INTRODUCTION AND SCOPE
- 3.1 Scope of Testing
- 3.2 Table of Contents, References, and Appendices
- 3.3 Glossary and Acronyms Used

SECTION 4—ORGANIZATIONAL ROLES AND RESPONSIBILITIES
- 4.1 Laboratory Organizational Structure
- 4.2 Responsibility and Authority

SECTION 5—QUALITY SYSTEMS
- 5.1 Quality Policy
- 5.2 Quality Manual

SECTION 6—DOCUMENT MANAGEMENT
- 6.1 Controlled Documents
  - 6.1.1 Document Changes to Controlled Documents
- 6.2 Obsolete Documents
- 6.3 Standard Operating Procedures
  - 6.3.1 Test Method SOPs

SECTION 7—REVIEW OF REQUESTS, TENDERS AND CONTRACTS
- 7.1 Procedure for the Review of Work Requests
- 7.2 Documentation of Review

SECTION 8—SUBCONTRACTING OF TESTS
SECTION 9—PURCHASING SERVICES AND SUPPLIES
SECTION 10—SERVICE TO THE CLIENT
- 10.1 Client Confidentiality

SECTION 11—COMPLAINTS
SECTION 12—CONTROL OF NON-CONFORMING WORK
SECTION 13—CORRECTIVE ACTION
- 13.1 Selection and Implementation of Corrective Actions
- 13.2 Monitoring of Corrective Action
- 13.3 Technical Corrective Action
- 13.4 Policy for Exceptionally Permitting Departures from Documented Policies and Procedures

SECTION 14—PREVENTIVE ACTION
SECTION 15—CONTROL OF RECORDS
- 15.1 Records Management and Storage
- 15.2 Legal Chain of Custody Records

SECTION 16—AUDITS AND MANAGEMENT REVIEW
- 16.1 Internal Audits
- 16.2 External Audits
- 16.3 Performance Audits
- 16.4 System Audits and Management Reviews

SECTION 17—PERSONNEL, TRAINING, AND DATA INTEGRITY
- 17.1 Job Descriptions
  - 17.1.1 Laboratory Director
  - 17.1.2 Technical Director(s)
  - 17.1.3 Quality Manager
- 17.2 Data Integrity and Ethics
- 17.3 Data Integrity and Ethics Training
- 17.4 General Training

SECTION 18—ACCOMMODATIONS AND ENVIRONMENTAL CONDITIONS

SECTION 19—TEST METHODS AND METHOD VALIDATION
- 19.1 Demonstration of Capability (DOC)
- 19.2 On-Going (or Continued) Proficiency
- 19.3 Initial Test Method Evaluation
  - 19.3.1 Limit of Detection (LOD)
  - 19.3.2 Limit of Quantitation
  - 19.3.3 Precision and Bias
  - 19.3.4 Selectivity
- 19.4 Estimation of Uncertainty
- 19.5 Laboratory-Developed or Non-Standard Method Validation
- 19.6 Control of Data

SECTION 20—EQUIPMENT
- 20.1 General Equipment Requirements
- 20.2 Support Equipment
  - 20.2.1 Support Equipment Maintenance
  - 20.2.2 Support Equipment Calibration
- 20.3 Analytical Equipment
  - 20.3.1 Maintenance for Analytical Equipment
  - 20.3.2 Initial Instrument Calibration
  - 20.3.3 Continuing Instrument Calibration
  - 20.3.4 Unacceptable Continuing Instrument Calibration Verifications

SECTION 21—MEASUREMENT TRACEABILITY
- 21.1 Reference Standards
- 21.2 Reference Materials
- 21.3 Transport and Storage of Reference Standards and Materials
- 21.4 Labeling of Reference Standards, Reagents, and Materials

SECTION 22—SAMPLE MANAGEMENT
- 22.1 Sample Receipt
- 22.2 Sample Acceptance

10 Laboratory Analyses

Yuncong Li, Renuka R. Mathur, and Lena Q. Ma

CONTENTS

10.1 INTRODUCTION

Although numerous methods for water quality analysis are currently available, many new analytical methods are being developed and old methods updated due to continual improvement of analytical technology. Often, selecting an analytical method from a plethora of new and existing methods is difficult, but very essential, since using an improper method can lead to inaccurate or false results. One purpose of this chapter is to explain the selection criteria of analytical methods for water quality monitoring programs. This is followed by a description of the analytical procedures for a selected group of water parameters. These procedures have been simplified from original descriptions of standard methods. It is the authors' intent to provide these simplified procedures to nonlaboratory professionals, including field sampling staff, so that they may gain insights on how water samples are analyzed in a laboratory. This chapter may also be used as a simple laboratory manual for laboratory technicians after they have read and understood the complete original methods. General information for each parameter is also presented to aid in interpreting the analysis and in referencing measured values, as compared to expected concentration ranges and numerical regulations.

10.2 SELECTION OF ANALYTICAL METHODS

10.2.1 Using Standard Methods

Standard methods refer to analytical methods that have been reviewed, validated, approved, and published by various standardization organizations or governmental agencies. Such methods generate defensible analytical data for scientific or legal purposes. Modification of these authoritative standard methods should be validated, documented, and, if possible, reviewed and approved by an accreditation organization. If a nonstandard method is used, it should be validated appropriately before use, and the results from the modified method should be comparable to those obtained using a standard method. Eaton et al. (2005) proposed a three-step process to validate new analytical methods, which includes (1) determination of single-operator precision and bias, (2) analysis of unknown samples, and (3) test of ruggedness.

The International Organization for Standardization (ISO) certifies standard methods for universal use. In the United States, most standard methods for water analysis have been developed by scientists affiliated with federal agencies, mainly the U.S. Environmental Protection Agency (USEPA), U.S. Geological Survey (USGS), and professional organizations, for example, the Association of Official Analytical Chemists (AOAC) International, American Society for Testing and Materials (ASTM) International, American Public Health Association (APHA), American Water Works Association, and Water Environment Federation. Some standard methods were developed by instrument manufacturers such as the method D6508 for fluoride (F) from Water Cooperation. The earliest standard methods were published by APHA in 1905, which were the Standard Methods (SM) for the Examination of Water and Wastewater. The latest version of this is the 21st edition, which was released in 2005 and includes over 350 testing methods; it is also available online (www.standardmethods.org; Eaton et al., 2005).

The USEPA has the obligation and authority to develop, evaluate, and certify analytical methods for drinking water, wastewater, and surface water in the United States. The USEPA's Forum on Environmental Measurements posts USEPA-approved analytical test methods for air, solid waste, and water at www.epa.gov/fem/methcollectns.htm. Recently, the USEPA issued the Methods Update Rule

(MUR) to approve 22 new methods for analysis of drinking water and wastewater (USEPA, 2007; 2008). It also deleted 105 USEPA methods, which were published in the USEPA's Methods for the Chemical Analysis of Water and Waste (USEPA, 1983b). In light of these changes, it is therefore necessary for a laboratory to review and update its analytical methods on a regular basis.

10.2.2 Meeting Method Performance Requirements

The National Water Quality Monitoring Network for U.S. Coastal Waters and Their Tributaries recently recommended laboratory performance requirements including detection limits for specific water quality parameters and stated that "if some areas within the sampling region or some times of year are consistently below the laboratory determined method detection limit, a better method with a lower detection limit needs to be used" (Table 10.1) (Caffrey et al., 2007). Therefore, method detection limit (MDL) is one of the criteria for selecting analytical methods.

TABLE 10.1
Laboratory Performance Requirements Recommended by the National Water Quality Monitoring Network for U.S. Coastal Waters and Their Tributaries

Parameters	Expected Ranges	Method Detection Limits (MDLs)
Chlorophyll *a*	0.01–150 μg/L	0.01 μg/L
Conductivity/salinity	0–1,000 mS/cm	1–100 μS/cm
Dissolved ammonium	0.007–0.50 mg N/L	0.007 mg N/L
Dissolved nitrate plus nitrite	0.007–10.0 mg N/L	0.007 mg N/L
Dissolved ortho phosphate	0.001–5.0 mg P/L	0.001 mg P/L
Dissolved organic carbon	0.22–50 mg C/L	0.22 mg C/L
Dissolved inorganic carbon	3–24 mg C/L	3 mg C/L
Dissolved oxygen	0–15 mg O_2/L	0.1 mg/L
Dissolved silicate	0.003–4.0 mg Si/L	0.003 mg Si/L
Particulate nitrogen	0.01–100 %	0.01%
Particulate phosphorus	0.005–5.0 mg P/L	0.005 mg P/L
pH	1–12 pH	0.01 pH
Particulate carbon	0.01–100%	0.01%
Photosynthetically active Radiation	0.01–10,000 μmol/m^2/s	0.01 μmol/m^2/s
Total dissolved nitrogen	0.001–10.0 mg N/L	0.001 mg N/L
Total nitrogen	0.03–15.0 mg N/L	0.03 mg N/L
Total dissolved phosphorus	0.01–5.0 mg P/L	0.01 mg P/L
Total phosphorus	0.01–10.0 mg P/L	0.01 mg P/L
Total suspended sediments	1–20,000 mg/L	10 mg/L

Source: Caffrey, J., T. Younos, G. Kohlhepp, D.M. Robertson, J. Sharp, and D. Whitall. 2007. Nutrient Requirements for the National Water Quality Monitoring Network for U.S. Coastal Waters and Their Tributaries. http://acwi.gov/monitoring/network/nutrients.pdf.

In order to achieve low detection limits, advanced instruments, well-trained personnel, and strictly controlled environmental conditions are often required. In addition, sample preparation should be completed meticulously. Often these stringent requirements increase both analytical costs and time. Therefore, clients frequently request simple, inexpensive, and quick methods or devices for water analyses. However, such compromises almost always decrease the quality of analyses, and especially the method detection limits. Many water quality testing kits or simple meters are cheap, easy to use, and available on the market. However, most are inaccurate (Li et al., 2006). Generally, handheld meters are either lacking or have deficient calibration functions. Of course, both portable pH meters and pH paper serve some purposes sufficiently, such as determining whether water is acidic or alkaline. A portable ion selective electrode for nitrate (NO_3–N), such as in the Horiba Cardy meter, has been reported to be useful for measuring NO_3–N in water. However, an instrument with such an electrode cannot be used for water quality analyses because the guaranteed lower detection limit is 62 mg/L, whereas the drinking water standard for NO_3–N in the United States is 10 mg/L. Different expensive and sophisticated instruments also have various instrument detection limits and precisions. For example, total phosphorus in water can be determined using a spectrophotometer, discrete analyzer, flow analyzer/autoanalyzer, ion chromatograph (IC), or ICP-AES (see Chapter 9). In general, the detection limit is the lowest with an autoanalyzer (~1 μg P/L), followed by spectrophotometer (~10 μg P/L), discrete analyzer (~20 μg P/L), IC (~50 μg P/L) and ICP-AES (~80 μg P/L). Other factors should also be considered when a method is selected. These factors include (1) selectivity, for example, the ability to identify interferences; (2) precision including repeatability, reproducibility, or random errors; (3) bias-systematic errors; (4) quantitation limits and ranges; (5) number of samples per hour; (6) skill required for operator; (7) cost and availability of equipment; (8) whether pretreatment is required; and (9) regulatory standards for specific parameters.

10.2.3 Considering Method Comparability

Many analytical methods were developed based on the best technology and knowledge at that time. Traditional techniques of colorimetric, titrimetric, and gravimetric methods without automation have been used for water analysis for a long time. Historic water quality data were most likely generated using one of these methods. Similar principles, but better performances (in terms of sensitivity, accuracy, precision, selectivity, robustness, and speed), have changed many analytical methods from a simple chemical procedure to an advanced instrumental operation. For example, NO_3-N in water was usually analyzed with either an electrode or a colorimetric method with a simple, manual spectrophotometer. These methods provide reliable assessment but have high MDL values because of low sensitivity of instruments and human error during operation. In the 1970s, autoanalyzers and flow analyzers were developed and became quite popular in the 1990s. Many standard methods were developed based on automated colorimetric methods. Following autoanalyzers, ion chromatography (IC) was accepted for NO_3-N analysis especially for drinking water because of the ease of use and minimal generation of laboratory wastes. Currently over 23 analytical methods are available for NO_3-N analysis based on the National Environmental Methods Index (NEMI, www.nemi.gov; Table 10.2).

Many analytical laboratories change methods as frequently as every few years based on instruments, costs, convenience, and other factors. The changes usually do not affect short term (a few years) water quality programs; very few laboratories will change analytical methods during the middle of a short term monitoring project. However, it may significantly affect long term monitoring programs such as the Everglades Water Quality Monitoring program, ongoing since 1977 (Hanlon et al., 2010). The method change contributes to data uncertainty, which is discussed in Chapter 12. The analytical method is one of the contributors for laboratory analysis uncertainty. One of the problems in using historic water quality data are inconsistencies of analytical methods within the same monitoring program and incompatibility with other monitoring programs. To address problems of method comparability, a multiagency Methods and Data Comparability Board (MDCB) created by the National Water Quality Monitoring Council recommends basic requirements for environmental analytical methods. Accordingly, information required for each chemical method should include (1) method identifier information (source, title, citation, and date); (2) applicability to analytes and media/matrices; (3) method summary (general procedural description) with keywords; (4) major interferences; (5) equipment (major instrumentation and/or critical apparatus and techniques); (6) performance data [detection level, bias and/or accuracy (e.g., percent recovery), precision, range, etc.]; (7) quality-control requirements (reference standards); (8) sample handling requirements (container, preservation, storage, holding times, etc.); and (9) sample preparation (filtering, dilution, homogenization, digestion, etc.; Keith et al., 2005). In order to assist comparison of analytical methods, the MDCB developed an online database of the National Environmental Methods Index (NEMI) in 2002. It has become one of the most useful tools for evaluating method compatibility. The searchable database provides comparisons of methods through the interactive Web site. The database provides a list of all or most available analytical methods for a specified analyte. The information for each method includes method source, name, detection level, detection level type, bias, precision, spike level, instrumentation, relative cost, and more. Answers for the most sensitive, the most precise, and the most cost-effective method for a specific analyte can be found through the database. An example of a search result for NO_3-N analysis is presented in Table 10.2.

10.3 ANALYTICAL PROCEDURES FOR SELECTED WATER QUALITY PARAMETERS

Selected standard analytical methods for the 20 most commonly monitored water quality parameters are described in this section. These methods are readily available either on the Web site of NEMI (www.nemi.gov) or USEPA method online collection (www.epa.gov/fem/methcollectns.htm). Most of these parameters have several USEPA-approved methods available but only one of these methods will be presented here, which is based on the simplicity, the requirement of instruments, and

TABLE 10.2
Analytical Methods for Nitrate Founded in the National Environmental Methods Index (NEMI)

Method[a]	Source	Method Name	Detection Level	Detection Level Type[b]	Bias[c]	Precision[d]	Spiking Level[e]	Instrument	Relative Cost[f]
300.0	USEPA	Inorganic anions by IC	2 μg/L	MDL	103% Rec (SL)	2% RSD (SL)	10 mg/L	IC	$$
300.1	USEPA	Anions in water by IC	8 μg/L	MDL	95% Rec (SL)	14% RSD (SL)	10 mg/L	IC	$$
352.1	USEPA	Nitrate by colorimetry	0.1 mg/L	RNGE	102% Rec (ML)	14% RSD (ML)	0.5 mg/L	Spectr	$
353.2	USEPA	Nitrate–nitrite nitrogen by colorimetry	N/A	RNGE	N/A	N/A	%	Spectr/auto	$$
353.3	USEPA	Nitrate–nitrite by cadmium reduction and colorimetry	N/A	RNGE	N/A	N/A	%	Spectr	$
353.4	USEPA	Nitrate–nitrate in estuarine and coastal waters by automated colorimetry	0.075 ug/L	MDL	106% Rec (SL)	1.5% RSD (SL)	140 μg/L	Spectr/auto	$
4110B	SM*	Anions in water by IC	2.7 μg/L	MDL	106% Rec (SL)	2.6% RSD (SL)	2.5 mg/L	IC	$$
4110C	SM	Anions in water by IC	17 μg/L	MDL	103% Rec (SL)	N/A RSD (SL)	8 mg/L	IC	$$
4500-NO3-D	SM	Nitrate electrode	0.14 mg/L	ADL	N/A	N/A	mg/L	ISE	$$$
4500-NO3-E	SM	Nitrate in water after cadmium reduction	0.01 mg/L	RNGE	99% Rec	14% RSD (ML)	.5 mg/L	Spectr	$
4500-NO3-F	SM	Nitrate by automated cadmium reduction	0.5 mg/L	RNGE	96% Rec	4 RSD	.29 mg/L	Spectr/auto	$
4500-NO3-H	SM	Nitrate by automated hydrazine reduction	0.01 mg/L	RNGE	101% Rec (SL)	N/A	2.97 mg/L	Spectr/auto	$
9056A	EPA	Anion chromatography	0.1 mg/L	RNGE	N/A	N/A	mg/L	IC	$$
993.3	AOAC	Inorganic anions in water	0.3 mg/L	RNGE	N/A	N/A	mg/L	IC	$$
A00243	SDI	Nitrate in water by colorimetric assay	0.5 mg/L		N/A	N/A	10 mg/L	IA	$

D4327	ASTM	Anions in water by IC	0.42 mg/L	RNGE	100% Rec	9.5% RSD (ML)	0.42 mg/L	IC	$$
D5996	ASTM	Anionic contaminants in high-purity water by on-Line IC	0.02 μg/L		N/A	N/A	μg/L	CIE-UV	$$$
D6508	ASTM	Anions in water by CIE-UV	0.82 mg/L	MDL	142.2% Rec (ML)	8.1	1.99 mg/L	IC	$$
I-2057	USGS	Anions, dissolved water, IC	0.05 mg/L	RL	N/A	8.3% RSD (SL)	0.12 mg/L	IC	$$
I-2058	USGS	Anions, dissolved water, IC	0.01 mg/L	RL	N/A	7.7% RSD (SL)	0.06 mg/L	Color_strip	$
MS310	DOE	Nitrate in water by colorimetric test	5 mg/L	RNGE	N/A	N/A	mg/L	Color	$
NO_3 EasyChem	Systea Scientific	Nitrate in water by colorimetry	11 μg/L	MDL	N/A	N/A	mg/L	Color	$
Nitrate via V(III) reduction	UC-Davis	Nitrate via manual vanadium reduction	2 ng/L	MDL	99.6% Rec	0.35 RSD	10 μg/L	Spectr	$

[a] Standard methods.

[b] The detection level (DL) can be (1) the point where a substance will be detected, but not quantified (e.g., method detection limit; MDL); (2) the lowest concentration at which the analyte can be accurately quantified (e.g., minimum level; ML); the lowest level of an analyte that can be determined by an instrument ignoring the possibility of analyte loss/contamination through sample pretreatment procedures (e.g., instrument detection limit, IDL); (3) the lowest level in the calibration range. "N/A" indicates bias data are not available or could not be found for the method.

[c] The systematic or persistent distortion of a measured value from its true value (this can occur during sampling design, the sampling process, or laboratory analysis).

[d] An expression of the reproducibility of measurements. Analytical precision measures the variability associated with two or more replicate analyses.

[e] The concentration of an analyte in sample used to determine precision and accuracy.

[f] Relative cost per procedure of a typical analytical measurement using the specified methods (i.e., the cost of analyzing a single sample). Additional considerations affect total project costs (e.g., labor and equipment/supplies for a typical sample preparation, QA/QC requirements to validate results reported, number of samples being analyzed, etc.).

Source: Searched at nemi.gov.

the preference of authors. Method selection for a specific laboratory or monitoring program should follow the recommendations provided in section 10.2. Several sections in the complete methods are omitted from this document, including the scope and applications, definitions, interferences, safety, sample collection, preservation and storage, quality control, laboratory performance assessment, analyte recovery assessment and data quality, calibration and standardization, method performance, pollution prevention, waste management, and references for USEPA methods and some information for SM methods. However, all information omitted from this section is important and laboratory technicians should read the information carefully at least once a year to avoid operation deviation from the standard methods. Some information omitted may briefly be included in the introduction section of each water quality parameter.

10.3.1 Alkalinity

10.3.1.1 Introduction

Alkalinity is the capability of water to neutralize acid, often referred to as *buffering capacity*, which is mainly due to the presence of bicarbonates, carbonates, and hydroxides in the water. It is expressed as mg/L of $CaCO_3$ with criteria often set as low as 20 mg/L for surface water to protect aquatic life and less than 500 mg/L for irrigation water.

10.3.1.2 Analytical Methods

According to the NEMI database, 10 methods from USEPA, USGS, SM, and Hach Company are available for water alkalinity determination. The USEPA approved methods listed in the MUR include SM 2320B, EPA 310.2, ASTM D1067-02B, ASTM D1067-92 B, and USGS I-1030-85. The procedure described below is based on SM 2320B for total alkalinity (Eaton et al., 2005). The precision (standard deviation) of the method is ~1 mg/L for alkalinity between 10–500 mg/L. Water samples should be refrigerated at ≤6°C and analyzed within 14 days after sampling.

10.3.1.3 Apparatus

An automatic potentiometric titrator, a 200 mL beaker, and a 1 L volumetric flask are needed.

10.3.1.4 Reagents

1. Sodium carbonate: Dry 3-5g Na_2CO_3 at 250°C for 4 h and cool in a desiccator.
2. Sodium carbonate solution (0.025 M): Dissolve 2.5 g oven-dried sodium carbonate in deionized or distilled (DI) water and make up to 1 L.
3. Standard hydrochloric acid (0.1 M): Add 8.3 mL concentrated HCl into 800 mL DI water and make up to 1 L with DI water.
4. Standard hydrochloric acid (0.02 M): Dilute 200 mL 0.1 M HCl solution to 1 L with DI water. Standardize it with 0.025 M Na_2CO_3 solution and calculate concentration.

10.3.1.5 Procedure

1. Calibrate the automatic titrator.
2. Add 100 mL water sample into a flask.
3. Titrate with standard acid in the increments of 0.5 mL or less, mix well, and record pH.
4. Continue titration to pH 4.5 or lower.
5. Plot titration curve for pH values versus volumes of acid used.

10.3.1.6 Calculation

$$\text{Total alkakinity, mg/L CaCO}_3 = \frac{(2B - C) \times M \times 50000}{\text{mL of sample}}$$

where:

B = mL acid to first recorded pH,
C = total mL acid to reach pH 0.3 unit lower, and
M = molality of acid.

10.3.1.7 Source for the Detailed Method

Standard Methods for the Examination of Water and Wastewater, 21st Edition, 2320 B—titration method for alkalinity, pages 2-27 to 2-29 (Eaton et al., 2005).

10.3.2 Aluminum

10.3.2.1 Introduction

Aluminum (Al, CAS No. 7429-90-5) exists in all water sources because of its abundance on the earth (~8% of the earth's crust). The average concentrations of Al are from <0.1 μg/L in groundwater, 54 μg/L in drinking water, 400 μg/L in streams, and 5 μg/L in seawater (Eaton et al., 2005; Marcovecchio et al., 2007). It is not an essential element for plants and humans. High concentration of Al (>5 mg/L) in irrigation water can be toxic to plants. The USEPA lists Al as one of the secondary drinking water contaminants, with a maximum contaminant level (MCL) of 50–200 μg/L. USEPA also suggests 750 μg/L as criteria maximum concentration (CMC) and 87 μg/L as the criterion continuous concentration (CCC) for fresh water to protect aquatic life and human health.

10.3.2.2 Analytical Methods

According to the NEMI database, 28 methods from USEPA, USGS, NOAA, AOAC, ASTM, Standard Methods, and Hach Company are available for Al determination. The detection levels range from 0.03 μg/L with inductively coupled plasma atomic mass spectrometry (ICP-MS, SM 3125) to 500 μg/L with flame atomic absorption spectroscopy (FAAS, ASTM D857). The USEPA approved methods for drinking water or wastewater include graphite furnace atomic absorption spectroscopy (GFAAS, EPA Method 200.9 and SM 3113B), ICP-AES (EPA Methods 200.5 and 200.7, and SM 3120 B) or ICP-MS (EPA Method 200.8).

TABLE 10.3
Wavelengths, Detection Limits, and Concentration Ranges for GFAAS Method

Element	Wavelength (nm) [a]	MDL (μg/L)[a]	Optimum Concentration (μg/L)[b]
Aluminum (Al)	309.3	7.8	20–200
Antimony (Sb)	217.6	0.8	20–300
Arsenic (As)	193.7	0.5	5–100
Beryllium (Be)	234.9	0.02	1–30
Cadmium (Cd)	228.8	0.05	0.5–10
Chromium (Cr)	357.9	0.1	5–100
Cobalt (Co)	242.5	0.7	5–100
Copper (Cu)	324.8	0.7	5–100
Iron (Fe)	248.3	—	5–100
Lead (Pb)	283.3	0.7	5–100
Manganese (Mn)	279.5	0.3	1–30
Nickel (Ni)	232.0	0.6	5–100
Selenium (Se)	196.0	0.6	5–100
Silver (Ag)	328.1	0.5	1–25
Thallium (Ti)	276.8	0.7	—
Tin (Sn)	286.3	1.7	—

[a] MDLs provided only as a guide to instrumental limits. The method detection limits are sample dependent and may vary as the sample matrix varies. Sourced from Creed, J.T., T.D. Martin, and J.W. O'Dell. 1994. Method 200.9—Determination of trace elements by stabilized temperature graphite furnace atomic absorption, Revision 2.2. Environmental; Monitoring Systems Laboratory, Office of Research and Development, U.S. Environmental Protection Agency, Cincinnati, OH.

[b] From Eaton, A.D., L.S. Clesceri, E.W. Rice, A.E. Greenberg, and M.A.H. Franson. 2005. *Standard Methods for the Examination of Water and Wastewater*, 21st ed. Washington D.C.: American Public Health Association.

Source: USEPA 200.9 Revision 2.2.

The procedure described below is based on EPA 200.9 Rev 2.2 (Creed et al., 1994). The method is also used for analysis of other trace elements and the optimum concentration ranges and MDLs for these elements are listed in Table 10.3. The MDL and optimum concentration range for Al is 7.8 μg/L and 20–200 μg/L, respectively. It is best to analyze for Al immediately after sample collection, or samples should be acidified with HNO_3 (pH<2), which can then be stored for up to six months.

10.3.2.3 Apparatus

1. GFAAS
2. Al hollow cathode lamp or electrodeless discharge lamp
3. Argon gas (high-purity grade, 99.99%) and hydrogen gas

4. Autosampler capable of adding matrix modifier solutions to the furnace
5. A temperature adjustable hot plate or a digestion block

10.3.2.4 Reagents

1. Hydrochloric acid (1+1): Add 500 mL concentrated HCl to 400 mL DI water and dilute to 1 L.
2. Hydrochloric acid (1+4): Add 200 mL concentrated HCl to 400 mL DI water and dilute to 1 L.
3. Nitric acid (1+1): Add 500 mL concentrated HNO_3 to 400 mL DI water and dilute to 1 L.
4. Nitric acid (1+5): Add 50 mL concentrated HNO_3 to 250 mL DI water.
5. Nitric acid (1+9): Add 10 mL concentrated HNO_3 to 90 mL DI water.
6. Concentrated ammonium hydroxide.
7. Tartaric acid, ACS reagent grade.
8. Matrix modifier, dissolve 300 mg palladium (Pd) powder in concentrated HNO_3 (1 mL of HNO_3, adding 0.1 mL of concentrated HCl if necessary). Dissolve 200 mg of $Mg(NO_3)_2$ in water. Pour the two solutions together and dilute to 100 mL with DI water.
9. Standard stock solutions may be purchased or prepared from ultra-high purity grade chemicals. All compounds must be dried for one hour at 105°C, unless otherwise specified. For preparing Al stock solution: Dissolve 1.000 g of Al metal in an acid mixture of 4.0 mL of (1+1) HCl and 1.0 mL of concentrated HNO_3 in a beaker. When dissolution is complete, transfer solution to a 1 L flask, add an additional 10.0 mL of (1+1) HCl, and dilute to 1 L with DI water.
10. *Preparation of calibration standards.* Fresh calibration standards should be prepared every two weeks. Dilute each of the stock standard solutions with acid solution to levels appropriate to the operating range of the instrument. The instrument calibration should be initially verified using a quality control sample.
11. *Blanks.* Four types of blanks are required for this method. A calibration blank is used to establish the analytical calibration curve, the laboratory reagent blank (LRB) is used to assess possible contamination from the sample preparation procedure and to assess spectral background, the laboratory fortified blank (LFB) is used to assess routine laboratory performance, and a rinse blank is used to flush the instrument autosampler uptake system.
12. *Instrument performance check (IPC) solution.* The IPC solution is used to periodically verify instrument performance during analysis. It should be prepared in the same acid mixture as the calibration standards and approximate the midpoint of the calibration curve. The IPC solution should be prepared from the same standard stock solutions used to prepare the calibration standards and stored in a FEP bottle.
13. *Quality control sample (QCS).* For initial and periodic verification of calibration standards and instrument performance, analysis of a QCS is required. The QCS must be obtained from an outside source different from the standard stock solutions and prepared in the same acid mixture as the

calibration standards. The concentration of the analytes in the QCS solution should be such that the resulting solution will provide an absorbance reading of approximately 0.1.

10.3.2.5 Procedure

10.3.2.5.1 Sample Pretreatment

1. For the determination of total recoverable Al in water samples other than drinking water, transfer a 100 mL sample to a 250 mL beaker (smaller sample size can be used).
2. Add 2 mL (1+1) nitric acid and 1.0 mL of (1+1) hydrochloric acid to the sample. Place the beaker on the hot plate and set at 85°C in a fume hood. The beaker should be covered with an elevated watch glass. Reduce the volume of the sample to about 20 mL by gently heating at 85°C. Do not boil.
3. Gently reflux the sample for 30 minutes. Allow the beaker to cool. Transfer the sample solution to a 50 mL volumetric flask, make to volume with DI water, and mix well.

10.3.2.5.2 Sample Analysis

1. After the warm-up period but before calibration, instrument stability must be verified by analyzing a standard solution with a concentration 20 times the instrument detection limit (IDL) five times. The resulting relative standard deviation (RSD) of absorbance signals must be <5%.
2. Calibrate the instrument using the calibration blank and calibration standards prepared at three or more concentrations within the usable linear dynamic range of Al.
3. An autosampler must be used to introduce all solutions into the graphite furnace.
4. Once the standard, sample, or QC solution plus the matrix modifier is injected, the furnace controller completes furnace cycles and cleanout period as programmed. Al signals must be integrated and collected as peak area measurements. Background absorbance and background corrected Al signals should be determined. Flush the autosampler solution uptake system with the rinse blank between each solution injected.
5. After completion of the initial requirements of this method, samples should be analyzed in the same operational manner used in the calibration routine.
6. Determined sample Al concentrations that are 90% or more of the upper limit of calibration must be diluted with acidified reagent water and reanalyzed.

10.3.2.6 Calculation

$$\mu g\ Al/L = C \times F,$$

where C = Al concentration as read directly from GFAAS (μg/L), and F = dilution factor.

10.3.2.7 Source of the Detailed Method

USEPA Method 200.9; Determination of trace elements by stabilized temperature graphite furnace atomic absorption, Revision 2.2. (Creed et al., 1994; http://www.epa.gov/waterscience/methods/method/files/200_9.pdf, accessed on December 12, 2009).

10.3.3 Ammonia

10.3.3.1 Introduction

Ammonia nitrogen (unionized ammonia, NH_3 and ammonium, NH_4^+, CAS No. 7664-41-7) in water may come from decomposition of organic matter, atmospheric deposition, fertilizers from cropland, and/or wastewater from urban and industrial areas. Ammonia nitrogen concentrations are less than 0.2 mg/L in unpolluted groundwater and can be as high as 12 mg/L in surface water (WHO, 2003). Ammonia does not affect human health within normal concentrations, but the unionized ammonia is toxic to fish and other aquatic life even at a concentration as low as 0.02 mg/L, which is the water quality standard set by most states in the United States.

10.3.3.2 Analytical Methods

According to the NEMI database, 13 methods from USEPA, USGS and Standard Methods are available for analysis of ammonia in water. Ammonia can be determined with titrimetric method (for concentration > 1 mg N/L), ammonia-selective electrode method (for concentrations between 0.03 and 1400 mg N/L), and phenol method (for concentrations between 0.0003 and 2 mg N/L), but only the automated phenol method (EPA 350.1) is USEPA-approved for water analysis (O'Dell, 1993a). Water samples should be placed in a refrigerator ≤ 6°C before analysis. Acidified samples with H_2SO_4 (pH < 2) can be stored for 28 days.

10.3.3.3 Apparatus

1. An autoanalyzer or a continuous-flow analyzer. The system manifold for ammonia can be found in O'Dell et al. (1994) or Eaton et al. (2005).
2. An all-glass distilling apparatus with an 800–1000 mL flask.

10.3.3.4 Reagents

1. Ammonia-free water: deionized (DI) water.
2. Sulfuric acid 2.5 M, air scrubber solution. Carefully add 139 mL of concentrated sulfuric acid to ~500 mL DI water, cool to room temperature, and dilute to 1 L.
3. Boric acid solution (20 g/L): Dissolve 20 g H_3BO_3 in DI water and dilute to 1 L.
4. Borate buffer: Add 88 mL of 0.1 M NaOH solution to 500 mL of 0.025 M sodium tetraborate solution (5.0 g anhydrous $Na_2B_4O_7$ or 9.5 g $Na_2B_4O_7C_{10}H_2O$ per L) and dilute to 1 L with DI water.
5. Sodium hydroxide, 1 M: Dissolve 40 g NaOH in DI water and dilute to 1 L.

6. Sodium phenolate: Dissolve 83 g phenol in 500 mL of DI water. In small increments with agitation, carefully add 32 g of NaOH. Cool flask under running water and dilute to 1 L.
7. Sodium hypochlorite solution: Dilute 250 mL of a bleach solution containing 5.25% NaOCl to 500 mL with DI water.
8. EDTA (disodium ethylenediamine-tetraacetate) (5%): Dissolve 50 g of EDTA and about six pellets of NaOH in 1 L of DI water.
9. Sodium nitroprusside (0.05%): Dissolve 0.5 g of sodium nitroprusside in 1 L of DI water.
10. Ammonia stock solution: Dissolve 3.819 g of anhydrous ammonium chloride, NH_4Cl, dried at 105°C, in DI water, and dilute to 1 L.
11. Ammonia standard solution A (10 mg N/L): Dilute 10.0 mL of stock solution to 1 L with DI water.
12. Standard solution B (1 mg N/L): Dilute 10 mL of standard solution A to 100 mL.

10.3.3.5 Procedure

1. Preparation of equipment: Add 500 mL of DI water to an 800 mL Kjeldahl flask. Steam out the distillation apparatus until the distillate shows no trace of ammonia.
2. Distillation: Transfer the sample, the pH of which has been adjusted to 9.5, to an 800 mL Kjeldahl flask and add 25 mL of the borate buffer. Distill 300 mL at the rate of 6–10 mL/min. into 50 mL of 2% boric acid contained in a 500 mL flask.
3. Since the intensity of the color used to quantify the concentration is pH dependent, the acid concentration of the wash water and the standard ammonia solutions should approximate that of the samples.
4. Set up ammonia manifold. Allow both colorimeter and recorder to warm up for 30 minutes. Obtain a stable baseline with all reagents, feeding water through a sample line.
5. Analyze ammonia standard and blank first. Arrange ammonia standards in sampler in order of decreasing concentration of nitrogen. Complete loading of sampler tray with unknown samples.
6. Switch sample line from distilled water to sampler and begin analysis.

10.3.3.6 Calculation

Prepare a calibration curve by plotting instrument response against standard concentration. Compute sample concentration by comparing sample response with the standard curve. Multiply answer by appropriate dilution factor. Report only those values that fall between the lowest and the highest calibration standards. Samples exceeding the highest standard should be diluted or reanalyzed. Report results in mg NH_3-N/L.

10.3.3.7 Source of the Detailed Method

USEPA Method 350.1: Determination of ammonia nitrogen by semiautomated colorimetry. Revision 2.0. (O'Dell, 1993a; http://www.epa.gov/waterscience/methods/method/files/350_1.pdf, accessed on December 12, 2009).

10.3.4 Arsenic

10.3.4.1 Introduction

Arsenic (As, CAS No. 7440-38-2) in water may come from dissolution of arsenic minerals, atmospheric deposition, and chemicals from pesticides and industrial uses. The average arsenic concentrations are ~3 μg/L in surface water, 1–2 μg/L in groundwater, and 4 μg/L in seawater (Boyd, 2000; Weiner, 2007). Inorganic arsenic is toxic and carcinogenic to humans and animals. The USEPA primary drinking water standard MCL for As is 10 μg/L and the maximum contaminant level goal (MCLG) is zero. USEPA also suggests 340 μg/L for criteria maximum concentration (CMC) and 150 μg/L for the criterion continuous concentration (CCC) for fresh water to protect aquatic life and human health.

10.3.4.2 Analytical Methods

According to the NEMI database, 27 methods from USEPA, USGS, NOAA, AOAC, ASTM, Standard Methods, and Hach Company are available for As determination. The detection levels range from 0.03 μg/L with GFAAS (EPA Method 1632) to 53 μg/L and ICP-AES (ASTM D1976). The USEPA approved methods for analysis of As in water are AAS (SM 3114B), GFAAS (EPA Method 200.9 and SM 3113B), ICP-AES (EPA Method 200.5), and ICP-MS (EPA Method 200.8). The GFAAS is a good choice if interference is not a problem. Otherwise, a hydride generation-AAS can be used. The ICP-AES is used to evaluate samples with high As concentrations (>50 μg/L) and the ICP-MS is often used for samples with low concentrations without chloride interference. The procedure described below is based on EPA Method 200.9 (Creed et al., 1994).

10.3.4.3 Apparatus (See Section 10.3.2.3)

10.3.4.4 Reagents (See Section 10.3.2.4, Except for the Following)

Preparing standard As solution for method 200.9: Arsenic stock solution, stock (1000 mg As/L): Dissolve 1.3200 g of As_2O_3 in 100 mL of DI water containing 10.0 mL concentrated NH_4OH. Warm the solution gently to help dissolution. Acidify the solution with 20 mL concentrated HNO_3 and dilute to 1 L.

10.3.4.5 Procedure (See Section 10.3.2.5)

10.3.4.6 Calculation (See Section 10.3.2.6)

10.3.4.7 Source of the Detailed Method (See Section 10.3.2.7)

10.3.5 Calcium

10.3.5.1 Introduction

Calcium (Ca, CAS No. 7440-70-2) exists in all water sources because of its abundance on the earth (~4.9% of the Earth's crust). The concentrations of Ca are usually

< 20 μg/L in groundwater but can be as high as 200 mg/L and ~400 mg/L in seawater. Calcium concentrations in surface water vary greatly. Dissolved Ca is one of the main contributors to hard water. Calcium is an essential element for plants and humans. There is no toxic effect directly linked to high concentration of Ca in water except that high concentration of Ca in drinking water may increase the risk of kidney stones. Therefore, USEPA does not specify the maximum acceptable concentration for Ca in drinking and surface waters.

10.3.5.2 Analytical Methods

The NEMI database lists 17 methods from USEPA, USGS, AOAC, and Standard Methods available for Ca determination. The detection levels range from 3 μg/L (SM 3111B with FAAS) to 500 μg/L (EPA 215.1 titrimetry method). The USEPA-approved methods for Ca analysis in drinking water and wastewater include FAAS (3114B), GFAAS (SM 3113B), ICP-AES (EPA Methods 200.5 and 200.7), ASTM International methods (D511-03A, D511-03B, D511-93A, D511-93B, and D6919-03), and Standard Methods (3111B and 3500-Ca). The procedure described below is based on EPA 200.7 Rev 4.4, which is the USEPA-approved method for drinking water, surface water, and wastewater (Martin et al., 1994). The method is also used for analysis of other elements and the upper limit concentrations and MDLs for these elements are listed in Table 10.4. The MDL and upper limit concentration for Ca is 30 μg/L and 100 mg/L, respectively. It is best to analyze Ca immediately after collection or samples should be acidified with HNO_3 (pH < 2), which can be stored for up to 6 months.

10.3.5.3 Apparatus

ICP-AES with (1) background-correction capability, (2) radio-frequency generator, and (3) a variable speed peristaltic pump.

10.3.5.4 Reagents

1. Hydrochloric acid (1+1): Add 500 mL concentrated HCl to 400 mL DI water and dilute to 1 L.
2. Hydrochloric acid (1+4): Add 200 mL concentrated HCl to 400 mL DI water and dilute to 1 L.
3. Hydrochloric acid (1+20): Add 10 mL concentrated HCl to 200 mL DI water.
4. Nitric acid (1+1): Add 500 mL concentrated HNO_3 to 400 mL DI water and dilute to 1 L.
5. Nitric acid (1+2): Add 100 mL concentrated HNO_3 to 200 mL DI water.
6. Nitric acid (1+5): Add 50 mL concentrated HNO_3 to 250 mL DI water.
7. Nitric acid (1+9): Add 10 mL concentrated HNO_3 to 90 mL DI water.
8. Concentrated ammonium hydroxide.
9. Tartaric acid, ACS reagent grade.
10. Hydrogen peroxide, 50%, stabilized certified reagent grade.
11. Standard stock solution (1000 mg Ca/L): Suspend 2.4980 g $CaCO_3$ (dried at 180°C for one hour before weighing) in water and dissolve cautiously with a minimum amount of (1+1) HNO_3. Add 10.0 mL concentrated HNO_3 and dilute to 1 L with DI water.

TABLE 10.4
Wavelengths, Detection Limits, and Up-Limit Concentrations for ICP-AES Method

Element	Wavelength (nm) [a]	MDL (μg/L)[a]	Upper Limit Concentration (mg/L)[b]
Aluminum (Al)	308.215	45	100
Antimony (Sb)	206.833	32	100
Arsenic (As)	193.759	53	100
Barium (Ba)	493.409	2.3	50
Beryllium (Be)	313.042	0.27	10
Boron (B)	249.678	5.7	50
Cadmium (Cd)	226.502	3.4	50
Calcium (Ca)	315.887	30	100
Cerium (Ce)	413.765	48	—
Chromium (Cr)	205.552	6.1	50
Cobalt (Co)	228.616	7.0	50
Copper (Cu)	324.754	5.4	50
Iron (Fe)	259.940	6.2	100
Lead (Pb)	220.353	42	100
Lithium (Li)	670.784	3.7	100
Magnesium (Mg)	279.079	30	100
Manganese (Mn)	257.610	1.4	50
Mercury (Hg)	194.227	2.5	—
Molybdenum (Mo)	203.844	12	100
Nickel (Ni)	231.604	15	50
Phosphorus (P)	241.914	76	—
Potassium (K)	766.491	700	100
Selenium (Se)	196.090	75	100
Silica (SiO_2)	251.611	26	100
Silver (Ag)	328.068	7.0	50
Sodium (Na)	588.995	29	100
Strontium (Sr)	421.552	0.77	50
Thallium (Tl)	190.864	40	100
Tin (Sn)	189.980	25	—
Titanium (Ti)	334.941	3.8	—
Vanadium (V)	292.402	7.5	50
Zinc (Zn)	213.856	1.8	100

[a] MDLs provided only as a guide to instrumental limits. The method detection limits are sample dependent and vary with sample matrix. From Martin, T.D., C.A. Brockhoff, J.T. Creed, and EMMC Methods Work Group. 1994. Method 200.7- determination of metals and trace elements in water and wastes by inductively coupled plasma-atomic emission spectrometry. Revision 4.4. Environmental Monitoring System Laboratory, Office of Research and Development, U.S. Environmental Protection Agency, Cincinnati, OH.

[b] The upper limit concentration is the upper limit of an effective calibration range. (From Eaton, A.D., L.S. Clesceri, E.W. Rice, A.E. Greenberg, and M.A.H. Franson. 2005. *Standard Methods for the Examination of Water and Wastewater*, 21st ed. Washington D.C.: American Public Health Association.)

Source: EPA 200.7 Revision 4,4.

12. If a sample will be analyzed for other elements, a mixed calibration standard solution should be prepared. For the analysis of total recoverable digested samples, prepare mixed calibration standard solutions by combining appropriate volumes of the stock solutions in 500 mL volumetric flasks containing 20 mL (1+1) HNO_3 and 20 mL (1+1) HCl and dilute to volume with water. Prior to preparing the mixed standards, each stock solution should be analyzed separately to determine possible spectral interferences or the presence of impurities. Care should be taken when preparing the mixed standards to ensure that the elements are compatible and stable together. To minimize the opportunity for contamination by the containers, it is recommended to transfer the mixed-standard solutions to acid-cleaned, new FEP fluorocarbon bottles for storage. Freshly mixed standards should be prepared, as needed.
13. Blanks: See Section 10.3.2.4. Reagents-13. LRB and LFB must be prepared through the same process as the samples.
14. Instrument Performance Check (IPC) Solution.
15. Quality Control Sample (QCS): The concentration of the analytes in the QCS solution should be ≥1 mg/L.
16. Spectral Interference Check (SIC) Solutions: When interelement corrections are applied, SIC solutions are needed containing concentrations of the interfering elements at levels that provide an adequate test of the correction factors.

10.3.5.5 Procedure

1. Sample preparation for total recoverable Ca:
   a. Transfer a 100 mL (±1 mL) acid–preserved sample to a 250 mL beaker. (Smaller sample size may be used.) Add 2 mL (1+1) nitric acid and 1.0 mL of (1+1) hydrochloric acid. Place the beaker on the hot plate for solution evaporation at a temperature less than 85°C in a fume hood. The beaker should be covered with an elevated watch glass. The process will take ~2 h for sample reducing to ~ 20 mL.
   b. Cover the lip of the beaker with a watch glass to reduce additional evaporation and gently reflux the sample for 30 minutes. Allow the beaker to cool, transfer the sample solution to a 50 mL volumetric flask, increase to 50 mL with DI water, stopper and mix. Allow any undissolved material to settle overnight, or centrifuge a portion of the prepared sample until clear.

2. Sample preparation for dissolved Ca:

   Transfer ~20 mL filtered and acid-preserved sample into a 50 mL centrifuge tube, add 0.4 mL (1+1) nitric acid, cap the tube, and mix.

3. Sample analysis
   a. Prior to daily calibration of the instrument, inspect the sample introduction system including the nebulizer, torch, injector tube, and uptake tubing. Clean the system when needed or on a daily basis.

b. After warming up the instrument for 30–60 min, complete required optical profiling or alignment specific to the instrument.
c. For initial and daily operations, calibrate the instrument according to the instrument manufacturer's recommended procedures, using mixed calibration standard solutions and the calibration blank. A peristaltic pump must be used to introduce all solutions to the nebulizer. To allow equilibrium to be reached in the plasma, aspirate all solutions for 30 seconds after reaching the plasma before beginning integration of the background corrected signal to accumulate data. When possible, use the average value of replicate integration periods of the signal to be correlated to the analyte concentration. Flush the system with the rinse blank for a minimum of 60 seconds between each standard. The calibration line should consist of a minimum of a calibration blank and a high standard. Replicates of the blank and highest standard provide an optimal distribution of calibration standards to minimize the confidence band for a straight-line calibration in a response region with uniform variance.

4. Samples should be analyzed in the same operational manner used in the calibration routine with the rinse blank also being used between all sample solutions, LFB, LFM, and check solutions.

10.3.5.6 Calculation

$$\text{mg Ca/L} = C \times F,$$

where
C = Ca concentration as read directly from ICP-AES (mg/L), and
F = dilution factor. For total recoverable analysis, C = 0.5.

10.3.5.7 Source of the Detailed Method

USEPA Method 200.7: determination of metals and trace elements in water and wastes by inductively coupled plasma-atomic emission spectrometry, Revision 4.4. (Martin et al., 1994; http://www.epa.gov/waterscience/methods/method/files/200_7.pdf, accessed on December 13, 2009).

10.3.6 Chloride

10.3.6.1 Introduction

Chloride (Cl, CAS No. 16887-00-6) is one of major anions in water. Concentrations of Cl are usually < 100 mg/L in groundwater and can be as high as 19,000 mg/L in seawater (Boyd, 2000). High Cl concentrations in the groundwater and surface water often result from saltwater intrusion, mineral dissolution, and industrial and domestic wastes. Chloride does not affect human health within normal concentrations, but increasing salinity may affect freshwater aquatic organisms. The USEPA sets 250 mg Cl/L for the national secondary drinking water standard and 860 mg Cl/L for the criteria maximum concentration (CMC) for surface water.

10.3.6.2 Analytical Methods

The NEMI database lists 18 analytical methods from AOAC, ASTM, USEPA, and USGS for analysis of Cl in water. Three of the USEPA methods (EPA 325.1, 325.2, and 325.3) listed in the NEMI have been recently withdrawn by USEPA MUR (USEPA, 2007) and should not be used. The detection levels range from 4 μg/L (EPA 300.1 and SM 4110B) to 1,000 μg/L (SM 4500-Cl-E). The procedure described below is based on EPA 300.1 (Hautman and Munch, 1997). The single laboratory precision (standard deviation) was reported from 0.06 to 0.44% for nine replicates of various water samples.

10.3.6.3 Apparatus

An ion chromatograph with an anion guard and analytical columns, an anion suppressor, and a conductivity detector.

10.3.6.4 Reagents

1. Eluent solution: Dissolve 1.91 g sodium carbonate in DI water and dilute to 2 L. Purge for 10 min with helium prior to use to remove dissolved gases.
2. Stock standard solutions (1 g/L): Dissolve 0.1649 g sodium chloride in water and dilute to 100 mL.

10.3.6.5 Procedure

1. Warm up the instrument and set operation condition: eluent flow = 0.4 mL/min and sample loop = 10 μL.
2. Sample preparation: Filter water samples through 0.45 μm membrane filter with a syringe.
3. Enter sample ID into the computer.
4. Run a minimum of three calibration standards and start injecting unknown samples manually or with an autosampler.

10.3.6.6 Calculation

Chloride concentration in a sample will be calculated by IC software.

10.3.6.7 Source of the Detailed Method

USEPA method 300.1: determination of inorganic anions in drinking water by ion chromatography (Hautman and Munch, 1997; http://www.epa.gov/waterscience/methods/method/files/300_1.pdf, accessed on December 13, 2009).

10.3.7 Coliforms and *Escherichia coli*

10.3.7.1 Introduction

The coliform bacteria including fecal coliform and *Escherichia coli* are present in the natural environment and in human and animal feces wastes. These bacteria are not generally health threats in themselves, but the presence of fecal coliform and *E. coli* indicates that the water has been contaminated and may be exposed to pathogens or disease-producing bacteria or viruses. Some disease-causing waterborne pathogens include *Giardia, Cryptosporidium, Salmonella,* Norwalk virus, bacterial gastroenteritis, and hepatitis A. The USEPA Total Coliform Rule (published 29 June 1989/effective 31 December 1990) for drinking water sets the health goal (MCLG) as zero per 100 mL of water and legal limits (MCL) at less than 5% for total coliform levels in drinking water. The USEPA sets water quality standard for *E. coli* in freshwater at 126 cfu/100 mL.

10.3.7.2 Analytical Methods

The NEMI database lists 13 methods for measuring total coliforms or *E. coli* from USEPA, Standard Methods, Hatch Company, IDEXX Laboratories, Micrology Laboratories, and CPI International. The USEPA-approved testing methods include EPA1604, SM 9221A,B,D, 9222A,B,C, and 9223 for total amount of coliforms and EPA1103.1, 1603, and 1604 for *E. coli* in drinking and wastewater. The procedure described below is based on EPA 1604 for total number of coliforms and *E. coli* in water by membrane filtration using a simultaneous detection technique (USEPA, 2002a). The single laboratory precision (coefficient of variation) ranged from 3.3% to 27.3% for *E. coli,* and from 2.5% to 5.1% for the six wastewater spiked samples tested by 19 laboratories. Water samples should be analyzed immediately after collection and the holding time should not exceed 30 hours.

10.3.7.3 Apparatus

1. An incubator set at 35°C ± 0.5°C, with approximately 90% humidity.
2. A stereoscopic microscope, with magnification of 10–15 ×, wide-field type and a microscope lamp producing diffuse light from cool, white fluorescent lamps adjusted to give maximum color.
3. Membrane filtration units, glass, plastic, or stainless steel.
4. A germicidal ultraviolet (254 nm) light box for sanitizing the filter funnels is desirable, but optional.
5. Line vacuum, electric vacuum pump, or aspirator is used as a vacuum source. In an emergency, a hand pump or a syringe can be used. Such vacuum-producing devices should be equipped with a check valve to prevent the return flow of air.
6. A vacuum filter flask, usually 1 L, with appropriate tubing. Filter manifolds to hold a number of filter bases are desirable, but optional.
7. A Bunsen or Fisher-type burner or electric incinerator unit.
8. A long wave ultraviolet lamp (366 nm), handheld 4-watt or 6-watt, or microscope attachment.
9. A water bath maintained at 50°C for tempering agar.

10.3.7.4 Reagents

1. Stock phosphate buffer solution: Add 34 g potassium dihydrogen phosphate (KH_2PO_4) into 500 mL water.
2. Stock buffer solution: Adjust the pH of the solution to 7.2 with 1 N NaOH, and bring volume to 1000 mL with water. Sterilize by filtration or autoclave for 15 minutes at 121°C.

3. $MgCl_2$ solution: Dissolve 38 g anhydrous $MgCl_2$ (or 81.1 g $MgCl_2 \cdot 6H_2O$) in 1 L DI water. Sterilize by filtration or autoclave for 15 min at 121°C. After sterilization of the stock buffer and $MgCl_2$ solutions, store in the refrigerator prior to use.
4. Working solution (final pH 7.0 ± 0.2): Add 1.25 mL phosphate buffer stock and 5 mL $MgCl_2$ stock into 1 L DI water. Mix well and dispense in appropriate amounts for dilutions in screw-cap dilution bottles or culture tubes, and/or into larger containers for use as rinse water. Autoclave at 121°C for 15 min.
5. MI agar composition:
   a. Proteose peptone #3 — 5 g
   b. Yeast extract — 3 g
   c. β-D-lactose — 1 g
   d. 4-methylumbelliferyl-β-D-galactopyranoside (MUGal) — 0.1 g
   e. (Final concentration 100 μg/mL)
   f. Indoxyl-β-D-glucuronide (IBDG) — 0.32 g
   g. (Final concentration 320 μg/mL)
   h. NaCl — 7.5 g
   i. K_2HPO_4 — 3.3 g
   j. KH_2PO_4 — 1.0 g
   k. Sodium lauryl sulfate — 0.2 g
   l. Sodium desoxycholate — 0.1 g
   m. Agar — 15 g
   n. DI water — 1 L
6. Cefsulodin Solution (1 mg/mL): Add 0.02 g of cefsulodin to 20 mL DI water, sterilize using a 0.22-μm syringe filter, and store in a sterile tube at 4°C until needed. Prepare fresh solution each time.
7. Autoclave the medium for 15 minutes at 121°C, and add 5 mL of the freshly prepared solution of Cefsulodin (5 μg/mL final concentration) per L of tempered agar medium. Pipette the medium into 9 × 50-mm Petri dishes (5 mL/plate). Solidify the medium in Petri dishes, and store plates at 4°C for up to 2 weeks. The final pH should be 7 ± 0.2.
8. MI broth: The composition of MI broth is the same as MI agar, but without the agar. The final pH of MI broth should be 7 ± 0.2. The broth is prepared and sterilized by the same methods described for MI agar except that absorbent pads are placed in 9 × 50 mm Petri dishes and saturated with 2–3 mL of MI broth containing 5 μg/mL final concentration of Cefsulodin. Alternately, the broth can be filter-sterilized. Excess broth is poured off before using the plates. Plates should be stored in the refrigerator and discarded after 96 hours.
9. Tryptic soy agar/trypticase soy agar (Difco 0369-17-6, BD 4311043, Oxoid CM 0129B, or equivalent) (TSA). The composition is:
   a. Tryptone — 15 g
   b. Soytone — 5 g
   c. NaCl — 5 g
   d. Agar — 15 g

Add the dry ingredients listed above to 1 L of DI water, and heat to boiling to dissolve the agar completely. Autoclave at 121°C for 15 min. Dispense the agar into 9 × 50-mm Petri dishes (5 mL/plate). Incubate the plates for 24–48 h at 35°C to check for contamination. Discard any plates with microbial growth. If more than 5% of the plates show contamination, discard all plates, and make new medium. Store plates at 4°C until needed. The final pH should be 7.3 ± 0.2.

10.3.7.5 Procedure

1. Filtration of samples: Place a membrane filter grid-side up on the porous plate of the filter base using a flamed forceps. Attach the funnel to the base of the filter unit.
2. Put ~30 mL of sterile dilution water in the bottom of the funnel.
3. Shake the sample container 25 times.
4. Measure an appropriate volume (100 mL for drinking water) or dilution of the sample with a sterile pipette or graduated cylinder, and pour it into the funnel. Turn on the vacuum and leave it on while rinsing the funnel twice with about 30 mL sterile dilution water.
5. Remove the funnel from the base of the filter unit. A germicidal ultraviolet (254 nm) light box can be used to hold and sanitize the funnel between filtrations. At least 2 min of exposure time is required for funnel decontamination.
6. Hold the membrane filter at its edge with a flamed forceps, and gently lift and place the filter grid-side up on the MI agar plate or MI broth pad plate. Slide the filter onto the agar or pad using a rolling action to avoid trapping air bubbles between the membrane filter and the underlying agar or absorbent pad. Run the tip of the forceps around the outside edge of the filter to make sure that the filter makes contact with the agar or pad.
7. Invert the agar Petri dish, and incubate the plate at 35°C for 24 hours. Pad plates with MI broth should be incubated grid-side up at 35°C for 24 h. If loose-lidded plates are used for MI agar or broth, the plates should be placed in a humid chamber.
8. Count all blue colonies on each MI plate under normal/ambient light, and record the results. This is the *E. coli* count. Positive results that occur in < 24 h are valid, but the results cannot be recorded as negative until the 24-h incubation period is complete.
9. Expose each MI plate to long wave ultraviolet light (366 nm), and count all fluorescent colonies (blue/green fluorescent *E. coli*, blue/white fluorescent total coliform other than *E. coli*, and blue/green with fluorescent edges [also *E. coli*]). Record the data.
10. Add any blue, non-fluorescent colonies (if any) found on the same plate to the TC count.

10.3.7.6 Calculation

$$E.coli/100\,\text{mL} = \frac{\text{Number of blue colonies}}{\text{Volume of sample filtered (mL)}} \times 100$$

$$\text{Total coliforms/100 mL} = \frac{\text{Number of } F \text{ colonies} + \text{Number of blue, } NF \text{ colonies}}{\text{Volume of sample filtered (mL)}} \times 100,$$

where F = fluorescent and NF = non-fluorescent colonies.

10.3.7.7 Source of the Detailed Method

USEPA method 1604: Total coliforms and *E. coli* in water by membrane filtration using a simultaneous detection technique (USEPA, 2002; http://www.epa.gov/waterscience/methods/method/biological/1604.pdf, accessed on December 13, 2009).

10.3.8 Copper

10.3.8.1 Introduction

Copper (Cu, CAS No. 7440-50-8) in water may come from corrosion of copper pipes, atmospheric deposition, and chemicals from fungicide and industrial uses. The average Cu concentrations are ~4-12 μg/L in surface water, <1 μg/L in groundwater, and 3 μg/L in seawater (Boyd, 2000; Eaton et al., 2005). Copper is one of the essential nutrients for plants, animals, and humans, but high Cu concentrations are toxic. The USEPA sets the secondary drinking water standard for Cu as 1 mg/L and also suggests 4.8 μg/L for the MCL and 3.1 μg/L for the criterion continuous concentration for saltwater.

10.3.8.2 Analytical Methods

The NEMI database lists 32 analytical methods for analysis of Cu in water. The detection levels range from 0.003 μg/L (SM 3125 with ICP-MS) to 10 μg/L (USGS I-1472-85 with ICP-AES and I-1270 with spectrometer). The USEPA–approved methods for Cu in drinking water and wastewater include AAS (SM 3111B), GFAAS (EPA 200.9 and SM 3113B), ICP-AES (EPA Methods 200.5 and 200.7), and ICP-MS (EPA Method 200.8). The ICP-AES and AAS methods are recommended because of less interference. The procedure described below is based on EPA Method 200.7 Rev 4.4 (ICP-AES), which is the USEPA–approved method for drinking water, surface water, and wastewater (Martin et al., 1994).

10.3.8.3 Apparatus (See Section 10.3.5.3)

10.3.8.4 Reagents (See Section 10.3.5.4 Except for the Following)

Standard stock solution (1 g Cu/L): Dissolve 1.000 g Cu metal in 50.0 mL (1+1) HNO_3 with heating to affect dissolution. Let solution cool and dilute to 1 L with DI water.

10.3.8.5 Procedure (See Section 10.3.5.5)

10.3.8.6 Calculation (See Section 10.3.5.6)

10.3.8.7 Source of the Detailed Method (See Section 10.3.5.7)

10.3.9 Fluoride

10.3.9.1 Introduction

Fluoride (F, CAS No. 16984-48-8) in water results from weathering of minerals, discharge from fertilizer and aluminum factories, phosphate mining, and additives in drinking water. Concentrations of F are usually < 1 mg/L and can be as high as 50 mg/L in groundwater, which often is related to F-containing minerals, fluorite or fluorite-apatite (Weiner, 2007). Drinking water with high F may cause bone disease and mottled teeth while low F concentrations in drinking water promote dental health. The Centers for Disease Control and Prevention recommended the optimal F concentration in drinking water at 0.7–1.2 mg/L. The EPA sets 4 mg F/L as the MCL for the national drinking water standard and 2 mg F/L as a secondary standard.

10.3.9.2 Analytical Methods

The NEMI database lists 18 methods for analysis of F in water. The detection levels range from 2 μg/L (SM 4110B with IC) to 300 μg/L (AOAC 993.3 with IC). The USEPA–approved methods for F analysis in drinking and wastewater are EPA (Methods 300.0 and 300.1) and SM (4110B, 4500-F, B, C, D, E), ASTM (D1179-04B, D1179-93B, D1179-99B, D4327-03, D4327-97), Bran+Luebbe (129-71W, 380-75WE), and Waters (D6508). EPA Methods 325.1, 340.1, 340.1, and 340.3 have been recently withdrawn by EPA MUR (USEPA, 2007). The procedure described below is based on EPA Method 300.1 (Hautman and Munch, 1997). The method detection limit of EPA Method 300.1 for F is 9 μg/L and the single laboratory precision (standard deviation) ranged from 5 to 10% for 9 replicates of various water samples.

10.3.9.3 Apparatus (See Section 10.3.6.3)

10.3.9.4 Reagents (See Section 10.3.6.4 Except for the Following)

Fluoride standard stock solution (1 g F/L): Dissolve 0.2210 g sodium fluoride (NaF) in water and dilute to 100 mL.

10.3.9.5 Procedure (See Section 10.3.6.5)

10.3.9.6 Calculation (See Section 10.3.6.6)

10.3.9.7 Source of the Detailed Method (See Section 10.3.6.7)

10.3.10 Iron

10.3.10.1 Introduction

Iron (Fe, CAS No. 7439-89-6) in water may come from dissolution of iron-containing minerals, organic matter decay, and human activities. The average Fe concentrations are ~ 0.7 mg/L in surface water, 0.1–10 mg/L in groundwater, and 0.01 mg/L in seawater (Boyd, 2000; Eaton et al., 2005). Iron is one of the essential nutrients for plants, animals, and humans. High Fe concentration may cause rusty color and metallic taste but is not considered toxic. The USEPA sets the secondary drinking water standard for Fe as 0.3 mg/L.

10.3.10.2 Analytical Methods

The NEMI database lists 22 methods for Fe analysis in water and the lowest detection levels are 1 μg/L (SM3113B with GFAAS and USGS I-2020-05 and I-4020-05 with ICP-MS). The USEPA–approved methods for Fe in drinking and wastewater include AAS (SM 3111B), GFAAS (EPA Method 200.9 and SM 3113B), and ICP-AES (EPA Methods 200.5 and 200.7). Both ICP-AES and AAS methods are adequate for Fe analysis in surface water and wastewater. GFAAS method may be used to determine water with low Fe concentrations. The procedure described below is based on SM 3111B, which is the USEPA-approved method for secondary drinking water contaminants (Easton et al., 2005). The method is also approved by USEPA for analysis of Cu, Mg, Mn, and Na in drinking water.

10.3.10.3 Apparatus

1. Flame atomic absorption spectrometer (FAAS) with Fe hollow cathode lamp and with adjustable, full range monochrometer, acetylene/air and acetylene/nitrous oxide burner heads.
2. Hot plate.
3. 125 mL Erlenmeyer flasks or Griffin beakers (acid washed and rinsed with water).

10.3.10.4 Reagents

1. Air: Clean compressed or commercially bottled air, devoid of oil, water, and other foreign substances.
2. Acetylene: Standard commercial grade with pressure > 689 Pa (100 psi).
3. Distilled deionized water (DDI).
4. Standard stock solution: Use commercially available 1000 mg/L standard solutions of Fe to prepare the stock solution. Use volumetric pipettes and flasks that have been rinsed with 1+1 Hydrochloric acid (HCl) and distilled water.
5. Standard working solution (100 mg/L): Pipette 10.0 ml of 1000 mg Fe/L standard stock solution into a 100-ml volumetric flask. Dilute to volume with DDI water, and mix well.
6. Calibration standards solution: Prepare the calibration standards from the above standard working solutions. Use volumetric pipettes and flasks rinsed in 1+1 HCl and distilled water. Fresh calibration standards should be prepared every two weeks, or as needed. Dilute the working standard solution to levels appropriate to the operating range of the instrument using the appropriate acid diluent (2% nitric acid). The element concentrations in each calibration solution should be sufficiently high to produce good measurement precision and to accurately define the slope of the response curve.
7. Concentrated nitric acid.
8. Nitric acid (1:1): Prepare a 1:1 dilution with DDI water by adding the concentrated acid to an equal volume of water.

9. Hydrochloric acid (1:1): Prepare a 1:1 solution of reagent grade hydrochloric acid and DDI.

10.3.10.5 Procedure

1. Sample preparation: For the determination of total metals the sample is acidified with 1:1 redistilled HNO_3 to a pH of less than 2 at the time of collection. The sample is not filtered before processing. Add 100 mL volume of well mixed sample appropriate for the expected level of metals to a Griffin beaker or 125 mL Erlenmeyer flask and add 3 ml of concentrated redistilled HNO_3. Heat the sample on a hot plate (85°C) and evaporate to near dryness (< 5 mL) cautiously, making certain that the sample does not boil. Cool the beaker and add another 5 ml portion of concentrated HNO_3. Cover the beaker/flask with a watch glass and return to the hot plate. Increase the temperature of the hot plate so that a gentle reflux action occurs. Continue heating, adding additional acid as necessary, until the digestion is complete (generally indicated when the digestate is light in color or does not change in appearance with continued refluxing). Again, evaporate to near dryness (< 5 mL) and cool the beaker. Add 10 mL of 1:1 HCl (5 mL/100 mL of final solution) and warm the beaker to dissolve any precipitate or residue resulting from evaporation. (If the sample is to be analyzed by the furnace procedure, substitute HNO_3 for 1:1 HCl and 15 mL of water.) Heat additional 15 min to dissolve any precipitate or residue. Cool and wash down the beaker walls and watch glass with distilled water. Filter the sample to remove silicate and other insoluble material that could clog the atomizer. Allow the beaker to cool. Quantitatively transfer the sample solution to a 100 mL volumetric flask. Add DDI water to bring the final volume of the sample solution to 100 mL.
2. Sample Analysis:
   a. Inspect the AA flame platform, the sample uptake system, and autosampler injector for any problems that would affect instrument performance. If necessary, clean the system and replace the flame platform.
   b. Prior to the use of this method, instrument operating conditions must be optimized. An analyst should follow the instructions provided by the manufacturer. Optimum conditions should provide the lowest reliable MDLs. Once optimum operating conditions have been determined, they should be recorded and available for daily reference.
   c. Configure the instrument: Because of differences among makes and models of spectrophotometers and furnace devices, consult the instrument manual for operations. Install hollow cathode lamp for Fe and set the wavelength to 248.3 nm. Set slit width according to manufacturer's suggested setting for Fe. Turn on the instrument and let energy source stabilize, generally about 30 min. Optimize the wavelength by adjusting wavelength dial until optimum energy gain is obtained. Align lamp in accordance with manufacturer's instructions. Install suitable burner head and adjust burner head position. Turn on air and adjust flow rate to give maximum sensitivity for the metal being measured. Turn on

acetylene, adjust flow rate to value specified by the manufacturer setting and ignite flame. Let flame stabilize (5-10 min). Aspirate a blank (DDI water or matrix solution) of same acid concentrations as the standards and the samples. Zero the instrument. Aspirate a standard and adjust nebulizer to get maximum sensitivity. Adjust burner position (vertically and horizontally) to get maximum response. Aspirate blank and auto zero the instrument. Aspirate a middle range standard and record the absorbance. The reading should be consistent with previous instrument optimization and calibration readings.

d. Standardization: At least 3 standards ranging in low, medium, and high concentrations must be used for standardization. First aspirate the method blank and auto zero the instrument. Then aspirate each standard and record the absorbance. Plot absorbance of standards versus their concentrations to prepare calibration curve. If the instrument is equipped with direct concentration readout then this step can be eliminated. After the warm up period, but before calibration, instrument stability must be demonstrated by analyzing a standard solution with a concentration 20 times the ML, a minimum of five times. The resulting relative standard deviation (RSD) of absorbance signals must be $< 5\%$. If the RSD is $> 5\%$, determine and correct the cause before calibrating the instrument.

e. Analysis: Aspirate water containing 1.5 ml concentrated nitric acid/L to rinse the nebulizer and zero instrument. Aspirate sample and determine the absorbance. Rinse the nebulizer between samples with DDI water.

10.3.10.6 Calculation

$$\text{mg Fe/L} = C \times F,$$

where C = Fe concentration as read directly from AAS (mg/L), and F = dilution factor.

10.3.10.7 Source of the Detailed Method

Standard Methods for the Examination of Water & Wastewater, 21st Edition, 3111 B–Direct air-acetylene flame method, pages 3-17 to 3-19 (Eaton et al., 2005; http://www.epa.gov/waterscience/methods/method/files/300_1.pdf, accessed on December 13, 2009).

10.3.11 Lead

10.3.11.1 Introduction

Lead (Pb, CAS No. 7439-92-1) as a contaminant in water may come from erosion of natural deposition, metal mining, corrosion of plumbing, leaded gasoline, coal mining, and commercial lead-containing products. The average Pb concentrations are ~3 μg/L in surface water, < 1 μg/L in groundwater, and 0.03 μg/L in seawater (Boyd, 2000; Eaton et al., 2005; Weiner, 2007). The USEPA primary drinking water standard sets the MCL for Pb at 0.015 μg/L and the maximum contaminant level goal as zero. USEPA also suggests 65 μg/L for criteria maximum concentration and

2.5 μg/L for the criterion continuous concentration for fresh water to protect aquatic life and human health.

10.3.11.2 Analytical Methods

The NEMI database lists 27 methods for Pb analysis in water and the detection levels range from 0.005 μg/L (SM3125 with ICP-MS) to 100 μg/L (USGS I-1399 and I-3399 with FAAS). The USEPA–approved analytical methods for Pb in drinking water and wastewater include GFAAS (EPA Method 200.9 and SM 3113B), ICP-AES (EPA Method 200.5), and ICP-MS (EPA Method 200.8). Both GFAAS and ICP-MS methods have low MDL, 0.7 μg/L for GFAAS. ICP-AES is often selected for samples with high Pb concentrations (> 42 μg/L). The procedure described below is based on EPA Method 200.9 (Creed et al., 1994).

10.3.11.3 Apparatus (See Section 10.3.2.3)

10.3.11.4 Reagents (See Section 10.3.2.4 Except for the Following)

Lead stock solution (1 g Pb/L): Dissolve 1.599 g $Pb(NO_3)_2$ in a minimum amount of (1+1) HNO_3. Add 20.0 mL (1+1) HNO_3 and dilute to a 1 L with water.

10.3.11.5 Procedure (See Section 10.3.2.5)

10.3.11.6 Calculation (See Section 10.3.2.6)

10.3.11.7 Source of the Detailed Method (See Section 10.3.2.7)

10.3.12 Magnesium

10.3.12.1 Introduction

Magnesium (Mg, CAS No. 7439-95-4), like Ca, is abundant on the earth (2.1% of the earth's crust) and one of the major cations in water. The average Mg concentrations are ~4 mg/L in surface water, > 5 mg/L in groundwater, and 1.3 g/L in seawater (Boyd, 2000; Eaton et al., 2005). Magnesium is one of the essential nutrients for plants, animals, and humans. The USEPA has no drinking water standard for Mg.

10.3.12.2 Analytical Methods

Magnesium can be determined with AAS (SM 3111B) and ICP-AES (EPA Methods 200.5 and 200.7). The procedure that follows is based on SM 3111B, which is the USEPA-approved method. The method is also approved by USEPA for analysis of Cu, Mg, Mn, and Na in drinking water.

10.3.12.3 Apparatus (See Section 10.3.10.3)

10.3.12.4 Reagents (See Section 10.3.10.4 Except for the Following)

Magnesium stock solution (100 mg Mg/L): Dissolve 0.1658 g MgO in a minimum amount of (1+1) HNO_3. Add 20.0 mL (1+1) HNO_3 and dilute to 1 L with DI water.

10.3.12.5 Procedure (See Section 10.3.10.5)

10.3.12.6 Calculation (See Section 10.3.10.6)

10.3.12.7 Source of the Detailed Method (See Section 10.3.10.7)

10.3.13 Manganese

10.3.13.1 Introduction

Manganese (Mn, CAS No. 7439-96-5) in water may come from sediment, rocks, mining, and industry waste. The average Mn concentrations are < 0.2 mg/L in surface water, < 0.1 mg/L in groundwater, and 0.002 mg/L in seawater (Boyd, 2000; Eaton et al., 2005; Weiner, 2007). Manganese is one of the essential micronutrients for plants, animals, and humans. There are no enforceable federal drinking water standards for Mn and the USEPA sets the secondary standard of Mn at 0.05 mg/L.

10.3.13.2 Analytical Methods

The NEMI database lists 22 methods for analysis of Mn in water from AOAC, ASTM, USEPA, and USGS and the detection levels range from 0.002 μg/L (SM 3125 with ICP-MS) to 42 μg/L (SM 3500-MN-B with spectrophotometer). The USEPA–approved methods for drinking water and wastewater for Mn analysis in water include AAS (SM 3111B), GFAAS (EPA Method 200.9; SM 3113B), ICP-AES (EPA Methods 200.5 and 200.7), and ICP-MS (EPA Method 200.8). The procedure described below is based on SM 3111B (Eaton et al., 2005).

10.3.13.3 Apparatus (See Section 10.3.10.3)

10.3.13.4 Reagents (See Section 10.3.10.4 Except for the Following)

Manganese stock solution (100 mg Mn/L): Dissolve 0.1000 g Mn metal in 10 mL concentrated HCl mixed with 1 mL concentrated HNO_3. Add 20.0 mL and dilute to 1 L with DI water.

10.3.13.5 Procedure (See Section 10.3.10.5)

10.3.13.6 Calculation (See Section 10.3.10.6)

10.3.13.7 Source of the Detailed Method (See Section 10.3.10.7)

10.3.14 Mercury

10.3.14.1 Introduction

Mercury (Hg, CAS No. 7439-97-6) as a contaminant in water may come from industry waste, mining, pesticides, coal, electrical equipment (batteries, lamps, and switches), and fossil-fuel combustion. Mercury concentrations are less than 0.005 μg/L in unpolluted surface water and less than 0.002 μg/L in unpolluted seawater (Fowler 1990; Gilmour and Henry 1991). The USEPA primary drinking water standard sets 0.002 mg/L for both the MCL and the maximum contaminant level goal. USEPA also suggests 1.4 μg/L for criteria maximum concentration and 0.77 μg/L for the criterion continuous concentration for mercury in fresh water to protect aquatic life and human health.

10.3.14.2 Analytical Methods

The NEMI database lists 11 methods for Hg analysis in water and the detection levels range from 0.0002 μg/L (EPA1631 with CVAFS) to 17 μg/L (EPA Method 6010C). The USEPA-approved analytical methods for Hg in drinking and waste water include CVAAS (EPA Method 245.1, SM 3113B), ICP-MS (EPA Method 200.8), and CVAFS (EPA Methods 245.7 and 1631E). The EPA 1631E with CVAFS is the method of choice for mercury with very low MDL and will be described below (USEPA, 2002b).

10.3.14.3 Apparatus

1. Cold vapor atomic fluorescence spectrometer (CVAFS).
2. Mercury purging system.
3. The dual-trap Hg(0) preconcentrating system.

10.3.14.4 Reagents

1. Hydrochloric acid: trace-metal purified reagent-grade HCl containing less than 5 pg/mL Hg.
2. Hydroxylamine hydrochloride: Dissolve 300 g of $NH_2OH{\cdot}HCl$ in DI water and bring to 1L.
3. Stannous chloride: Bring 200 g of $SnCl_2{\cdot}2H_2O$ and 100 mL concentrated HCl to 1 L with DI water. Purge overnight with mercury-free N_2 and tightly capped.
4. Bromine monochloride (BrCl): In a fume hood, dissolve 27 g of KBr in 2.5 L of low-Hg HCl. Slowly add 38 g KBrO3 to the acid while stirring. Loosely cap the bottle, and allow to stir another hour before tightening the lid.
5. Stock mercury standard: NIST-certified 10,000-ppm aqueous Hg solution (NIST-3133).
6. Secondary Hg standard (1.00 μg/mL): Add 0.5 L of DI water and 5 mL of BrCl solution to a 1-L flask. Add 0.1 mL of the stock mercury standard to the flask and dilute to 1 L with DI water.
7. Working Hg Standard A (10.0 ng/mL): Dilute 1.00 mL of the secondary Hg standard to 100 mL with DI water containing 0.5% by volume BrCl solution.
8. Working Hg Standard B (0.10 ng/mL)—Dilute 0.10 mL of the secondary Hg standard to 1 L with DI water containing 0.5% by volume BrCl solution.
9. Nitrogen—Grade 4.5 nitrogen that has been further purified by the removal of Hg using a gold-coated sand trap.
10. Argon—Grade 5.0 argon that has been further purified by the removal of Hg.

10.3.14.5 Procedure

10.3.14.5.1 Sample Preparation

1. Transfer 100 mL of the water sample into a sample container. If BrCl was not added as a preservative, add 0.5-1 mL or more BrCl solution.
2. Matrix spikes and matrix spike duplicates for every 10 samples.

10.3.14.5.2 Hg Reduction and Purging

1. Hg reduction and purging for the bubbler system: Add 0.2–0.25 mL of NH_2OH solution to the BrCl-oxidized sample in the sample bottle. Cap the bottle and swirl the sample. The yellow color will disappear, indicating the destruction of the BrCl. Allow the sample to react for 5 min with periodic swirling to be sure that no traces of halogens remain. Connect a fresh trap to the bubbler, pour the reduced sample into the bubbler, add 0.5 mL of $SnCl_2$ solution, and purge the sample onto a gold trap with N_2 at 350 ± 50 mL/min for 20 min.
2. Hg reduction and purging for the flow-injection system: Add 0.2–0.25 mL of NH_2OH solution to the BrCl-oxidized sample in the sample bottle or in the autosampler tube. Cap the bottle and swirl the sample. The yellow color will disappear. Allow the sample to react for 5 minutes with periodic swirling. Pour the sample solution into an autosampler vial and place the vial in the rack.

10.3.14.5.3 Desorption of Hg from the Gold Trap

1. Remove the sample trap from the bubbler, place the Nichrome wire coil around the trap, and connect the trap into the analyzer train between the incoming Hg-free argon and the second gold-coated (analytical) sand trap.
2. Pass argon through the sample and analytical traps at a flow rate of approximately 30 mL/min for approximately 2 min.
3. Apply power to the coil around the sample trap for 3 min to thermally desorb the Hg (as Hg(0)) from the sample trap onto the analytical trap.
4. After the 3-min desorption time, turn off the power to the Nichrome coil, and cool the sample trap using the cooling fan.
5. Turn on the chart recorder or other data acquisition device to start data collection, and apply power to the Nichrome wire coil around the analytical trap. Heat the analytical trap for 3 min.
6. Stop data collection, turn off the power to the Nichrome coil, and cool the analytical trap to room temperature using the cooling fan.
7. Place the next sample trap in line and proceed with analysis of the next sample.

10.3.14.6 Calculation

$$ngHg / L = \left(\frac{As - Abb}{CFm \times V} \right)$$

10.3.14.7 Source of the Detailed Method

EPA Method 1631, Revision E: Mercury in water by oxidation, purge and trap, and cold vapor atomic fluorescence spectrometry (USEPA, 2002; http://www.epa.gov/waterscience/methods/method/mercury/1631.pdf, accessed on January 30, 2010).

10.3.15 Nitrate

10.3.15.1 Introduction

Nitrate (NO_3- CAS No. 14797-55-8) in water may come from fertilizers, industrial wastes, animal waste, sewage, atmospheric deposition, and decomposition of organic matter. Nitrate concentrations are usually <2 mg/L in natural groundwater and can be over 100 mg/L in contaminated water. Nitrate toxicity is often linked to methemoglobinemia, commonly called "blue baby syndrome." The USEPA primary drinking water standard for NO_3-N is 10 mg /L.

10.3.15.2 Analytical Methods

The NEMI database lists 23 methods for analysis of NO_3-N in water from AOAC, ASTM, DOE, USEPA, Strategic Diagnostics Inc., Standard Methods, Systea Scientific, UC-Davis, and USGS (Table 10.2). The detection levels range from 2 μg/L (EPA300 with IC) to 5 mg/L (DOE MS310 with color strip). The USEPA–approved methods for analysis of NO_3-N in drinking water include ASTM (D3867-90A, D386790-B, D4327-03, and 4327-97), ATIorion (601), EPA (300.0 and 353.2), SM (4110B, 4500-NO_3-E, D, E, and F), and Waters (B1011 and D6508). The procedure described below is based on EPA 300.1 (Hautman and Munch, 1997).

10.3.15.3 Apparatus (See Section 10.3.6.3)

10.3.15.4 Reagents (See Section 10.3.6.4 Except for the Following)

Nitrate standard stock solution (1 g F/L): Dissolve 0.6068 g sodium nitrate ($NaNO_3$) in water and dilute to 100 mL.

10.3.15.5 Procedure (See Section 10.3.6.5)

10.3.15.6 Calculation (See Section 10.3.6.6)

10.3.15.7 Source of the Detailed Method (See Section 10.3.6.7)

10.3.16 Orthophosphate

10.3.16.1 Introduction

Phosphate (PO_4-P, CAS No. 98059-61-1) in water may come from fertilizers, industrial wastes, animal waste, sewage, atmospheric deposition, and decomposition of organic matter. Phosphate concentrations are usually < 0.03 mg/L in uncontaminated surface water and can be over 0.1 mg/L in contaminated water. High PO_4-P in water often causes algal blooming and eutrophication, which has been a serious problem in aquatic systems worldwide.

10.3.16.2 Analytical Methods

The NEMI database lists 20 methods for analysis of phosphate in water from AOAC, ASTM, DOE, USEPA, Hach Co., SM, and USGS. The detection levels range from 0.7 μg/L (EPA Method 365.5 with automated spectrophotometer) to 690 μg/L (ASTM Method D4327 with IC). The procedure described below is based on EPA

Method 365.1 (O'Dell, 1993b). The applicable range of the method is 0.01–1.0 mg P/L. Approximately 20 to 30 samples per hour can be analyzed.

10.3.16.3 Apparatus

An autoanalyzer or a continuous-flow analyzer. The system manifold for phosphate can be found in O'Dell et al. (1994).

10.3.16.4 Reagents

1. Sulfuric acid solution, 2.5 M: Slowly add 70 mL of concentrated H_2SO_4 to ~400 mL of water. Cool to room temperature and dilute to 500 mL with water.
2. Antimony potassium tartrate solution: Weigh 0.3 g $K(SbO)C_4H_4O_6\ Cl{\cdot}2H_2O$ and dissolve in 50 mL water and dilute to 100 mL. Store at 4°C in a dark, glass-stoppered bottle.
3. Ammonium molybdate solution: Dissolve 4 g $(NH_4)_6Mo_7O_{24}{\cdot}4H_2O$ in 100 mL water. Store in a plastic bottle at 4°C.
4. Ascorbic acid (0.1 M): Dissolve 1.8 g of ascorbic acid in 100 mL of DI water.
5. Combined reagent: Mix the above reagents in the following proportions for 100 mL of the mixed reagent: 50 mL of 2.5 M H_2SO_4, 5 mL of $K(SbO)C_4H_4O_6\ Cl{\cdot}2H_2O$ solution, 15 mL of $(NH_4)_6Mo_7O_{24}{\cdot}4H_2O$ solution.
6. Phenolphthalein indicator solution (5 g/L): Dissolve 0.5 g of phenolphthalein in a solution of 50 mL of isopropyl alcohol and 50 mL of DI water.
7. Stock phosphorus solution (100 mg P/L): Dissolve 0.4393 g of pre-dried (105°C for 1 hour) potassium phosphate monobasic KH_2PO_4 (CASRN 7778-77-0) in DI water and dilute to 1L.
8. Standard phosphorus solution (10 mg P/L): Dilute 10.0 mL of stock solution to 100 mL with DI water.
9. Standard phosphorus solution (1 mg P/L): Dilute 10.0 mL of standard solution to 100 mL with DI water.

10.3.16.5 Procedure

1. Add 1 drop of phenolphthalein indicator solution to ~50 mL of sample. If a red color develops, add H_2SO_4 solution dropwise to just discharge the color. Acid samples must be neutralized with 1 M sodium hydroxide (40 g NaOH/L).
2. Set up manifold and allow system to equilibrate as required. Obtain a stable baseline with all reagents, feeding reagent water through the sample line.
3. Place standards in sampler in order of decreasing concentration. Complete filling of sampler tray.
4. Switch sample line from water to sampler and begin analysis.
5. Run at least three standards, covering the desired range, and a blank by pipetting and diluting suitable volumes of working standard solutions into 100 mL volumetric flasks. Suggested ranges include 0.00–0.10 mg/L and 0.20–1.00 mg/L.

10.3.16.6 Calculation

Prepare a calibration curve by plotting instrument response against standard concentration. Compute sample concentration by comparing sample response with the standard curve. Multiply answer by appropriate dilution factor. Report only those values that fall between the lowest and the highest calibration standards. Samples exceeding the highest standard should be diluted or reanalyzed. Report results in mg P/L.

10.3.16.7 Source of the Detailed Method

USEPA Method 365.1: Determination of phosphorus by semi-automated colorimetry, Revision 2.0 (O'Dell, 1993b; http://www.epa.gov/waterscience/methods/method/files/365_3.pdf, accessed on December 12, 2009).

10.3.17 Pesticide—Diuron

10.3.17.1 Introduction

Diuron ($C_9H_{10}Cl_2N_2O$, CAS No 330-54-1) is a substituted urea herbicide used to control a wide variety of weeds for noncrop areas and cropland; it is one of the most commonly used herbicides. It has high water solubility (~36 mg/L), long half-life (~372-1290 d) and a low adsorption coefficient (KOC = ~499; PANNA, 2009). The California Department of Pesticide Regulation suggested that pesticides with K_{OC} <1,900, and water solubility > 3 mg/L have potential to contaminate groundwater and surface water. The U.S. Geological Survey found diuron in ~20% of the rivers and streams the agency sampled in its national monitoring program (USGS, 2006). U.S. Drinking Water Equivalent Level for diuron is 70 μg/L and Canada's drinking water maximum acceptable concentration is 150 μg/L.

10.3.17.2 Analytical Methods

Diuron in water can be determined using high performance liquid chromatography (HPLC) with an ultraviolet detector (EPA Methods 531.1 and 632). The method detection limit of EPA Method 632 for Diuron is 0.009 μg/L and the single operator precision (standard deviation) was reported as 1%–5% for water samples spiked with Diuron. The procedure described below is based on EPA Method 1604. The method is also used for the determination of other carbamate and urea pesticides including Aminocarb, Barban, Carbaryl, Carbofuran, Chlorpropham, Fenuron, Fluometuron, Linuron, Methiocarb, Methomyl, Mexacarbate, Monuron, Neburon, Oxamyl, Propham, Propoxur, Siduron, and Swep. Water samples are extracted within a week and the extracts can be stored in a refrigerator (≤6°C) up to 40 days before analysis.

10.3.17.3 Apparatus

1. HPLC system with an ultraviolet detector capable of monitoring at 254 nm and 280 nm.
2. Analytical column: 30 cm long by 4 mm ID stainless steel C8 column or equivalent.
3. Guard column: C8 or equivalent.

4. Water bath: Heated, with concentric ring cover, capable of temperature control (±2°C).
5. Filtration apparatus: As needed to filter chromatographic solvents prior to HPLC.

10.3.17.4 Reagents

1. Acetone, acetonitrile, hexane, methylene chloride, and methanol in pesticide-quality or equivalent.
2. Ethyl ether: Nanograde, redistilled in glass if necessary.
3. Sodium sulfate: ACS, granular, anhydrous.
4. Florisil: PR grade (60/100 mesh).
5. Acetic acid: Glacial.
6. Stock standard solutions (1.00 μg/μL): Stock standard solutions may be prepared from pure standard materials or purchased as certified solutions. Weigh 10.00 mg of pure material. Dissolve the material in pesticide-quality acetonitrile or methanol and dilute to 10 mL. Transfer the stock standard solutions into TFE-fluorocarbon-sealed screw-cap vials. Store at 4°C and protect from light.

10.3.17.5 Procedure

10.3.17.5.1 Sample Extraction

1. Pour the sample into a 2 L separatory funnel. Add 60 mL methylene chloride to the sample, seal, and shake 30 seconds to rinse the inner walls.
2. Transfer the solvent to the separatory funnel and extract the sample by shaking the funnel for 2 minutes with periodic venting to release excess pressure. Allow the organic layer to separate from the water phase for a minimum of 10 min.
3. Add a second 60 mL volume of methylene chloride to the sample bottle and repeat the extraction procedure a second time, combining the extracts in the flask.
4. Pass a measured fraction or all of the combined extract through a drying column containing about 10 cm of anhydrous sodium sulfate and collect the extract in a 500 mL flask. Attach the flask to the rotary evaporator and partially immerse in the 50°C water bath. Concentrate the extract to approximately 5 mL.
5. Add 50 mL of hexane, methanol, or acetonitrile to the flask and concentrate the solvent extract as before. When the apparent volume of liquid reaches approximately 5 mL, remove the 500-mL round-bottom flask from the rotary evaporator and transfer the concentrated extract to a 10-mL volumetric flask, quantitatively washing with 2 mL of solvent. Adjust the volume to 10 mL.

10.3.17.5.2 Instrument Calibration

The instrument should be calibrated daily using either the external or internal standard technique.

1. External standard calibration: Prepare calibration standards at a minimum of three concentration levels by adding accurately measured volumes of one or more stock standards to a volumetric flask and diluting to volume with acetonitrile or methanol. One of the external standards should be representative of a concentration near, but above, the method detection limit. The other concentrations should correspond to the range of concentrations expected in the sample concentrates or should define the working range of the detector. Using injections of 10 μL of each calibration standard, tabulate peak height or area responses against the mass injected.
2. Internal standard calibration: Prepare calibration standards at a minimum of three concentration levels for each parameter of interest by adding volumes of one or more stock standards to a volumetric flask. To each calibration standard, add a known constant amount of one or more internal standards, and dilute to volume with acetonitrile or methanol. One of the standards should be representative of a concentration near, but above, the method detection limit. The other concentrations should correspond to the range of concentrations expected in the sample concentrates, or should define the working range of the detector. Using injections of 10 μL of each calibration standard tabulate the peak height or area responses against the concentration for each compound and internal standard.

10.3.17.6 Calculation

If the external standard calibration procedure is used, calculate the amount of material injected from the peak response using the calibration curve or calibration factor. The concentration in the sample can be calculated as follows:

$$\text{Concentration}\,(\mu\text{g/L}) = \frac{\text{A Vt}}{\text{Vi Vs}},$$

where

A = Amount of material injected, in ng,
Vi = Volume of extract injected, in μL,
Vt = Volume of total extract, in μL, and
Vs = Volume of water extracted, in mL.

If the internal standard calibration procedure was used, calculate the concentration in the sample as follows:

$$\text{Concentration}\,(\mu\text{g/L}) = \frac{\text{As Is}}{\text{Ais RF Vo}},$$

where

As = Response for parameter to be measured,
Ais = Response for the internal standard,
I = Amount of internal standard added to each extract, in μg,
Vo = Volume of water extracted, in L, and
RF = Calculate response factors.

10.3.17.7 Source of the Detailed Method

USEPA Method 632: The Determination of Carbamate and Urea Pesticides in Municipal and Industrial Wastewater (USEPA, 1999; http://www.epa.gov/waterscience/methods/method/files/632.pdf, accessed on December 29, 2009).

10.3.18 pH

10.3.18.1 Introduction

Water pH indicates water's acidity and is measured by hydrogen activity in the water. When hydrogen activity is less than 10^{-7} mol/L, the water is acidic (pH < 7). Basic water has hydrogen activity greater than 10^{-7} mol/L and pH greater than 7. Water pH is a water quality parameter tested frequently. High pH causes a bitter taste and low pH water will corrode or dissolve metals. A pH range of 6.0 to 9.0 is needed for healthy ecosystems. Sudden pH change often indicates chemical pollution.

10.3.18.2 Analytical Methods

The NEMI database lists 20 methods for water pH measurements from AOAC, ASTM, USEPA, Hach Co, SM, and USGS. The detection limits range from 0.001 pH unit (SM 4500-H-B) to 1 pH unit (USGS NFM 634C). The procedure of EPA Method 150.1 is described in this section (USEPA, 1982). The method precision is 0.1 pH unit in the range of pH 6–8.

10.3.18.3 Apparatus

A pH meter, a glass electrode, a reference electrode or combination electrodes, and a temperature sensor.

10.3.18.4 Reagents

1. Primary standard buffers (pH 4, 7, and 10).
2. Secondary buffers may be prepared or purchased.

10.3.18.5 Procedure

1. Follow the manufacturer's recommendation for operation and installation of the system.
2. The electrode should be calibrated at a minimum of two points that represent expected pH of samples.
3. After rinsing and gently wiping the electrodes, if necessary, immerse them into the sample and stir at a constant rate.
4. Record sample pH and temperature. Repeat measurement on successive volumes of sample until values differ by less than 0.1 pH units.

10.3.18.6 Calculation

pH meters read directly in pH units. Report pH to the nearest 0.1 unit and temperature to the nearest degree centigrade.

10.3.18.7 Source of the Detailed Method

USEPA 150.1 pH (Electrometric) (USEPA, 1982; http://www.nemi.gov/apex/f?p=237:38:2056482884271689::::P38_METHOD_ID:4685, accessed on December 13, 2009).

10.3.19 Specific Conductivity

10.3.19.1 Introduction

Specific conductance (SC) or electrical conductivity (EC) is the ability of water to conduct an electrical current. It is affected by the presence of dissolved salts such as chloride, sulfate, sodium, calcium, and others, which may come from natural deposition, industrial wastes, fertilizer, and other sources. Specific conductance ranges from 0 to 1,300 μS/cm in freshwater, ranges from 1,301 to 28,800 μS/cm in brackish water, and > 28,800 μS/cm in sea water. Rapid changes in SC may indicate salt water intrusion or water pollution. Currently there is no federal regulation on SC.

10.3.19.2 Analytical Methods

The NEMI database lists seven methods for SC measurements from ASTM, USEPA, SM, and USGS, with method detection levels being ~10 μS/cm. The procedure of EPA Method 120.1 is described in this section (USEPA, 1982).

10.3.19.3 Apparatus

1. Conductivity meter
2. Conductivity cell
3. Thermometer

10.3.19.4 Reagents

Standard potassium chloride solutions (0.01 M): Dissolve 0.7456 g of pre-dried (2 h at 105°C) KCl in distilled water and dilute to 1 L.

10.3.19.5 Procedure

Follow the direction of the manufacturer for operating the instrument. Allow samples to come to room temperature (23 to 27°C) if possible. Determine the temperature of samples within 0.5°C. If the temperature of the samples is not at 25°C, make temperature correction.

10.3.19.6 Source of the Detailed Method

USEPA Method 120.1: Conductance (Specific Conductance, umhos at 25°C) (USEPA, 1983a; http://www.epa.gov/waterscience/methods/method/files/120_1.pdf, accessed on December 12, 2009).

10.3.20 Turbidity

10.3.20.1 Introduction

Turbidity measures water clarity, which is closely related to the amount of particle matter that is suspended in water. The suspended materials include clay, silt, finely divided organic and inorganic matter, soluble colored organic compounds, plankton, and microscopic organisms. It is one of the indicators used to assess water quality. The USEPA has not set a water quality standard for turbidity. The World Health Organization stated that the turbidity of drinking water shouldn't be more than 5 NTU, and should ideally be below 1 NTU.

10.3.20.2 Analytical Methods

The NEMI lists 8 methods for measuring water turbidity and the detection levels are from 0 (EPA Method 180.1) to 1 NTU (ASAM Method D1889). The procedure of EPA Method 180.1 is described in this section (O'Dell, 1993c).

10.3.20.3 Apparatus

The turbidimeter should consist of a nephelometer, with light source for illuminating the sample, and one or more photo-electric detectors with a readout device to indicate the intensity of light scattered at right angles to the path of the incident light.

10.3.20.4 Reagents

1. Stock standard suspension (Formazin):
   a. Dissolve 1.00 g hydrazine sulfate, $(NH_2)_2.H_2SO_4$ in DI water and dilute to 100 mL.
   b. Dissolve 10.00 g hexamethylenetetramine in DI water and dilute to 100 mL.
   c. In a 100 mL volumetric flask, mix 5.0 mL of each solution. Allow to stand 24 h at 25 ± 3°C, then dilute to the mark with DI water.

2. Primary calibration standards: Mix and dilute 10.00 mL of stock standard suspension to 100 mL with DI water. The turbidity of this suspension is defined as 40 NTU.
3. Secondary standards may be acceptable as a daily calibration check, but must be monitored on a routine basis for deterioration and replaced as required.

10.3.20.5 Procedure

Allow samples to come to room temperature before analysis. Mix the sample to thoroughly disperse the solids. Wait until air bubbles disappear, then pour the sample into the turbidimeter tube. Read the turbidity directly from the instrument scale or from the appropriate calibration curve.

10.3.20.6 Source of the Detailed Method

USEPA Method 180.1: The determination of turbidity by nephelometry (O'Dell, 1993c; http://www.epa.gov/waterscience/methods/method/files/180_1.pdf, accessed on December 12, 2009).

10.4 SUMMARY

A number of analytical methods are available for assessing water quality parameters. It is often not easy to select an appropriate method for a specific analytical problem. Comparison of results obtained with a standard method is an important criterion to select an appropriate method. A new method developed for a specific purpose has to be validated and calibrated with a standard method, or certified by an accreditation organization. Even when a standard method is used, uncertainties may arise during chemical analysis. Following quality control procedures of the selected method is important for reliable chemical analyses. Method compatibility should be considered when a new method or new instrument is used for a long-term monitoring program.

REFERENCES

Boyd, C.E. 2000. *Water Quality*. Boston: Kluwer Academic Publishers.

Caffrey, J., T. Younos, G. Kohlhepp, D.M. Robertson, J. Sharp, and D. Whitall. 2007. Nutrient Requirements for the National Water Quality Monitoring Network for U.S. Coastal Waters and Their Tributaries. http://acwi.gov/monitoring/network/nutrients.pdf.

Coastal Waters and their Tributaries. Nutrients Workgroup, Advisory Committee on Water Information, and the National Water Quality Monitoring Council. http://acwi.gov/monitoring/network/nutrients.pdf (accessed December 12, 2009).

Creed, J.T., T.D. Martin, and J.W. O'Dell. 1994. Method 200.9—Determination of trace elements by stabilized temperature graphite furnace atomic absorption, Revision 2.2. Environmental Monitoring Systems Laboratory, Office of Research and Development, U.S. Environmental Protection Agency, Cincinnati, OH.

Eaton, A.D., L.S. Clesceri, E.W. Rice, A.E. Greenberg, and M.A.H. Franson. 2005. *Standard Methods for the Examination of Water and Wastewater*, 21st ed. Washington D.C.: American Public Health Association.

Fowler, S.W. 1990. Critical review of selected heavy metal and chlorinated hydrocarbon concentrations in marine environment. *Mar Environ* Res 29:1–64.

Gilmour, C.C., and E. A. Henry. 1991. Mercury methylation in aquatic systems affected by acid deposition. *Environ. Pollution* 71(2–4):131–169.

Hanlon, E.A., X.H. Fan, B. Gu, K. Migliaccio, Y.C. Li, and T.W. Dreschel. 2010. Water quality trends at inflows to Everglades National Park, 1977–2005. *J. Environ. Qual.* (in press).

Hautman, D.P., and D.J. Munch. 1997. Method 300.1—determination of inorganic anions in drinking water by ion chromatography, version 1.0. National Exposure Research Laboratory Office of Research and Development, U.S. Environmental Protection Agency, Cincinnati, OH.

Keith, L.H., H.J. Brass, D.J. Sullivan, J.A. Boiani, and K. T. Alben. 2005. An introduction to the National Environmental Methods Index. *Environ. Sci. Technol.* 39 (8):173A–176A. DOI: 10.1021/es053241l.

Li, Y.C., T. Olczyk, and K. Migliaccio. 2006. County faculty in-service training for water sampling and chemical analysis. *Proc. Fla. State Hort. Soc.* 119: 249–254.

Marcovecchio, J.E., S.E. Botte, and R.H. Freije. 2007. Heavy metals, major metals, trace elements. In *Handbook of Water Analysis*, ed. L.M.L. Nollet, 275–311. Boca Raton: Taylor & Francis.

Martin, T.D., C.A. Brockhoff, J.T. Creed, and EMMC Methods Work Group. 1994. Method 200.7—determination of metals and trace elements in water and wastes by inductively coupled plasma-atomic emission spectrometry. Revision 4.4. Environmental Monitoring System Laboratory, Office of Research and Development, U.S. Environmental Protection Agency, Cincinnati, OH.

O'Dell, J.W. 1993a. Method 350.1—Determination of ammonia nitrogen by semi-automated colorimetry. Revision 2.0. Environmental Monitoring System Laboratory, Office of Research and Development, U.S. Environmental Protection Agency, Cincinnati, OH.

O'Dell, J.W. 1993b. Method 365.1—Determination of phosphorus by semi-automated colorimetry. Revision 2.0. Environmental Monitoring System Laboratory, Office of Research and Development, U.S. Environmental Protection Agency, Cincinnati, OH.

O'Dell, J.W. 1993c. Method 180.1—Determination of turbidity by nephelometry. Revision 2.0. Environmental Monitoring System Laboratory, Office of Research and Development, U.S. Environmental Protection Agency, Cincinnati, OH.

O'Dell, J.W., B.B. Potter, L.B. Lobring, and T.D. Martin. 1994. EPA Method 245.1—Determination of mercury in water by cold vapor atomic absorption spectrometry, Revision 3.0. Environmental Monitoring System Laboratory, Office of Research and Development, U.S. Environmental Protection Agency, Cincinnati, OH.

PANNA (Pesticide Action Network North America). 2009. PAN Pesticides Database—Chemicals. http://www.pesticideinfo.org/ (accessed December 6, 2009).

USEPA (U.S. Environmental Protection Agency). 1982. Method 150.1: pH (Electrometric). Washington D.C.: Environmental Protection Agency.

USEPA. 1983a. Method 120.1: Conductance (Specific Conductance, umhos at 25°C). Washington D.C.: Environmental Protection Agency.

USEPA. 1983b. *Methods for Chemical Analysis of Water and Wastes.* Alexandria: The National Technical Information Service (NTIS).

USEPA. 1999. Method 632: The Determination of Carbamate and Urea Pesticides in Municipal and Industrial Wastewater. Washington D.C.: Environmental Protection Agency.

USEPA. 2002a. Total coliforms and *Escherichia Coli* in water by membrane filtration using a simultaneous detection technique. EPA-821-R-02-024. Washington D.C.: Environmental Protection Agency.

USEPA. 2002b. Method 1631, Revision E: Mercury in water by oxidation, purge and trap, and cold vapor atomic fluorescence spectrometry. EPA-821-R-02-019. Washington D.C.: Environmental Protection Agency.

USEPA. 2007. Guidelines establishing test procedures for the analysis of pollutants under the Clean Water Act; national primary drinking water regulations; and national secondary drinking water regulations; analysis and sampling procedures; final rule. 40 CFR Part 122, 136. Federal Register, Vol. 72, No. 47. Washington D.C.: Environmental Protection Agency.

USEPA. 2008. Expedited approval of alternative test procedures for analysis of contaminants under the Safe Drinking Water Act: Analysis and sampling procedures. 40 CFR Part 141. Federal Register, Vol. 73, No. 107. Washington D.C.: Environmental Protection Agency.

USGS (U.S. Geological Survey). 2006. Pesticides in the Nation's Streams and Ground Water. Reston: Office of Communication, U.S. Department of the Interior, U.S. Geological Survey.

Weiner, E.R. 2007. *Application of Environmental Aquatic Chemistry.* Boca Raton: CRC Press.

WHO (World Health Organization). 2003. Ammonia in drinking-water, background document for development of WHO guidelines for drinking-water quality. Geneva: World Health Organization.

11 Sampling and Analysis of Emerging Pollutants

David A. Alvarez and Tammy L. Jones-Lepp

CONTENTS

11.1 INTRODUCTION

Historically, environmental monitoring programs have tended to focus on organic chemicals, particularly those that are known to resist degradation, bioaccumulate in the fatty tissues of organisms, and have a known adverse toxicological effect. The Stockholm Convention on Persistent Organic Pollutants (http://chm.pops.int) identified several classes of chemicals of environmental concern—chlorinated pesticides, polychlorinated biphenyls, polychlorinated dioxins, and furans—and later developed policy criteria leading to the worldwide limitation or ban on the use of a dozen chemicals in these classes (UNEP, 2005). These chemicals and others that fit the described criteria are typically referred to as persistent organic pollutants (POPs).

Recently, it has been recognized that risks to aquatic and terrestrial organisms, including humans, are not limited to chemicals fitting the classical POP definition. An examination of the complex mixtures of chemicals present in natural water reveals the presence of organic chemicals covering a wide range of water solubilities and environmental half-lives. Many of these chemicals have been termed *emerging contaminants* (ECs) by the scientific community.

"Emerging contaminants" is a phrase commonly used to broadly classify chemicals that do not fall under standard monitoring and regulatory programs but may be candidates for future regulation once more is known about their toxicity and health effects (Glassmeyer, 2007). The term *emerging* can be misinterpreted as an indication that the chemical's presence in the environment is new, when in fact it means the chemical has recently gained the interest of scientific and regulatory communities. Chemicals such as polybrominated diphenyl ether (PBDE) flame retardants, musk fragrances, and pharmaceuticals have been present in the environment since their first use decades ago (Garrison et al., 1976; Hignite and Azarnoff, 1977; Yamagishi et al., 1981; de Wit, 2002), but only recently have they been considered and measured due to advances in monitoring techniques and the increased understanding of their toxicological impact. Other chemicals, such as nanomaterials, can truly be defined as emerging, that is, "new." Although nanomaterials have been present in research laboratories since the early 1980s, it has only been since the early 2000s that nanomaterials have been produced in sufficient quantities for consumer use (Englert, 2007). Some of these new nanomaterials may become a concern as the probability is high for their continual release into the aquatic environment via multiple consumer applications such as nanosilver disinfectants released into washing machines, water purifiers, and athletic socks and nano-titanium dioxide in cosmetics and sunblocks (Woodrow Wilson International Center for Scholars Nanotechnology Project Inventories, 2009).

Effluents, treated and nontreated, from wastewater treatment plants (WWTPs) and industrial complexes, leaking septic tanks, rural and urban surface runoff, and improper disposal of wastes are all common sources of ECs. ECs commonly include complex mixtures of new generation pesticides, antibiotics, prescription and nonprescription drugs (human and veterinary), personal-care products, and household and industrial compounds such as antimicrobials, fragrances, surfactants, and fire retardants (Alvarez et al., 2005). The fate of such contaminants in WWTPs is largely unknown; however, the limited data that does exist suggests that many of these chemicals survive treatment and some others are transformed back into their biologically active form via deconjugation of metabolites (Desbrow et al., 1998; Halling-Sørensen et al., 1998; Daughton and Ternes, 1999).

The plethora of ECs in the environment is highlighted by Kolpin et al. (2002), who found at least one EC in 80% of the 139 streams sampled across the United States. Rowe et al. (2004) reported that at least one EC was present in 76% of shallow urban wells sampled in the Great and Little Miami River Basins in Ohio and found that the number of ECs detected increased with increasing urban land use. Urban streams are impacted by EC contamination due to the concentration of people and potential point sources; however, surface and groundwater systems in rural areas can also be at risk due to less efficient waste treatment systems and nonpoint source contamination from agricultural practices (Barnes et al., 2008; Focazio et al., 2008). Widespread use of pesticides and land application of manure from large animal feeding operations are common contributors of anthropogenic contaminants to rural water systems (Boxall et al., 2003; Sarmah et al., 2006; Burkholder et al., 2007).

Diminishing fresh water supplies has prompted a "use and reuse" practice where water is often used, treated, and released back into a reservoir or river, before being reused again as drinking water by the same or downstream communities (Drewes et al., 2002; USEPA, 2004; Radjenović et al., 2008). The pathways for removal of ECs from wastewater streams are poorly understood and, as a result, many ECs survive conventional water treatment processes and persist in drinking water supplies (Stackelberg et al., 2007; Benotti et al., 2009). Gibbs et al. (2007) found that 52 of 98 ECs remained unaltered in chlorinated drinking water 10 days after treatment.

Releases of ECs into the environment, albeit at trace (parts per billion and parts per trillion) concentrations, have the potential to cause adverse biological effects across a range of species (Daughton and Ternes, 1999; Sumpter and Johnson, 2005). Several common ECs are known or suspected to alter the endocrine function in fish, resulting in impaired reproductive function, feminization or masculinization of the opposite sex, and other anomalies (Sumpter and Johnson, 2005). Pharmaceuticals designed for human or veterinary use have a specific biological mode of action; however, the impact on nontarget species is rarely known. Since ECs are released into the environment as complex mixtures, and not single compounds, the possibility exists for synergistic or antagonistic interactions resulting in unexpected biological effects. The concentrations of ECs in water supplies are likely to be below any level of direct risk to humans; however, indirect risks are likely, such as the presence of antibiotics in the environment resulting in the development of antibiotic-resistant strains of bacteria that could become a serious threat to human health (Schwartz et al., 2003; Kümmerer, 2004; Josephson, 2006; Schwartz et al., 2006).

The first step in understanding the potential biological impact of ECs in the environment is to identify and quantify the types of ECs that are present. To do so, innovative sampling methodologies need to be coupled with analytical techniques that can confirm the identity of targeted and unknown chemicals at trace concentrations in complex environmental samples. This chapter will discuss common techniques that can be used to address the issues of sampling and analysis of ECs, such as pharmaceuticals, hormones, personal care products, perfluorinated chemicals, and nanomaterials in water.

11.2 SAMPLING METHODS

11.2.1 Development of a Sampling Plan

Obtaining a sample of the matrix of interest is an often-overlooked but vital component of any environmental monitoring program. Failure to properly collect a sample can invalidate any results subsequently obtained. The sample should be representative of the original environmental matrix (air, water, sediment, biota, etc.) and be free of any contamination arising during sample collection and transport to the analytical facility. The collection of a representative sample starts in the office or laboratory with the training of personnel and formulation of a sampling plan, moves to the field for the actual sampling, and ends with the delivery of the sample to the laboratory.

A successful sampling strategy must begin with a thorough plan and established protocols. Questions that need to be addressed while planning the sampling trip include (1) selection of the sampling method to obtain a representative sample, (2) determination of the sample quantity needed to meet the minimum quantitation limits of the analytical method, (3) identification of quality control (QC) measures to be taken to address any bias introduced by the sample collection, (4) identification of safety measures that need to be taken, and (5) determination of sampling objectives. The study plan must define the chemicals to be assayed in the sample and sample size requirements of the analytical methods. Different extraction and processing procedures may be needed to isolate targeted chemical classes from each other and potential interferences, resulting in larger sample size requirements. If sample size is limited, then alterations to the processing methods or changes to the overall study design may be needed. If possible, reconnaissance trips to sampling sites will greatly aid in the determination of the logistical needs of the sampling plan.

Documentation of the sampling trip is critical as observations and measurements made in the field are often necessary for the integrity of the sample and can be instrumental in the final interpretation of the chemical analyses. Depending on the study design and properties of the targeted chemicals, water quality parameters such as temperature, flow, pH, turbidity, etc., may need to be measured. The field log should include sample collection procedures, location of the sampling sites on maps, global positioning system (GPS) coordinates or other data to identify the site(s), date and time samples were collected, types of QC that were used, and names of the personnel involved in the sample collection. Additional information on weather conditions during sampling, visible point sources of contamination, and surrounding land use can be useful during the final interpretation of the data. Photographs of the sampling sites are often helpful, especially if the project officer or the scientists interpreting the data and writing the report are not familiar with the location.

The sample collection plan becomes a balancing game between the numbers of samples that can be taken, defined by sample availability and funding, and the amount of uncertainty that can be tolerated by the study objectives. When collecting samples, regardless of the matrix, the amount of uncertainty associated with the sampling decreases with increasing number of samples. Sample collection can follow a judgmental, systematic, or random pattern approach (Keith, 1991; Radtke, 2005). A judgmental approach focuses the sampling points around a predetermined spot such as a known point source. A systematic approach involves taking samples from locations identified by a consistent grid pattern. The random approach has no defined locations for sample collection and requires a high number of samples to be taken, but generally results in the lowest uncertainty.

Regardless of the type of sample matrix method used, issues of sample preservation, storage conditions and time, and delivery methods must be resolved. Samples should be collected with equipment made of stainless steel, aluminum, glass, or fluorocarbon polymers. Materials made of polyethylene, rubber, Tygon, or other plastics should be avoided due to the potential for these materials to absorb or desorb targeted chemicals from/into the collected sample. Since plasticizers and flame retardants are commonly targeted ECs, plastics should not be used as they may contain high levels of these chemicals from the manufacturing process. The need for sample preservation, which can vary among chemical classes, often requires the addition of chemicals to water samples, but this is generally not recommended for most ECs. If elevated levels of residual chlorine are present in a water sample, sodium thiosulfate is often added to prevent the formation of chlorinated by-products (Keith, 1991; USEPA, 2007a,b). To prevent alteration, samples are transported chilled (< 4–6°C, and overnight if by carrier) to the laboratory. ECs potentially sensitive to UV radiation require use of amber bottles to prevent photodegradation.

11.2.2 Traditional Sampling Techniques

Water is an extremely heterogeneous matrix both spatially and temporally (Keith, 1990). The mixing and distribution of waterborne chemicals in a water body are controlled by the hydrodynamics of the water, the sorption partition coefficients of the chemicals, and the amount of organic matter (suspended sediment, colloids, and dissolved organic carbon) present. Stratification due to changes in temperature, water movement, and water composition can occur in lakes and oceans resulting in dramatic changes in chemical concentrations with depth (Keith, 1990). Because episodic events from surface runoff, spills, and other point source contamination can result in isolated and/or short-lived chemical pulses in the water, sampling sites and methods must be carefully selected.

11.2.2.1 Surface Water

The most common method for collecting surface water samples is taking grab or spot samples. This may involve taking a single sample or a composite sample representative of a width- and depth-integrated profile. Collecting a sample by hand directly into the shipping sample container is the easiest method, especially in small, wadeable streams. In deeper water such as lakes and reservoirs, samples are often taken using bailers or thief samplers (Lane et al., 2003). Common samplers include the Kemmerer, Van Dorn, and double check-valve bailer designs (Figure 11.1), all of which consist of a tube or bottle that collects the water sample. The sample is constrained by caps or check valves that close upon being released by a messenger (a weight or other object released along a tether line from the surface). These types of samplers are useful for collecting discrete samples from specific depths.

Depth-integrated samplers generally fall into two categories: hand-held samplers used in wadeable streams and cable-and-reel samplers for nonwadeable bodies of water (Lane et al., 2003). These samplers are designed to accumulate a representative water sample as the sampler is guided across a vertical cross-section of the water body. Depth-integrated samplers often are a torpedo-shaped device that maintains a horizontal orientation as it is raised and lowered in the water column (Figure 11.1). Water enters through a small port in the nose and is collected in a container inside the sampler.

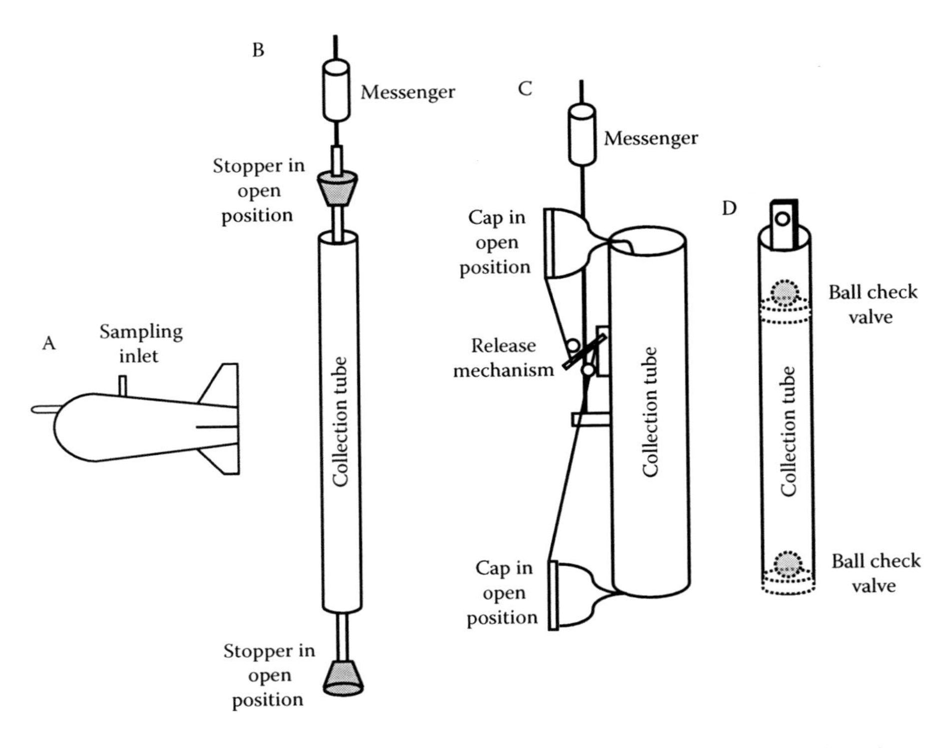

FIGURE 11.1 Commonly used grab and depth-integrating samplers for collection of surface water samples. A. depth-integrating sampler; B. Kemmerer sampler; C. Van Dorn sampler; D. double-check-valve bailer.

Automated sampling systems are often used in remote locations (ephemeral, small streams, storm drains, effluent discharges) where the presence of water may be intermittent, and to collect composite samples over time. They can be programmed to take samples at predetermined intervals or be started by an external sensor such as a flow meter or depth gauge. A basic system consists of a pump to draw water into a collection vessel, while more sophisticated systems can collect multiple samples, have refrigerated storage chambers, and can transmit and receive programming and data via land-line or cellular phone connections (see also Section 5.7.3.2, Automatic Sampling, in Chapter 5).

11.2.2.2 Groundwater

Groundwater samples are generally collected from existing supply wells or monitoring wells. The sampling methods vary depending on water depth and well size. Because monitoring wells are generally small, sampling is less frequent. Automated samplers as discussed above are often used due to their ease of use at multiple sites. Portable peristaltic pumps can also be used to obtain groundwater from monitoring wells. Because supply wells for domestic, industrial, and agricultural use often require routine monitoring, large-capacity pumps and autosampling systems are often permanently installed. Additional details on groundwater sampling are provided in Chapter 6.

11.2.2.3 Soil and Sediment Pore Water

Pore water samples are an important component in assessing toxicity to benthic invertebrates and understanding the potential trophic transfer of contaminants (Winger and Lasier, 1991; Ankley and Schubauer-Berigan, 1994). Pore water can also be a marker of chemicals that may be released into the overlying water column. Pore water samples can be collected in situ using passive sampling devices (Section 11.2.3.3) or in the laboratory. Collection of pore water from sediment samples in the laboratory can be achieved by centrifugation, squeezing, and vacuum filtration (Bufflap and Allen, 1995; Angelidis, 1997). Centrifugation involves placing a soil/sediment sample in a centrifuge tube and then centrifuging until the soil/sediment forms a pellet in the bottom of the tube. The supernatant is then decanted and filtered prior to further processing or analysis. The squeezing method uses pressurized systems with either a diaphragm or piston to compress the sediment and release the interstitial water. Vacuum filtration can be performed in the laboratory or as an in situ active sampling method that involves a sediment probe made of porous plastic, ceramic, or other material that is placed in the sediment. The probe is attached via a length of tubing to a syringe, hand-operated or with an automatic vacuum pump that withdraws the pore water from the sediment. Since pore water samples collected by vacuum filtration are not exposed to air, pore water characteristics are retained and loss of volatile chemicals is minimized (Winger and Lasier, 1991). Additional details on pore water sampling of soil and sediment are provided in Chapter 7.

11.2.3 Time-Integrated (Passive) Sampling Techniques

Because time-weighted average (TWA) concentrations of chemicals are commonly used to determine exposure, they are a fundamental part of an ecological risk assessment process for chemical stressors (Huckins et al., 2006). Since grab samples only represent the concentration of chemicals at the instant of sampling, TWA exposure is difficult to accurately estimate even with repetitive sampling. Episodic events are often missed with routine grab sampling schedules. In addition, the detection of trace concentrations of ECs can be problematic as standard methods are designed to handle small (<5 L) volumes of water. Passive sampling devices provide an alternative to grab sampling, overcoming many of the inherent limitations of those traditional techniques.

Successful use of personal passive monitors or dosimeters in determining TWA concentrations of chemicals to measure exposure in the workplace (Fowler, 1982) has contributed to the application of the same principle to dissolved organic contaminants in water (Huckins et al., 2006). Integrative or equilibrium passive samplers can be used depending on their design, the exposure time in the field, and the properties of the targeted chemicals. Integrative samplers are characterized by having an infinite sink for the retention of sampled chemicals, providing a higher degree of assurance that episodic changes of chemical concentrations in the water will not be missed. The use of an integrative sampler is essential for the determination of TWA concentrations. Equilibrium samplers are characterized by having low capacity for retaining chemicals and high chemical loss rates. Although simplicity in the uptake models makes equilibrium samplers an attractive option, one of the difficulties encountered is assessing whether equilibrium—which can be affected by temperature, water flow, and biofouling—has been reached or not (Huckins et al., 2006).

11.2.3.1 Surface Water

The major use of passive sampling devices outside of occupational-exposure monitoring for human health and safety in the workplace is in surface water applications. A growing number of passive samplers have been developed for sampling organic chemicals in water. These samplers include, but are not limited to, semipermeable membrane devices (SPMD), polar organic chemical integrative samplers (POCIS), Chemcatchers, polyethylene strips, polymers on glass, and solid-phase microextraction (SPME) devices (Namieśnik et al., 2005). The SPMD and POCIS are two of the most widely used passive samplers for measuring ECs in surface water (Figure 11.2). SPMDs consist of a layflat low density polyethylene membrane tube containing a neutral lipid such as triolein (Huckins et al., 2006). The POCIS consists of a solid phase sorbent or mixture of sorbents contained between two sheets of a microporous polyethersulfone membrane (Alvarez et al., 2004, 2007). SPMDs sample chemicals with moderate to high (>3) octanol-to-water partition coefficients (K_{ow}s) due to the affinity of these hydrophobic chemicals to partition into the lipid and hydrophobic membrane of the sampler. Chemicals with log K_{ow}s <3 are sampled using the POCIS, which has a hydrophilic membrane and modified adsorbents to remove polar organics from the water. By using the two samplers *in concert*, a wide range of organic chemicals can be measured.

A generic processing scheme for SPMDs and POCIS (Figure 11.3) begins with collecting the passive sampler used in the field or laboratory and storing it at <0°C in a solvent-rinsed airtight container, such as a metal can, for transport to the laboratory

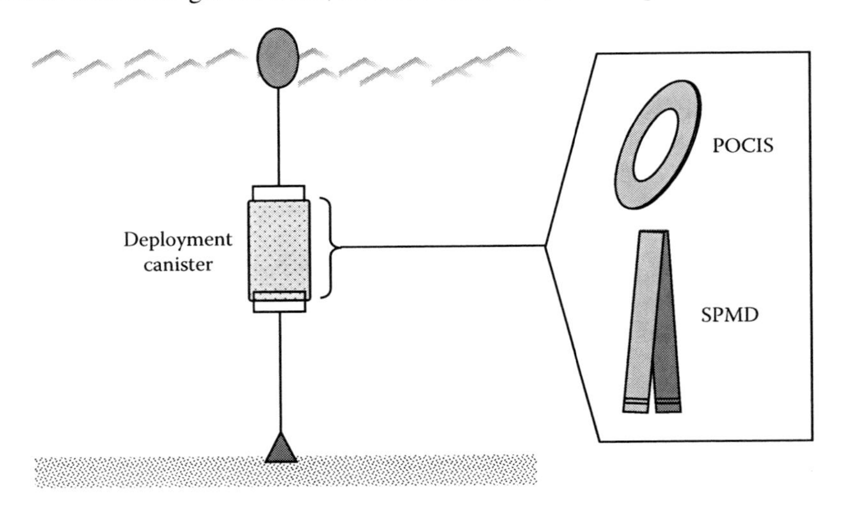

FIGURE 11.2 A typical apparatus for suspending passive samplers in the water column. Polar organic chemical integrative samplers (POCIS) and semipermeable membrane devices (SPMDs) are commonly housed in this type of protective canister.

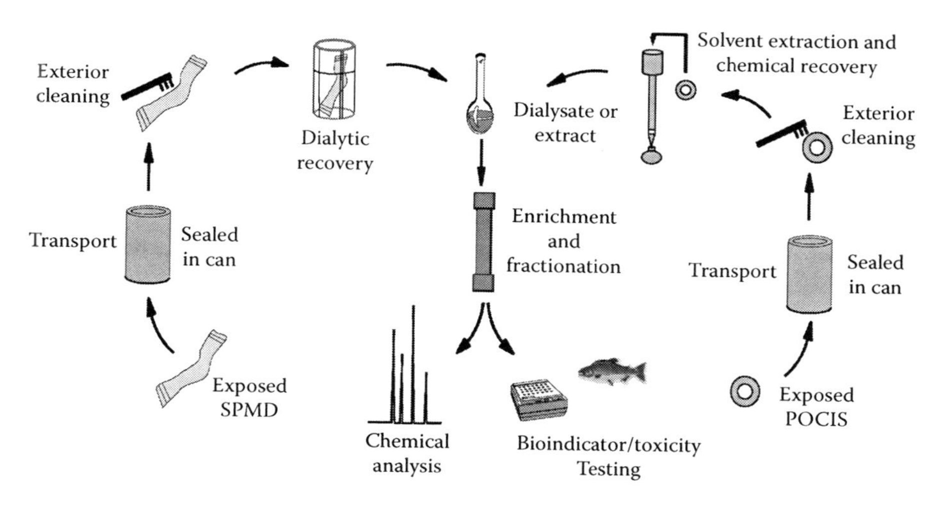

FIGURE 11.3 General schematic for the processing, analysis, and/or biological testing of passive samplers.

and storage until processing. At the onset of processing, the exterior of the sampler is gently cleaned with a soft toothbrush and running water to remove any particulate matter on the surface that may fall into the sample on opening. Chemical residues are recovered from the sampler by extraction using a suitable method such as dialysis for the SPMD or solvent extraction of the sorbent for the POCIS. Depending on the requirements of the analytical method, the extract can undergo further enrichment and fractionation to isolate the targeted chemicals from potential interferences. The extracts are then available for analysis using common analytical instrumentation, for bioassay or toxicity testing, or for dosing experiments to determine effects on organisms.

SPMDs and POCIS are commonly used for measuring levels of ECs in surface water. Leiker et al. (2009) determined levels of methyl triclosan in Las Vegas Wash, a channel receiving treated WWTP effluents from the city of Las Vegas, Nevada, using SPMDs. Trace water concentrations of PBDEs have been measured using SPMDs in the Columbia River (Washington State) and off the Dutch coast (Booij et al., 2002; Morace, 2006). POCIS have been used in numerous surface water monitoring studies to assess pharmaceuticals and other ECs in WWTP effluents (Jones-Lepp et al., 2004; Alvarez et al., 2005, 2009; MacLeod et al., 2007; Mills et al., 2007). Chemicals such as antibiotics, fragrances, plasticizers, and surfactants were commonly found in these studies. Comparisons between POCIS and traditional grab sampling techniques have shown that the latter can miss the sporadic or low level occurrence of ECs and that TWA data are less variable and easier to interpret than data obtained using repetitive grab samples (Alvarez et al., 2005; Vermeirssen et al., 2006).

11.2.3.2 Groundwater

Passive samplers, which have a minimal effect on water circulation and preserve stratification of water within a well, have an advantage over active sampling techniques (Vrana et al., 2005). Samplers based on diffusion have been used for the monitoring of volatile organic compounds (VOCs) since the early 1990s (Vroblesky et al., 1991). While most groundwater passive samplers have focused on VOCs, semi-volatile organic compounds have been sampled using SPMDs (Vrana et al., 2005; Huckins et al., 2006). The use of any passive sampler such as the SPMD, which has a large capacity and high sampling or clearance rates, can be limited in systems with low groundwater flow. If the exchange volume of the well is less than the clearance volume of the SPMD, chemicals can potentially be depleted, changing the equilibrium between the sediment and water (Vrana et al., 2005). This can be avoided by using smaller SPMDs or choosing a different passive sampler that has a clearance volume less than the groundwater recharge rate.

11.2.3.3 Soil and Pore Water

Collection of pore water samples in situ avoids possible alteration during collection, shipment, storage, and processing of whole sediment samples. A dialysis sampler gives more accurate estimates of the pore water concentrations than centrifugation because sediment–water interactions can result in altered chemical measurements (Angelidis, 1997). The most common passive sampler for pore water is a dialysis system occasionally referred to as Peepers or the Hesslein In-situ Pore Water Sampler (Hesslein, 1976). Peepers that are based on the diffusion of chemicals across a membrane are equilibrium samplers, whose efficiency is determined by the equilibration time and the diffusion coefficient for a chemical, temperature, and by sediment porosity. For more information on peepers, see Section 7.5.2 in Chapter 7.

The development of the solid-phase microextraction device (SPME), which is an equilibrium sampler consisting of a coated fiber housed in a syringe body, provides a new means of collecting an in situ sample of organic chemicals in pore water. The fiber is plunged into the sediment where it reaches equilibrium with the pore water and then is retracted into the syringe body (Ouyang and Pawliszyn, 2007; Maruya et al., 2009). Specially outfitted gas chromatographs can allow the SPME fiber to be inserted into an injector where the sampled chemicals are recovered via thermal desorption and directly analyzed.

11.2.4 Quality Control (QC)

Bias in the form of variability and sample contamination, which is present in every sample, can be identified by the use of appropriate QC measures. Common types of QC samples include replicates, blanks, and fortified samples (spikes). Identical conditions (i.e., sampling devices, containers, and protocols) must be used for both the field and QC samples. The selection of the matrix for blank and spiked QC samples must be nearly identical to the field sample matrix but free of the chemicals of interest in the study.

Three types of blanks are commonly used: field, trip, and equipment blanks. Field blanks are exposed to the ambient air during the sampling process to measure any potential contamination. Generally, these blanks consist of analyte-free water, freshly prepared passive samplers, or some other surrogate matrix. In contrast, trip blanks, which are not exposed to the air, accompany the field samples from the sampling site to the laboratory to assess contamination during shipping, handling, and storage. Equipment blanks are rinses of the sampling equipment (e.g., buckets, bailers, autosamplers, etc.) that are collected to determine how adequately the equipment was decontaminated between uses. Steps to minimize sample contamination include thoroughly cleaning the sampling equipment, reducing exposure time to ambient air, and avoiding contact with or consumption of personal-use products and medications, which may contain the chemicals of interest.

Quality control spikes can include field-spiked samples where known quantities of targeted chemicals are added to collected samples to identify field, transportation, and matrix effects (Keith, 1991). If there is not sufficient sample available in the field, a surrogate matrix can be used for these spike samples. Budgetary constraints can limit the amount of QC that is used; however, it should not limit the types of samples that are collected. As an alternative, the field blanks can be analyzed and the remaining QC samples archived unless problems are identified in the field blanks (Keith, 1991).

11.3 SAMPLE PREPARATION, EXTRACTION, CLEANUP, AND ANALYSIS

One of the challenges facing the analytical chemistry community is the development of robust and standardized analytical methods and technologies that can easily be transferred to laboratories worldwide. While today's analysts can detect pg L^{-1} and ng L^{-1} concentrations of numerous ECs (e.g., PFOS, PFOAs, pharmaceuticals, nonyl- and alkyl-phenolethoxylates, steroids, hormones, and their metabolites) in various water matrices (e.g., surface waters, wastewaters, groundwater), proper analytical methods must still be followed. Table 11.1 provides a summary of the methods discussed in this section.

11.3.1 PREPARATION, EXTRACTION, AND CLEANUP

Concentrations of ECs found in the water samples are typically below the μg L^{-1} range, making extraction, pre-concentration, and cleanup prior to detection an important step. Solid phase extraction (SPE) is one of the most widely reported methods for isolating ECs from environmental aqueous samples. SPE was developed as an alternative to liquid–liquid extraction (LLE), which is labor intensive, difficult to automate, and requires large portions of high-purity solvents, such as methylene chloride. Nevertheless, LLE has been used to extract ECs containing hydroxyl groups (e.g., bisphenol A, nonylphenol ethoxylates, alkylphenol ethoxylates, and most steroids and hormones) from water. This process often requires derivatization of the hydroxyl groups prior to extraction using agents such as *N*-methyl-*N*-(*tert*-butyldimethylsilyl), trifluoroacetamide (MTBSTFA), bis(trimethylsilyl) trifluoroacetamide (BSTFA), and *N*-methyl-*N*-(trimethylsilyl) trifluoroacetamide (MSTFA), and less frequently diazomethane (Kelly, 2000; Moeder et al., 2000; Mol et al., 2000; Öllers et al., 2001; Ternes et al., 2002; Liu et al., 2004). Most of the LLE methods described below

TABLE 11.1
Review of Extraction and Detection Methods for the Analysis of Select Emerging Contaminants

Analyte(s) Class(es)	Aqueous Matrix	Extraction Method[a]	Detection Method[b]	Reference
Pharmaceuticals, hormones, illicit drugs, surfactants, plasticizers, pesticides, personal care products	Surface water, wastewater	SPMD POCIS	GC-MS LC-MS LC-MS/MS	Alvarez et al., 2005 Alvarez et al., 2007 Alvarez et al., 2009 Jones-Lepp et al., 2004
Nonyl-, octyl-phenols Bisphenol A, 17β-estradiol, 17α-ethynylestradiol	Surface water	LLE	GC-MS w/derivatization	Mol et al., 2000
Alkylphenol ethoxylate nonionic surfactants, flame retardants, plasticizers, fecal sterols, disinfectants	Surface water, stormwater overflows, domestic and industrial wastewater	CLLE	GC-MS	Zaugg et al., 2007
Nonylphenols, bisphenol A, *p*-tert-octylphenol, nonylphenol menoethoxylate, nonylphenol diethoxylate	Surface water, wastewater	LLE	GC-MS	ASTM D 7065-06
Nonylphenol, nonylphenol ethoxylate, nonylphenol diethoxylate, octylphenol	Surface water, wastewater, sea water	SPE (C_{18})	LC-MS/MS	ASTM D 7485-09
Iodo-disinfection byproducts	Drinking water	LLE	GC/NCI-MS w/derivatization	Richardson et al., 2008
C_{60} and C_{70} fullerenes, [6,6]-phenyl C_{61}-butyric acid methyl ester	Surface water	LLE	HPLC-UV	Bouchard and Ma, 2008
Steroids, hormones	Surface water	SPE (styrenedivinylbenzene) field sampler	HPLC-fluorescence and radioimmunoassay	Snyder et al., 1999
Parabens, alkylphenols, phenylphenol, bisphenol A	Surface water, wastewater	SPE (HLB)	LC-MS/MS	Jonkers et al., 2009
Musks, synthetic musks (e.g., tonalide, galaxolide)	Surface water, wastewater	SPE[polystyrene/ poly (methyl methacrylate)]	GC-MS	Osemwengie and Steinberg, 2001

Nine neutral pharmaceuticals (e.g., diazepam, caffeine, glibenclamide, omeprazole)	Surface water	SPE (C_{18})	LC-MS/MS	Ternes et al., 2001
Carbamazepine, ibuprofen, diclofenac, ketoprofen, naproxen, clofibric acid, triazines, acetamides, phenoxy acids	Drinking water, surface water, wastewater	SPE (HLB)	GC-MS (two step analysis, derivatization for acidic compounds)	Öllers et al., 2001
Estrogens, progestrogens	Surface water, sediments	SPE (HLB, C18, polydivinylbenzene resin-GP)	LC-DAD-MS	de Alda and Barceló, 2001
95 compounds: veterinary and human antibiotics, prescription drugs, nonprescription drugs, phthalates, insecticides, nonylphenols, polynuclear aromatic hydrocarbons	Surface water	LLE SPE (HLB)	GC-MS LC-MS	Kolpin et al., 2002
7 basic pharmaceuticals and 11 acidic drugs: carbamazepine, aspirin, caffeine, gemfibrozil, naproxen	Surface water, wastewater	SPE (HLB)	GC-MS w/derivatization	Togola and Budzinski, 2007
21 pharmaceuticals: corticosteroids (cortisone, dexamethasone, hydrocortisone, prednisone); β-blockers (atenolol, metoprolol, propanolol)	Wastewater (influent, effluent)	SPE (MCX)	LC-MS/MS	Piram et al., 2008
Antibiotics: macrolides, sulfonamides, tetracyclines, trimethoprim, chloramphenicol, penicillins	Surface water	Lypholization and SPE (C_{18})	LC-MS/MS	Hirsch et al., 1998
Ciprofloxacin, enrofloxacin, tetracyclines	Surface water, well water, wastewater	SPE (HLB)	LC-MS (SIM)	Reverté et al., 2003
Erythromycin-H_2O, roxithromycin, tylosin	Surface water, CAFO wastewater	SPE (HLB)	LC-MS/MS	Yang and Carlson, 2003
13 antibiotics: fluoroquinolones, sulfonamides, tetracyclines, macrolides	Surface water	SPE (HLB)	LC-MS/MS	Batt and Aga, 2005
Tetracyclines, sulfonamides, macrolides, ionophore polyethers	Surface water	SPE (HLB)	LC-MS/MS	Kim and Carlson, 2006

(continued)

TABLE 11.1 (continued)
Review of Extraction and Detection Methods for the Analysis of Select Emerging Contaminants

Analyte(s) Class(es)	Aqueous Matrix	Extraction Method[a]	Detection Method[b]	Reference
Azithromycin	Wastewater	LLE	LC-MS	Koch et al., 2005
Azithromycin, roxithromycin, clarithromycin, methamphetamine, MDMA, urobilin	Surface water, wastewater	SPE (HLB)	LC-MS/MS	Jones-Lepp, 2006
Azithromycin, roxithromycin, clarithromycin, methamphetamine, MDMA	Surface water, wastewater (influent, effluent)	SPE (HLB)	LC-MS/MS	Loganathan et al., 2009
Perfluorooctanesulfonates (PFOSs), perfluorooctanoates (PFOAs)	Surface water, wastewater	SPE (C18)	LC-MS	Tseng et al., 2006
PFOSs, PFOAs	Wastewater	SPE (HLB)	LC-MS/MS	Loganathan et al., 2007
PFOSs, PFOAs, steroids, hormones	Surface water	SPE (HLB)	LC-MS/MS	Loos et al., 2009
PFOSs, PFOAs	Surface water	SPE (HLB)	LC-MS/MS	Gros et al., 2009
Sulfonamides	Surface waters	LLLME	HPLC/UV	Liu and Huang, 2008
Sulfonamides, macrolides, trimethoprim	Surface water	SPME	LC-MS/MS	McClure and Wong, 2007
Estrogens: diethylstilbestrol, estrone, 17β-estradiol, 17α-ethynylestradiol	Reservoir water, drinking water	PC-HFME	GC-MS w/derivatization	Basheer et al., 2005
diethylstilbestrol, estrone, 17β-estradiol, estriol	Wastewater	MIP	HPLC/UV-vis	Meng et al., 2005
17β-estradiol	River water	MIP	LC-MS	Watabe et al., 2006
8 β-blockers: atenolol, sotalol, pindolol, timolol, metoprolol, carazolol, propranolol, betaxolol	Wastewater	MIP	LC-MS/MS	Gros et al., 2008

[a] Extraction methods: SPMD—semipermeable membrane device, POCIS—polar organic chemical integrative sampler, LLE—liquid-liquid extraction, CLLE—continuous liquid-liquid extraction, SPE—solid phase extraction, HLB—hydrophilic lipophilic blend, MCX—mixed mode cation exchange, LLLME—liquid-liquid-liquid microextraction, SPME—solid phase microextraction, PC-HFME—polymer coated hollow fiber microextraction, MIP—molecularly imprinted polymers.

[b] Detection methods: GC-MS—gas chromatography mass spectrometry, LC-MS—liquid chromatography mass spectrometry, LC-MS/MS—liquid chromatography tandem mass spectrometry, GC/NCI-MS—gas chromatography negative chemical ionization mass spectrometry, HPLC-UV—high performance liquid chromatography ultraviolent detection, LC-DAD-MS—liquid chromatography diode array detection coupled with mass spectrometry, SIM—selection ion monitoring.

use derivatization before gas chromatography-mass spectrometry (GC-MS) analysis. Information on GC and MS can be found in Section 9.3.3 of Chapter 9.

11.3.1.1 Liquid–Liquid Extraction

Mol et al. (2000) proposed a specific LLE procedure for several nonyl- and octyl-phenols, 4-*tert*-butylbenzoic acid, bisphenol A, 17β-estradiol, and 17α-ethynylestradiol where water samples are acidified and extracted with two portions of ethyl acetate. The extracts are then reduced in volume and derivatized prior to analysis by gas chromatography-mass spectrometry (GC-MS). Zaugg et al. (2007) described a continuous LLE (CLLE) procedure for extracting several classes of ECs (e.g., alkylphenol ethoxylate nonionic surfactants, flame retardants, plasticizers, fecal sterols, and disinfectants) from surface and storm-sewer overflow water samples that are not pH-adjusted or filtered. This method is different from traditional LLE methods in that it uses smaller amounts of methylene chloride and shorter extraction times (6 vs. 24 hrs). The extracts are then concentrated and analyzed by GC-MS.

The American Society for Testing and Materials (ASTM, 2007) published a standard test method for the determination of nonylphenols, bisphenol A, *p-tert*-octylphenol, nonylphenol monoethoxylate, and nonylphenol diethoxylate in environmental waters using LLE as an extraction method with subsequent detection by GC-MS. This method calls for acidified water samples, placed into a LLE along with methylene chloride, and extracting the water sample for 18 to 24 hours. After the extraction is complete the extracts are concentrated, dried over anhydrous sodium sulfate, and subsequently analyzed by GC-MS.

Richardson et al. (2008) reported a LLE-gas chromatography-negative chemical ionization mass spectrometry (GC/NCI-MS) method for a newly recognized class of ECs, namely, iodo-disinfection byproducts. One L of drinking water is adjusted to pH <0.5, 350 g sodium sulfate is added, and the sample is extracted with methyl tert-butyl ether (MTBE). The iodo-acids partition into the organic phase and are derivatized with the addition of diazomethane, thus converting the iodo-acids into methylated iodo-acids prior to GC/NCI-MS analysis.

Nanomaterials are defined as carbon- or metallic-based, dendrimers, and bio-inorganic composites with particle sizes in the nm range ($1–100 \times 10^{-9}$ m). Nanomaterials can be considered as new chemicals because their physicochemical properties are very different at this extremely small scale. They have relatively large specific surface areas and at the very low end of the scale, quantum effects can override their general physicochemical properties (Motzer, 2008). While the number of consumer products containing nanomaterials is soaring, there are few methods for their detection in the environment and correspondingly few papers published on the subject. The first method was developed to detect naturally occurring C_{60} and C_{70} fullerenes in geologic samples (Heymann et al., 1995) and could probably be adapted to water samples. Their method requires the sample to be slurried by sonication for 4 hr with adequate amounts of toluene before extraction on a Soxhlet apparatus. A preparatory liquid chromatography column (19×300 nm) coupled to a photo diode array (PDA) detector facilitated the separation of the fullerenes from the organic solution using a methanol and toluene mobile phase with a flow rate of 10 mL min^{-1}. Extraction efficiencies were >90% for C_{60} and C_{70}. The second method, reported by Bouchard and Ma (2008), is a simple LLE procedure for extracting C_{60}, C_{70}, and [6, 6]-phenyl C_{61}-butyric acid methyl ester (PCBM) from environmental waters. In this method, water samples are stirred for 13 days and then allowed to settle for 1 hr before sampling. An aliquot was collected from which three subaliquots were extracted with toluene. An aliquot of toluene phase was analyzed by HPLC-UV. One consideration when trying to extract fullerenes from environmental waters is that at neutral pH, fullerenes are negatively charged, thereby facilitating their formation of stable colloidal suspensions, but negating their ability to partition into organic solvents for extraction. Therefore, something needs to be added to the water suspensions to facilitate the break-up of the nano-colloidal suspensions. Bouchard and Ma (2008) showed that the addition of $Mg(ClO_4)$ destabilized the colloidal suspensions, thereby enhancing the partitioning of the carbon-based nanomaterials into the toluene.

11.3.1.2 Solid Phase Extraction

Because many hydrophilic ECs do not partition into an organic solvent from the aqueous phase, resulting in poor extraction efficiencies, SPE rather than LLE should be used. SPE offers lower solvent consumption, shorter processing times, automation options, and simpler procedures than LLE. Since direct sampling in the field is an option for SPE, the need for transport and storage of large sample volumes of water to the laboratory can be avoided (Osemwengie and Steinberg, 2001; Primus et al., 2001). Field-portable SPE can reduce the possibility of degradation of target analytes during sample holding times after sample collection.

Solid-phase extraction is commercially available in three basic formats: thin flat discs (47 and 90 mm), small cylindrical cartridges (usually < 6 mL reservoirs), and 96-well plates. Each type of format can employ a wide variety of sorbents such as silica based (e.g., C_{18}), hydrophilic lipophilic balanced (HLB), mixed cation exchange (MCX), and mixed anionic exchange (MAX). SPE sorbents are selected for their ability to retain the ECs of interest, based upon a variety of physico-chemical properties of both the SPE phase and the analytes (e.g., pK_a and K_{ow}). For example, C_{18} is used as a universal extraction sorbent, with a pH range from 2 to 8, and its retention mechanism is primarily governed by hydrophobic interactions between the analytes and the carbonaceous moieties of the C_{18} alkyl chains (Poole, 2003). Other less commonly used SPE sorbents include weak cation-exchange (WCX), weak anionic-exchange (WAX), strong MAX, anion or cation exchange sorbents without mixed mode sorbents, and C_8 (Benito-Peña et al., 2006; Kasprzyk-Hordern et al., 2007). The ion exchange cartridges are useful not only for extraction and concentration but also for sample cleanup. For example, SPEs can be used to separate humic and fulvic acids from basic ECs, or separate neutral lipids from charged analytes. EC extraction is completed by first cleaning and conditioning the SPE cartridge with the solvent that will be used for the extraction solvent (e.g., methanol), followed by a neutral solvent of the same composition as that of the sample (e.g., water). Once the cartridge is prepared, 0.1 to 2 L of sample is passed through the SPE cartridge, at approximately 7 to 10 mL min^{-1}, using either gravity, vacuum-induced, or syringe-push flow, after which the cartridge is dried for a varying amount of time, and finally extracted using various solvents or solvent mixtures dependent on the p*K*a's and polarities of the analytes of interest (Poole, 2003).

Although LLE can be used for a variety of hydrophilic ECs as discussed before, much of the current work uses SPE for the extraction of these compounds from water. Snyder et al. (1999) were one of the first to report the use of SPE for hormones and steroids detection in source waters. What makes their method interesting is that it is an in situ (field) extraction technique utilizing 90-mm styrenedivinylbenzene SPE discs, which allow for very large volumes (5 L) of water to be extracted. A stainless steel mesh filter at the head of the tubing prevents large particles from entering with the SPE disc encompassed between two glass fiber filters. Once the extraction is complete, the SPE discs are removed, frozen, and shipped to the laboratory for recovery of the hormones and steroids. The resulting organic extract is analyzed using a HPLC-fluorescence detector and radioimmunoassay (Snyder et al., 1999). A more recent analytic procedure using SPE (HLB sorbent) and LC-MS/MS was developed for parabens, alkylphenolic compounds, phenylphenol, and bisphenol A (Jonkers et al., 2009). Water samples at neutral pH were extracted using SPE (HLB sorbent) cartridges. The cartridges were dried and the analytes were eluted using 3 mL methyl-*tert*-butyl ether:2-propanol (1:1) followed by 3 mL methanol. The eluent was evaporated to approximately 250 μL before adding another 250 μL methanol:water (1:1) to bring the final volume to 500 μL before analysis by LC-MS/MS.

Osemwengie and Steinberg (2001) also used an in situ SPE extraction technique for concentrating natural musks and synthetic musks (e.g., tonalide, galaxolide, cashmeran, versalide) from natural waters. A proprietary sorbent (a mix of polystyrene and poly[methyl methacrylate]) packed between polyethylene frits was used. After extraction of ~60 L of water, the cartridges were returned to the laboratory for extraction and clean-up, using gel permeation chromatography, and analysis by GC-MS.

Ternes et al. (2001) used SPE C_{18} to extract nine neutral pharmaceuticals (e.g., diazepam, caffeine, glibenclamide, omeprazole, phenylbutazone) from water. Briefly, this method calls for the extraction of 500 mL of filtered, pH adjusted (7.0 to 7.5) water, and subsequent elution with 3×1 mL of methanol. The extracts were further concentrated to 20 μL, brought back up to 1 mL with a phosphate buffer, and stored at < −20°C until analysis by LC-electrospray-triple stage quadrupole mass spectrometry (LC-ESI-QqQ MS).

These early SPE papers used a variety of familiar sorbents (C_{18}, polyvinylstyrenes), but recently several proprietary sorbents have been developed that are better suited for the emerging contaminants. Since 2004, the most frequently used SPE sorbents used for extracting ECs from water matrices are HLB and MCX sorbents. De Alda and Barceló (2001), who were among the first to report using the HLB-type sorbent, compared on- versus offline SPE extraction, and the recoveries of several estrogens, progestrogens, and their synthetic counterparts (e.g., ethynyl estradiol, diethylstilbestrol, norethindrone, levonorgestrel) from three types of SPE sorbents, namely, HLB, C_{18}, and a polydivinyl benzene resin-GP (general phase) cartridge. Each type of sorbent has its merits dependent upon the amount of interfering substances in the water samples and the limit of detection (LOD) required (de Alda and Barceló, 2001). Öllers et al. (2001) proposed a method for the simultaneous extraction of neutral and acidic pharmaceuticals and a few pesticides from water utilizing HLB cartridges. Their methodology involves filtration of a 1 L water sample adjusted to a pH of 3, followed by sample enrichment onto the cartridge, and elution of the analytes with 6 mL of 50:50 ethyl acetate and acetone mixture. Neutral compounds were assayed by GC-MS, followed by the addition of diazomethane to derivatize the acidic pharmaceuticals before a second GC-MS analysis.

Kolpin et al. (2002) describes five different methodologies utilizing combinations of SPE (HLB cartridges) and LLE (using methylene chloride as the extraction solvent) with subsequent analyses either by LC-MS or GC-MS (with derivatization of the acidic compounds before analysis) to characterize 95 ECs in US streams. Togola and Budzinski (2007) developed two extraction methods using both HLB and C_{18} sorbents for 18 different pharmaceuticals (7 basic compounds and 11 acidic drugs) including carbamazepine, aspirin, caffeine, gemfibrozil, and naproxen. However, they later further refined their method to using only HLB sorbent (Togola and Budzinski, 2008). After the same preextraction procedures (filtering, pH adjustment < 2), the waters were extracted at a rate of 12 to 15 mL min^{-1}, the cartridge is dried for 1 hr under N_2, before extraction with 3 mL ethyl acetate, 3 mL ethyl acetate:acetone (50:50, v/v), and 3 mL ethyl acetate:acetone:ammonium hydroxide (48:48:2 v/v/v). The extracts were evaporated and taken up in 100 μL ethyl acetate, and MSTFA is added to derivatize the acidic compounds (e.g., aspirin, ibuprofen, diclofenac, naproxen, gemfibrozil, and clofibric acid) before analysis by GC-MS. An optimized method using SPE (MCX sorbent) was developed for 21 pharmaceuticals from corticosteroids (e.g., cortisone, dexamethasone, hydrocortisone, prednisone) and β-blockers (e.g., atenolol, metoprolol, propanolol) classes by Piram et al. (2008). In this method, 400 mL water is acidified with formic acid before loading onto MCX cartridges. The corticosteroids are eluted with 1 mL methanol:water:formic acid (70:30:0.1, v/v/v) and the β-blockers are eluted in a second stage with methanol:ammonia (95:5, v/v). The subsequent eluants are evaporated to dryness and taken up in acetonitrile:water (25:75, v/v) before analysis by LC-MS/MS.

Antibiotics [e.g., fluoroquinolones (FQs), macrolides (MCs), sulfonamides (SAs), tetracyclines (TCs)] are EC classes of interest due to their possible adverse effect on the environment by promoting the development of antibiotic-resistant bacteria in waters receiving wastewater effluents (Miyabara et al., 1995; Schwartz et al., 2003; Schwartz et al., 2006). Hirsch et al. (1998) were among the first to describe the extraction and detection of multiple classes of antibiotics (e.g., MCs, SAs, TCs, trimethoprim, chloramphenicol, and penicillins) in water. Their early methodology compared a lyophilization procedure with SPE C_{18} sorbent. The resulting extract was analyzed by LC-MS/MS (using a QqQ). Other researchers have successfully reported the use of SPE sorbents for recovery of antibiotics from water. Reverté et al. (2003) describe a SPE (HLB sorbent) extraction method for the recovery of ciprofloxacin, enrofloxacin, and 4 TCs from water. In their method, 100 to 250 mL water samples are pH adjusted to <3, the analytes were eluted with 5 mL of methanol, the eluate was evaporated to dryness, and re-dissolved in methanol:water (50:50, v/v) before analysis by selected-ion monitoring (SIM) LC-MS. A watershed scale field study was conducted by Yang and Carlson (2003) to determine contamination occurring due to TCs and SAs used in animal production to treat and prevent disease and promote growth. The compounds were found in manure and waste lagoons from confined animal feed operations (CAFOs). Because TCs are unstable in acid solutions, the pH of the waters is adjusted to < 3 just immediately before extraction with

5 mL methanol (1% trifluoroacetic acid) to remove the TCs and SAs. Separate water samples are extracted at neutral and pH < 3 to recover the SAs. Subsequently, all eluants are evaporated to 50 μL before analysis by SIM LC-MS. Batt and Aga (2005) describe a SPE methodology using HLB sorbents to extract 13 antibiotics of various classes (FQs, SAs, TCs, MCs) from water, which is initially adjusted to pH <3 and then Na_2EDTA added to chelate metal ions competing with the sorbent, followed by extraction of the analytes from the SPE sorbent with 10 mL acetonitrile. The eluate is reconstituted in 1 mL of deionized water, before analysis by LC-MS/MS. A SPE method to enrich 4 different classes (TCs, SAs, MCs, and ionophore polyethers [IPs]) of 19 veterinary antibiotics from water samples was developed by Kim and Carlson (2006). TCs, SAs, and MCs are used as both human and veterinary drugs to treat disease and prevent infection, while the IPs are used to promote growth and efficiency of feed conversion in animal production. Their methodology is a modification of that of Yang and Carlson (2003) with optimization of the SPE method for the IPs. As a result no pH adjustments or additives to the methanol and water used for cartridge conditioning or eluting solvent (methanol) are required.

One of the most widely used human antibiotics in the United States is azithromycin (annual sales in 2007 were $1.3 billion dollars, equivalent to over 45 million prescriptions; see http://drugtopics.modernmedicine.com/drugtopics/data/articlestandard/drugtopics/102008/500218/article.pdf). Only a few methods have been published on its extraction and detection. Koch et al. (2005) used methyl-*tert*-butyl ether added to 10 mL of water prior to vortexing and centrifugation. The supernatant is transferred to a glass tube, dried, reconstituted in mobile phase, and subsequently analyzed by LC-MS/MS. Jones-Lepp (2006) published a SPE method (HLB sorbent) in which 500 mL of water sample is acidified to pH <3, passed through the HLB cartridge before extracting the analytes with either methanol (1% acetic acid) or a methanol:MTBE (10:90, v/v) mixture, and analysis (and analyzed) by LC-MS/MS. Focazio et al. (2008) added this compound to their list of analytes being monitored in a large survey of U.S. waters, using the methodologies reported by Kolpin et al. (2002), while Loganathan et al. (2009) used a modification of the method developed by Jones-Lepp (2006).

Other classes of ECs are the perfluorinated surfactants including perfluorooctanesulfonate (PFOS) and perfluorooctanoate (PFOA). Analogous to the persistence of many historic contaminants such as polychlorinated biphenyls (PCBs), perfluorinated compounds are ubiquitous in the environment throughout the world (Giesy and Kannan, 2001) due to their multiple uses as surfactants and surface protectors in a variety of consumer goods. Tseng et al. (2006) report an optimized SPE method in which a water sample at pH 3 is extracted using a C18 SPE cartridge prior to LC-MS analysis. Loganathan et al. (2007) used a SPE (HLB sorbent) and LC-MS/MS methodology to detect PFOS and PFOA in wastewater.

A recent European Union (EU) survey of a variety of ECs in European river waters used a simple SPE (HLB sorbent) extraction procedure followed by LC-MS/MS detection for the analysis of a variety of ECs compromising pharmaceuticals, PFOS, PFOA, steroids, and hormones (Loos et al., 2009). A 400 mL unfiltered water sample is passed through SPE and the analytes eluted with 6 mL of methanol. It is assumed that the eluent was further concentrated before LC-MS/MS analysis.

Isotopically labeled compounds were used to correct for extraction losses inherent in the method. Gros et al. (2009) reported a simplified SPE (HLB sorbent) extraction followed by a more sophisticated analytical detection approach using LC-quadrupole-linear ion trap mass spectrometry (LC-QtrapMS) and automated library searching for the detection and identification of 73 pharmaceutical residues (covering a wide range of pharmaceutical classes) in both surface water and wastewaters. Their SPE methodology was optimized by comparing both MCX and HLB type sorbents, and a combination thereof with and without sample acidification and with and without the addition of Na_2EDTA. They concluded that the optimum conditions were no acidification, Na_2EDTA addition, and HLB sorbent.

11.3.1.3 Other Extraction Techniques

Other extraction techniques including two types of microextraction techniques have been used. The first involves equilibrium liquid–liquid–liquid microextraction (LLLME) rather than exhaustive LLE to extract SAs from small volumes (µL) of water (McClure and Wong, 2007). Unlike in solid-phase microextraction (SPME), the extract phase of LLLME does not come into contact with the sample solution. Instead, LLLME uses a disposable polypropylene hollow fiber to extract SAs into a few µL of an organic phase and subsequently into another phase before analysis. The risk of carryover and cross contamination is essentially eliminated due to the disposable nature of the sampling apparatus. The second method is an SPME technique that uses hollow fibers to extract compounds from an aqueous sample by absorption in the case of liquid coatings, or adsorption in the case of solid coatings, and is similar to LLLME. Moeder et al. (2000) were among the first to use SPMD fiber coatings, and the resultant SPME extracts were derivatized prior to GC-MS analysis. Basheer et al. (2005) describe a modified SPME procedure, termed polymer-coated hollow fiber microextraction (PC-HFME), in which SPME fibers were coated with a new polymer having a large number of functional groups (-OH) more compatible with polar compounds, such as the estrogens. Using PC-HFME, they extracted diethylstilbestrol, estrone, 17β-estradiol and 17α-ethynylestradiol from spiked reservoir and tap water samples. The extracts were derivatized and analyzed using GC-MS. Some obvious advantages of SPME are small sample size and solvent volume while disadvantages include interferents (e.g., surfactants, humic and fulvic acids) competing for limited bonding sites and extended equilibrium times necessary for ensuring representative extraction efficiencies.

On the horizon is a novel extraction technique utilizing molecularly imprinted polymers (MIPs) that are target class specific imprinted with specificity for either a single analyte or a class of analytes. Once only in the realm of the research laboratory, there are now several commercially available MIP sorbents. Meng et al. (2005) developed a reusable (up to 5 extractions) nonspecific MIP to extract 17β-estradiol, diethylstilbestrol, estriol, and estrone from wastewater but a limitation is the difficulty in completely removing the target analytes from the MIP template. This is especially problematic at the low levels at which most pharmaceuticals and hormones are found in the environment (ng L^{-1}), making accurate quantitation of the target compound difficult. This problem was solved by Watabe et al. (2006) who developed a MIP template to extract only 17ß-estradiol (E2) from river water. The MIP template used a similarly structured analog of 17β–estradiol, namely, 6-ketoestradiol (KE2), which has a different chromatographic retention time than that of 17β-estradiol. Gros et al. (2008) developed a method that uses a commercially available MIP template (MIP Technologies, Lund, Sweden) to selectively extract eight β-blockers from wastewater. Comparing MIP and SPE (HLB) extracts, they found that while recoveries were similar, the MIP extract yielded a lower overall detection limit due to the specificity of the MIP template.

11.3.2 Detection Techniques

As discussed above, most detection techniques for ECs are based on mass spectrometry, which has become the preferred method in environmental analysis due to the inherent complexity of most environmental samples. For example, early attempts at measuring estrogens in the environment used HPLC-fluorescence detection, but numerous interferences made identification of the targeted estrogens difficult (Snyder et al., 1999). They later utilized the mass accuracy and specificity of a mass specific detector for the same analytes with the additional benefit of being able to characterize other pharmaceuticals in the same lake water matrix (Vanderford et al., 2003).

A variety of mass spectrometers (quadrupole, ion traps [ITMS], time-of-flight [TOF], triple quadrupole [QqQ], magnetic sector, and orbitrap) are now used as detectors coupled to either GCs or LCs. Selection of the type of mass analyzer for environmental analyses depends on the separation technique used (GC or LC), information needed, mass accuracy necessary, and specificity dictated by regulation. A better understanding of mass spectrometry and its application to environmental analysis can be gained from McLafferty (1980), Busch et al. (1988), Barceló (1996), Grayson (2002), and Herbert and Johnstone (2003).

11.4 ANALYTICAL DIFFICULTIES

Environmental samples, especially surface water samples containing WWTP effluents, can be extremely complex. Even with state-of-the-art mass spectrometers, positive identification of chemicals can be difficult to nearly impossible to achieve. Problems of coeluting chemicals, chemicals with common mass-to-charge ratios, and matrix effects such as ion suppression and shifting retention times can all lead to misidentification of compounds. Jones-Lepp et al. (2004) observed shifting retention times during the LC-MS analysis of the illicit drugs methamphetamine and methylenedioxymethamphetamine (MDMA or Ecstasy) in POCIS extracts. Identification and quantitation of the drugs was made possible by the use of collision-induced dissociation (CID) and the method of standard addition to the extracts. Azithromycin has also been shown to share a common product ion with some surfactants requiring CID to prevent misidentification.

Sample cleanup is often essential in isolating chemicals of interest from the rest of the sample. Methods for isolating ECs in environmental samples are limited but generally involve modifications to common techniques such as SPE, LLE, and dialysis, among others. Coextracted chemicals in environmental samples can be structurally similar to those of interest, making their removal difficult. For example, steroidal hormones share similar ring systems with many naturally occurring sterols (e.g., cholesterol). Many standard cleanup methods are not applicable to EC analyses as the background interferences they were designed to remove are now part of some EC chemical lists. Besides the sample cleanup problems, laboratory and field contamination issues are different for ECs than for historic contaminants such as pesticides. Soaps, deodorants, cleaning supplies, insect repellants, plasticizers from computer cases, foams, and many other items can all be sources of the chemicals that are part of many EC monitoring programs. Knowledge of a chemical's use and good laboratory practices are essential in preventing accidental contamination of samples.

In addition to the analytical difficulties posed by the complexity of environmental samples, the availability of authentic pure standard materials is limited. Many proprietary chemicals, degradation products, and metabolites of ECs are only available from the original manufacturer or through custom synthesis. Surrogate chemicals, such as isotopically labeled analogs of targeted ECs and certified reference materials to be used in QC programs, are not available for many ECs. As demand for these materials and potential for new regulatory action increase, additional ECs will become available from commercial sources to be used in environmental monitoring studies.

11.5 CONCLUSIONS

The field of emerging contaminants research is ever-changing as new chemicals are developed and new environment threats are recognized. Pharmaceuticals, personal care products, natural and synthetic hormones, plasticizers, and flame retardants are currently the center of attention due to their constant release into surface-, ground-, and ultimately drinking water.

Water sample collection methods for these ECs are similar to most common sampling techniques. Sample preservation or special handling is generally not required but the use of products containing these ECs during collection must be avoided to prevent contamination. Grab samples have the advantages of being easy to collect and relatively inexpensive. Passive sampling techniques are now favored in EC monitoring studies due to their ability to concentrate trace levels of ECs, catch EC pulses into the environment, and selectively sample dissolved chemicals (not bound to particulate matter).

Because of improvements in EC detection, interest in and understanding of ECs in the environment has greatly increased. The advent of reasonably priced sophisticated mass spectrometry systems coupled to gas or liquid chromatographs has allowed a greater number of laboratories to gain the needed instrumentation to undertake EC analyses. As knowledge of these ECs increases and new regulations are implemented, sampling and analytical methods for ECs will become commonplace, but the cycle will continue as new classes of ECs are identified.

NOTICE

The United States Environmental Protection Agency through its Office of Research and Development collaborated in the research described here under to United States

Geological Survey. It has been subjected to Agency review and approved for publication. Any use of trade, product, or firm names is for descriptive purposes only and does not imply endorsement by the U.S. Government.

REFERENCES

Alvarez, D.A., J.D. Petty, J.N. Huckins, et al. 2004. Development of a passive, in situ, integrative sampler for hydrophilic organic contaminants in aquatic environments. *Environ Toxicol Chem* 23:1640–1648.

Alvarez, D.A., P.E. Stackelberg, J.D. Petty, et al. 2005. Comparison of a novel passive sampler to standard water-column sampling for organic contaminants associated with wastewater effluents entering a New Jersey stream. *Chemosphere* 61:610–622.

Alvarez, D.A., J.N. Huckins, J.D. Petty, et al. 2007. Tool for monitoring hydrophilic contaminants in water: Polar organic chemical integrative sampler (POCIS). In *Passive Sampling Techniques in Environmental Monitoring. Comprehensive Analytical Chemistry*, ed. R. Greenwood, G. Mills, B. Vrana, 171–197. Amsterdam, The Netherlands: Elsevier.

Alvarez, D.A., W.L. Cranor, S.D. Perkins, et al. 2009. Reproductive health of bass in the Potomac, USA drainage: Part 2. Seasonal occurrence of persistent and emerging organic contaminants. *Environ Toxicol Chem* 28:1084–1095.

American Society of Standard Testing and Materials International (ASTM). 2007. Standard test method for determination of nonylphenol, bisphenol A, p-tert-octylphenol, nonylphenol monoethoxylate and nonylphenol diethoxylate in environmental waters by gas chromatography mass spectrometry. ASTM D 7065-06.

Angelidis, T.N. 1997. Comparison of sediment pore water sampling for specific parameters using two techniques. *Water Air Soil Pollut* 99:179–185.

Ankley, G.T., and M.K. Schubauer-Berigan. 1994. Comparison of techniques for the isolation of sediment pore water for toxicity testing. *Arch Environ Contam Toxicol* 27:507–512.

Barceló, D. 1996. *Applications of LC-MS in Environmental Chemistry*, Amsterdam: Elsevier.

Barnes, K.K., D.W. Kolpin, E.T. Furlong, S.D. Zaugg, M.T., Meyer, and L.B. Barber. 2008. A national reconnaissance of pharmaceuticals and other organic wastewater contaminants in the United States–I) Ground water. *Sci Total Environ* 402:192–200.

Basheer, C., A. Jayaraman, M.K. Kee, S. Valiyaveettil, and H.K. Lee 2005. Polymer-coated hollow-fiber microextraction of estrogens in water samples with analysis by gas chromatography-mass spectrometry. *J. Chromatogr. A* 1100:137–143.

Batt, A.L., and D.S. Aga 2005. Simultaneous analysis of multiple classes of antibiotics by ion trap LC/MS/MS for assessing surface water and groundwater contamination. *Anal. Chem.* 77:2940–2947.

Benito-Peña, E., A.I. Partal-Rodera, M.E. León-González, and M.C. Moreno-Bondi. 2006. Evaluation of mixed mode solid phase extraction cartridges for the preconcentration of beta-lactam antibiotics in wastewater using liquid chromatography with UV-DAD detection. *Anal. Chim. Acta* 556:415–422.

Benotti, M.J., R.A. Trenholm, B.J. Vanderford, J.C. Holady, B.D. Standford, and S.A. Snyder. 2009. Pharmaceuticals and endocrine disrupting compounds in U.S. drinking water. *Environ Sci Technol* 43:597–603.

Booij, K., B.N. Zegers, and J.P. Boon. 2002. Levels of some polybrominated diphenyl ether (PBDE) flame retardants along the Dutch coast as derived from their accumulation in SPMDs and blue mussels (*Mytilus edulis*). *Chemosphere* 46:683–688.

Bouchard, D., and X. Ma. 2008. Extraction and high-performance liquid chromatographic analysis of C60, C70, and [6,6]-phenyl C61-butyric acid methyl ester in synthetic and natural waters. *J Chromatogr A* 1203:153–159.

Boxall, A.B.A., D.W. Kolpin, B. Halling-Sorensen, and J. Tolls. 2003. Are veterinary medicines causing environmental risks? *Environ Sci Technol* 37:286A–294A.

Bufflap, S.E., and H.E. Allen. 1995. Comparison of pore water sampling techniques for trace metals. *Water Res* 29:2051–2054.

Burkholder, J., B. Libra, P. Weyer, S. Heathcote, et al. 2007. Impacts of waste from concentrated animal feeding operations (CAFOs) on water quality. *Environ Health Perspect* 115: 308–312.

Busch, K., G. Glish, and S. McLuckey. 1988. *Mass Spectrometry/Mass Spectrometry: Techniques and Applications of Tandem Mass Spectrometry*. New York: VCH Publishers.

Daughton, C., and T. Ternes. 1999. Pharmaceuticals and personal care products in the environment: Agents of subtle change? *Environ Health Perspect* 107:907–938.

de Alda, M.J.L., and D. Barceló. 2001. Use of solid-phase extraction in various of its modalities for sample preparation in the determination of estrogens and progestogens in sediment and water. *J Chromatogr A* 938:145–153.

Desbrow, C., E. Routledge, G. Brighty, J. Sumpter, and M. Waldock. 1998. Identification of estrogenic chemicals in stw effluent. 1. Chemical fractionation and in vitro biological screening. *Environ Sci Technol* 32:1549–1558.

de Wit, C.A. 2002. An overview of brominated flame retardants in the environment. *Chemosphere* 46:583–624.

Drewes, J.E., T. Heberer, and K. Reddersen. 2002. Fate of pharmaceuticals during indirect potable reuse. *Water Sci Technol* 46:73–80.

Englert, B.C. 2007. Nanomaterials and the environment: Uses, methods, and measurement. *J Environ Monitor* 9:1154–1161.

Focazio, M.J., D.W. Kolpin, K.K. Barnes, et al. 2008. A national reconnaissance for pharmaceuticals and other organic wastewater contaminants in the United States–II) Untreated drinking water sources. *Sci Total Environ* 402:201–216.

Fowler, W.K. 1982. Fundamentals of passive vapor sampling. *Am Lab* 14:80–87.

Garrison, A.W., J.D. Pope, and F.R. Allen. 1976. GC/MS analysis of organic compounds in domestic wastewaters. In *Identification and Analysis of Organic Pollutants in Water*, ed. C.H. Keith, 517–556. Ann Arbor, MI: Ann Arbor Science Publishers.

Gibbs, J., P.E. Stackelberg, E.T. Furlong, M.T. Meyer, S.D. Zaugg, and R.L. Lippincott. 2007. Persistence of pharmaceuticals and other organic compounds in chlorinated drinking water as a function of time. *Sci Total Environ* 373:240–249.

Giesy, J.P., and K. Kannan. 2001. Perfluorochemical surfactants in the environment. *Environ Sci Technol* 35:1339–1345.

Glassmeyer, S.T. 2007. The cycle of emerging contaminants. *Water Res Impact* 9:5–7.

Grayson, M. 2002. *Environmental Distress in Measuring Mass: From Positive Rays to Proteins*. Philadelphia: Chemical Heritage Press.

Gros, M., M. Petrović, and D. Barceló. 2009. Tracing pharmaceutical residues of different therapeutic classes in environmental waters using liquid chromatography/quadrupole-linear ion trap mass spectrometry and automated library searching. *Anal Chem* 81:898–912.

Gros, M., T.M. Pizzolato, M. Petrović, M.J.L. de Alda, and D. Barceló. 2008. Trace level determination of β-blockers in waste waters by highly selective molecularly imprinted polymers extraction followed by liquid chromatography-quadrupole-linear ion trap mass spectrometry. *J Chromatogr A* 1189:374–384.

Halling-Sørensen, B., S. Nors Nielsen, P. Lanzley, F. Ingerslev, H. Holten Lützhøft, and S. Jørgensen. 1998. Occurrence, fate and effects of pharmaceuticals substances in the environment—a review. *Chemosphere* 36:357–393.

Herbert, C., and R. Johnstone. 2003. *Mass Spectrometry Basics*. Boca Raton: CRC Press.

Hesslein, R.H. 1976. An in situ sampler for close interval pore water studies. *Limnol Oceanogr* 21:912–914.

Heymann, D., L.P.F. Chibante, and R.E. Smalley. 1995. Determination of C60 and C70 fullerenes in geologic materials by high-performance liquid chromatography. *J Chromatogr A* 689:157–163.

Hignite, C., and D.L. Azarnoff. 1977. Drugs and drug metabolites as environmental contaminants: Chlorophenoxyisobutyrate and salicylic acid in sewage water effluent. *Life Sci* 20:337–341.

Hirsch, R., T. Ternes, K. Haberer, A. Mehlich, F. Ballwanz, and K.L. Kratz. 1998. Determination of antibiotics in different water compartments via liquid chromatography-electrospray tandem mass spectrometry. *J Chromatogr A* 815:213–223.

Huckins, J.N., J.D. Petty, and K. Booij. 2006. *Monitors of Organic Chemicals in the Environment—Semipermeable Membrane Devices*. New York: Springer.

Jones-Lepp, T.L. 2006. Chemical markers of human waste contamination: Analysis of urobilin and pharmaceuticals in source waters. *J Environ Monit* 8:472–478.

Jones-Lepp, T.L., D.A. Alvarez, J.D. Petty, and J.N. Huckins. 2004. Polar organic chemical integrative sampling (POCIS) and LC-ES/ITMS for assessing selected prescription and illicit drugs treated sewage effluent. *Arch Environ Contam Toxicol* 47:427–439.

Jonkers, N., H-P.E. Kohler, A. Dammshäuser, and W. Giger. 2009. Mass flows of endocrine disruptors in the Glatt River during varying weather conditions. *Environ Pollut* 157:714–723.

Josephson, J. 2006. The microbial resistome. *Environ Sci Technol* 40:6531–6534.

Kasprzyk-Hordern, B., R.M. Dinsdale, and A.J. Guwy. 2007. Multi-residue method for the determination of basic/neutral pharmaceuticals and illicit drugs in surface water by solid-phase extraction and ultra performance liquid chromatography-positive electrospray ionization tandem mass spectrometry. *J Chromatogr A* 1161:132–145.

Keith, L.H., 1990. Environmental sampling: A summary. *Environ Sci Technol* 24:610–617.

Keith, L.H., 1991. *Environmental Sampling and Analysis: A Practical Guide*. Boca Raton, FL: CRC.

Kelly, C. 2000. Analysis of steroids in environmental water samples using solid-phase extraction and ion-trap gas chromatography-mass spectrometry and gas chromatography-tandem mass spectrometry. *J Chromatogr A* 872:309–314.

Kim, S.-C., and K. Carlson. 2006. Quantification of human and veterinary antibiotics in water and sediment using SPE/LC/MS/MS. *Anal Bioanal Chem* 387:1301–1315.

Koch, D.E., A. Bhandari, L. Close, and R.P. Hunter. 2005. Azithromycin extraction from municipal wastewater and quantitation using liquid chromatography/mass spectrometry. *J Chromatogr A* 1074:17–22.

Kolpin, D.W., E.T. Furlong, M.T. Meyer, et al. 2002. Pharmaceuticals, hormones, and other organic wastewater contaminants in U.S. streams, 1999–2000—A national reconnaissance. *Environ Sci Technol* 36:1202–1211.

Kümmerer, K. 2004. Resistance in the environment. *J Antimicrob Chemother* 54:311–320.

Lane, S.L., S. Flanagan, and F.D. Wilde. 2003. Selection of equipment for water sampling (ver. 2.0): U.S. Geological Survey Techniques of Water-Resources Investigations, book 9, chap. A2. http://pubs.water.usgs.gov/twri9A2/ (accessed May 2, 2008).

Leiker, T.J., S.R. Abney, S.L. Goodbred, and M.R. Rosen. 2009. Identification of methyl triclosan and halogenated analogues in male common carp (*Cyprinus carpio*) from Las Vegas Bay and semipermeable membrane devices from Las Vegas Wash, Nevada. *Sci Total Environ* 407:2102–2114.

Liu, C., and S. Huang. 2008. Application of liquid–liquid–liquid microextraction and high-performance liquid-chromatography for the determination of sulfonamides in water. *Anal Chim Acta* 612:37–43.

Liu, R., J.L. Zhou, and A. Wilding. 2004. Microwave-assisted extraction followed by gas chromatography-mass spectrometry for the determination of endocrine disrupting chemicals in river sediments. *J Chromatogr A* 1038:19–26.

Loganathan, B.G., K.S. Sajwan, E. Sinclair, K.S. Kumar, and K. Kannan. 2007. Perfluoroalkyl sulfonates and perfluorocarboxylates in two wastewater treatment facilities in Kentucky and Georgia. *Water Res* 41:4611–4620.

Loganathan, B.G., M. Phillips, H. Mowery, and T.L. Jones-Lepp. 2009. Contamination profiles and mass loadings of select macrolide antibiotics and illicit drugs from a small urban wastewater treatment plant. *Chemosphere* 75:70–77.

Loos, R., B.M. Gawlik, G. Locoro, E. Rimaviciute, S. Contini, and G. Bidoglia. 2009. EU-wide survey of polar organic persistent pollutants in European river waters. *Environ Pollut* 157:561–568.

MacLeod, S.L., E.L. McClure, and C.S. Wong. 2007. Laboratory calibration and field deployment of the polar organic chemical integrative sampler for pharmaceuticals and personal care products in wastewater and surface water. *Environ Toxicol Chem* 26:2517–2529.

Maruya, K.A., E.Y. Zeng, D. Tsukada, and S.M. Bay. 2009. A passive sampler based on solid-phase microextraction for quantifying hydrophobic organic contaminants in sediment pore water. *Environ Toxicol Chem* 28:733–740.

McClure, E.L., and C. Wong. 2007. Solid phase microextraction of macrolide, trimethoprim, and sulfonamide antibiotics in wastewaters. *J Chromatogr A* 1169:53–62.

McLafferty, F. 1980. *Interpretation of Mass Spectra,* 3rd ed. Mill Valley: University Science Books.

Meng, Z., W. Chen, and A. Mulchandani. 2005. Removal of estrogenic pollutants from contaminated water using molecularly imprinted polymers. *Environ Sci Technol* 39:8958–8962.

Mills, G.A., B. Vrana, I. Allan, D.A. Alvarez, J.N. Huckins, and R. Greenwood. 2007. Trends in monitoring pharmaceuticals and personal-care products in the aquatic environment by use of passive sampling devices. *Anal Bioanal Chem* 387:1153–1157.

Miyabara, Y., M. Imoto, S. Arai, J. Suzuki, and S. Suzuki. 1995. Distribution of antibiotic resistant *Staphylococcus aureus* in river water. *Environ Sci* 8:171–179.

Moeder, M., S. Schrader, M. Winkler, and P. Popp. 2000. Solid phase microextraction-gas chromatgography-mass spectrometry of biologically active substances in water samples. *J Chromatogr A* 873:95–106.

Mol, H., S. Sunarto, and O. Steijger. 2000. Determination of endocrine disruptors in water after derivatization with *N*-methyl-*N*-(*tert*-butyldimethyltriflouroacetamide) using gas chromatography with mass spectrometric detection. *J Chromatogr A* 879:97–112.

Morace, J.L. 2006. Water-quality data, Columbia River Esturary, 2004–05. U.S. Geological Survey Data Series 213, 18 pp.

Motzer, W.E. 2008. Monograph for California Groundwater Resources Association of California, http://grac.org/Nanomaterials_and_Water_Resources.pdf (accessed March 17, 2009).

Namieśnik, J., B. Zabiegala, A. Kot-Wasik, M. Partyke, and A. Wasik. 2005. Passive sampling and/or extraction techniques in environmental analysis: A review. *Anal Bioanal Chem* 381:279–301.

Öllers, S., H. Singer, P. Fässler, and S. Müller. 2001. Simultaneous quantification of neutral and acidic pharmaceuticals and pesticides at the low-ng L^{-1} level in surface and waste water. *J Chromatogr A* 911:225–234.

Osemwengie, L.I., and S. Steinberg. 2001. On-site solid-phase extraction and laboratory analysis of ultra-trace synthetic musks in municipal sewage effluent using gas chromatography-mass spectrometry in the full-scan mode. *J Chromatogr A* 932:107–118.

Ouyand, G., and J. Pawliszyn. 2007. Passive sampling devices for measuring organic compounds in soils and sediments. In *Passive Sampling Techniques. Comprehensive Analytical Chemistry*, R. Greenwood, G. Mills, B. Vrana, eds., 379–390. Amsterdam, The Netherlands: Elsevier.

Piram, A., A. Salvador, J-Y. Gauvrit, P. Lanteri, and R. Faure. 2008. Development and optimization of a single extraction procedure for the LC/MS/MS analysis of two pharmaceutical classes residues in sewage treatment plant. *Talanta* 74:1463–1475.

Poole, C. 2003. New trends in solid-phase extraction. *Trends Anal Chem* 22:362–373.

Primus, T.M., D.J. Kohler, M. Avery, P. Bolich, M.O. Way, and J.J. Johnston. 2001. Novel field sampling procedure for the determination of methiocarb residues in surface waters from rice fields. *J Agric Food Chem* 49:5706–5709.

Radjenović, J., M. Petrović, F. Ventura, and D. Barceló. 2008. Rejection of pharmaceuticals in nanofiltration and reverse osmosis membrane drinking water treatment. *Trends Anal Chem* 26:1132–1144.

Radtke, D.B. 2005. Bottom-material samples: U.S. Geological Survey Techniques of Water-Resources Investigations, book 9, chap. A8. http://pubs.water.usgs.gov/twri9A8/ (accessed May 2, 2008).

Reverté, S., F. Borrull, E. Pocurull, and R.M. Marcé. 2003. Determination of antibiotic compounds in water by solid-phase extraction-high-performance liquid chromatography-(electrospray) mass spectrometry. *J Chromatogr A* 1010:225–232.

Richardson, S. 2008. Environmental mass spectrometry: Emerging contaminants and current issues. *Anal Chem* 80:4373–4402.

Richardson, S.D., F. Fasano, J.J. Ellington, et al. 2008. Occurrence and mammalian cell toxicity of iodinated disinfection byproducts in drinking water. *Environ Sci Technol* 42:8330–8338.

Rowe, G.L., Jr., D.C. Reutter, D.L. Runkle, J.A. Hambrook, S.D. Janosy, and L.H. Hwang. 2004. Water quality in the Great and Little Miami River basins: U.S. Geological Survey Circular 1229, 40 pp.

Sarmah, A.K., M.T. Meyer, and A.B.A. Boxall. 2006. A global perspective on the use, sales, exposure pathways, occurrence, fate and effects of veterinary antibiotics (VAs) in the environment. *Chemosphere* 65:725–759.

Schwartz, T., W. Kohnen, B. Jansen, and U. Obst. 2003. Detection of antibiotic-resistant bacteria and their resistance genes in wastewater, surface water, and drinking water biofilms. *FEMS Microbiol Ecol* 43:325–335.

Schwartz, T., H. Volkmann, S. Kirchen, et al. 2006. Real-time PCR detection of *Pseudomonas aeruginosa* in clinical and municipal wastewater and genotyping of the ciprofloxacin-resistant isolates. *FEMS Microbiol Ecol* 57:158–167.

Snyder, S., T.L. Keith, D.A. Verbrugge, et al. 1999. Analytical methods for detection of selected estrogenic compounds in aqueous mixtures. *Environ Sci Technol* 33:2814–2820.

Stackelberg, P.E., J. Gibs, E.T. Furlong, M.T. Meyer, S.D. Zaugg, and R.L. Lippincott. 2007. Efficiency of conventional drinking-water-treatment processes in removal of pharmaceuticals and other organic compounds. *Sci Total Environ* 377:255–272.

Sumpter, J.P., and A.C. Johnson. 2005. Lessons from endocrine disruption and their application to other issues concerning trace organics in the aquatic environment. *Environ Sci Technol* 39:4321–4332.

Ternes, T., M. Bonerz, and T. Schmidt. 2001. Determination of neutral pharmaceuticals in wastewater and rivers by liquid chromatography-electrospray tandem mass spectrometry. *J Chromatogr A* 938:175–185.

Ternes, T., H. Andersen, D. Gilberg, and M. Bonerz. 2002. Determination of estrogens in sludge and sediments by liquid extraction and GC/MS/MS. *Anal Chem* 74:3498–3504.

Togola, A., and H. Budzinksi. 2007. Analytical development for analysis of pharmaceuticals in water samples by SPE and GC-MS. *Anal Bioanal Chem* 388:627–635.

Togola, A., and H. Budzinski. 2008. Multi-residue analysis of pharmaceutical compounds in aqueous samples. *J Chromatogr A* 1177:150–158.

Tseng, C.-L., L. Li-Lian, C.-M. Chen, and W.-H. Ding. 2006. Analysis of perfluorooctanesulfonate and related fluorochemicals in water and biological tissue samples by liquid chromatography-ion trap mass spectrometry. *J Chromatogr A* 1105:119–126.

UNEP (United Nations Environment Programme). 2005. Ridding the World of POPs: A Guide to the Stockholm Convention on Persistent Organic Pollutants. http://chm.pops.int/Portals/0/Repository/CHM-general/UNEP-POPS-CHM-GUID-RIDDING.English.PDF (assessed March 4, 2009).

USEPA (U.S. Environmental Protection Agency). 2004. Guidelines for water reuse. US EPA Office of Technology Transfer and Regulatory Support. EPA/625/R-04/108.

USEPA. 2007a. Method 1694—Pharmaceuticals and personal care products in water, soil, sediment, and biosolids by HPLC/MS/MS. http://www.epa.gov/waterscience/methods/method/other.html (accessed May 2, 2008).

USEPA. 2007b. Method 1698—Steroids and hormones in water, soil, sediment, and biosolids by HRGC/HRMS. http://www.epa.gov/waterscience/methods/method/other.html (accessed May 2, 2008).

Vanderford, B., R.A. Pearson, D.J. Rexing, and S.A. Snyder. 2003. Analysis of endocrine disruptors, pharmaceuticals, and personal care products in water using liquid chromatography/tandem mass spectrometry. *Anal Chem* 75:6265–6274.

Vermeirssen, E.L.M., M.J-F. Suter, and P. Burkhardt-Holm. 2006. Estrogenicity patterns in the Swiss Midland River Lützelmurg in relation to treated domestic sewage effluent discharges and hydrology. *Environ Toxicol Chem* 25:2413–2422.

Vrana, B., H. Paschke, A. Paschke, P. Popp, and G. Schüürmann. 2005. Performance of semipermeable membrane devices for sampling of organic contaminants in groundwater. *J Environ Monit* 7:500–508.

Vroblesky, D.A., M.M. Lorah, and S.P. Trimble. 1991. Mapping zones of contaminated-ground-water discharge using creek-bottom-sediment vapors, Aberdeen Proving Ground, Maryland. *Ground Water* 29:7–12.

Watabe, Y., T. Kubo, T. Nishikawa, T. Fujita, K. Kaya, and K. Hosoya. 2006. Fully automated liquid chromatography-mass spectrometry determination of 17β-estradiol in river water. *J Chromatogr A* 1120:252–259.

Winger, P.V., and P.J. Lasier. 1991. A Vacuum-operated pore-water extractor for estuarine and freshwater sediments. *Arch Environ Contam Toxicol* 21:321–324.

Woodrow Wilson International Center for Scholars Nanotechnology Project Inventories. 2009. www.nanotechproject.org/inventories/consumer/analysis_draft/ (accessed March 17, 2009).

Yamagishi, T., T. Miyazaki, S. Horii, and S. Kaneko. 1981. Identification of musk xylene and musk ketone in freshwater fish collected from the Tama River, Tokyo. *Bull Environ Contam Toxicol* 26:656–662.

Yang, S., and K. Carlson. 2003. Evolution of antibiotic occurrence in a river through pristine urban and agricultural landscapes. *Water Res* 37:4645–4656.

Zaugg, S.D., S.G. Smith, and M.P. Schroeder. 2007. Determination of wastewater compounds in whole water by continuous liquid-liquid extraction and capillary-column gas chromatography/mass spectrometry. Chap. 4, Section B, Book 5, United States Geological Survey.

12 Uncertainty in Measured Water Quality Data

Daren Harmel, Patricia Smith, and Kati W. Migliaccio

CONTENTS

12.1 INTRODUCTION

Water quality assessment, management, and regulation continue to rely on measured water quality data, in spite of advanced modeling capabilities (Silberstein, 2006). However, very little information is available on one very important component of the measured data—the inherent measurement uncertainty. Although all measurements are in fact uncertain to some degree, the uncertainty in measured data is rarely estimated and thus typically is ignored.

By ignoring data uncertainty, the numerous benefits of such information are often not realized (Brown et al., 2005). Specifically, monitoring projects are not optimized in terms of cost-effectiveness and data quality because measurement uncertainty and alternatives to reduce uncertainty are not included in project design and implementation (Beven, 2006a; Harmel et al., 2006b; Rode and Suhr, 2007). Science-based decision-making and stakeholder support are not fully achieved because measurement uncertainty is rarely estimated and not adequately communicated to researchers, public interest groups, regulators, and elected officials (Collins et al., 2000; Bonta and Cleland, 2003; Reckhow, 2003; Nature, 2005; Beven, 2006a; Pappenberger and Beven, 2006). Similarly, model-based predictions are not optimized because the uncertainty of measured data, which drive model calibration and validation, is typically unknown and/or not considered (Reckhow, 1994; Kavetski et al., 2002; Pappenberger and Beven, 2006; Beven, 2006b; Shirmohammadi et al., 2006; Harmel and Smith, 2007).

To facilitate uncertainty estimation and enhanced understanding, this chapter presents recent scientific advances related to uncertainty in measured water quality data. It nullifies previously used technical justifications for ignoring measurement uncertainty by

- Summarizing current scientific understanding
- Describing a user-friendly uncertainty estimation method
- Presenting uncertainty estimates for measured data

12.1.1 Chapter Scope

This chapter addresses uncertainty in measured discharge and constituent (sediment, dissolved and particulate nitrogen [N] and phosphorus [P]) concentration data collected at the field and small watershed scale. Uncertainty in pathogen/bacteria data is only touched upon in this chapter; see McCarthy et al. (2008) for more information. A detailed summary of uncertainty in basin-scale water quality data is provided by Rode and Suhr (2007). Similarly, uncertainty introduced by watershed spatial characteristics and the number and location of sampling sites, which is better addressed by watershed models and geospatial-statistical tools, is outside the scope of this chapter and thus not discussed. For information on these topics, see Isaaks and Srivastava (1989), Rouhani et al. (1996), Deutsch and Journel (1998), Hunsaker et al. (2001), and Haan (2002).

12.2 SOURCES OF UNCERTAINTY

The typical data collection procedures (monitoring methods) for water quality data have been classified into four categories: discharge measurement, sample collection, sample preservation and storage, and laboratory analysis (Harmel et al., 2006a). It is by procedures in these categories that uncertainty—defined as random statistical variation affected by appropriate and accepted application of each procedure—is introduced into measured data. Uncertainty can also be contributed by missing values, assumptions made to estimate missing values, and mistakes in data management and reporting; therefore, a fifth procedural category "data processing and management" was also established (Harmel et al., 2009). Together these five categories were developed to "characterize the dispersion of the values that could reasonably be attributed to the particular quantity subject to measurement" as described by ISO (1993a) in EA (1999).

The subsequent sections briefly describe sources of uncertainty without discussing specific alternatives to reduce uncertainty. These alternatives are briefly discussed in Chapter 5—Surface Water Sampling in Small Streams and Canals and reviewed more comprehensively in Pelletier (1988), Herschy (1995), Kotlash and Chessman (1998), USGS (1999), Jarvie et al. (2002), Meyer (2002), Harmel et al. (2006a, 2006b), and Rode and Suhr (2007).

12.2.1 Discharge Measurement

Although discharge (flow) is not a water quality constituent, measured discharge provides valuable information related to constituent transport. The uncertainty associated with discharge measurement is well understood but rarely presented. Important exceptions are instantaneous discharge measurements and annual station discharge records published by the United States Geological Survey (Novak, 1985). For most discharge measurements, stage (depth) measurement is a major source of uncertainty (Herschy, 1995). The uncertainty in stage measurement is determined by stage sensor accuracy and channel bed conditions (Pelletier, 1988; Sauer and Meyer, 1992). The presence/absence of a stilling well, which reduces the influence of wind and water turbulence, also affects stage measurement. Instantaneous velocity measurements, usually taken in vertical cross-section segments, also introduce uncertainty related to meter accuracy, flow direction, and measurement location (Young, 1950; Carter and Anderson, 1963; Rantz et al., 1982; Hipolito and Leoureiro, 1988; Pelletier, 1988; Sauer and Meyer, 1992). For continuous in-stream velocity measurement, the uncertainty depends on stage measurement, channel stability, meter accuracy over the entire cross-sectional flow area, and the variability of water chemical and physical characteristics (McIntyre and Marshall, 2008). When discharge is determined from an established stage-discharge relationship, the uncertainty is determined to a large extent by the presence/absence of a hydraulic control structure (flume or weir), channel stability, range and accuracy of measured flows used to develop the relationship, and frequency of relationship adjustment (Dickinson, 1967; Buchanan and Somers, 1976; Brakensiek et al., 1979; Kennedy, 1984; Pelletier, 1988; Carter and Davidian, 1989; Sauer and Meyer, 1992; Herschy, 1995; Schmidt, 2002; Slade, 2004).

12.2.2 Sample Collection

The uncertainty introduced by sample collection can be the dominant source in environmental investigations (Ramsey, 1998), but relatively little information on uncertainty related to water quality sample collection was available until recently. The potential for sample collection to introduce substantial uncertainty, especially for particulate-associated constituents, results from cross-sectional and temporal concentration variability (Martin et al., 1992; Ging, 1999; Harmel and King, 2005). How well this variability is captured is affected by sample frequency, constituent type, and collection method (grab, integrated, or automated) and ultimately determines the uncertainty introduced by baseflow and storm sampling (Martin et al., 1992; Ging, 1999; USGS, 1999; Robertson and Roerish, 1999; King and Harmel, 2003; Harmel et al., 2003; Harmel and King, 2005; Miller et al., 2007; Rode and Suhr, 2007). In addition, the definition of storm occurrence as it relates to the determination of sample collection timing also introduces uncertainty in storm water data (Harmel et al., 2002).

12.2.3 Sample Preservation and Storage

The uncertainty contributed by sample preservation and storage regularly receives considerable quality assurance attention (Lambert et al., 1992; Kotlash and Chessman, 1998; Jarvie et al., 2002). Research indicates that nutrient forms and concentrations can be altered during the interval between sample collection and analysis and that water chemical and biological characteristics affect the degree and rate of alteration (Fitzgerald and Faust, 1967; Johnson et al., 1975; Lambert et al., 1992; Robards et al., 1994; Haygarth et al., 1995; Jarvie et al., 2002). Thus, relatively low ambient concentrations and transformation potentials can introduce substantial uncertainty in dissolved and total nutrient concentrations (Kotlash and Chessman, 1998; Jarvie et al., 2002; Meyer, 2002). In contrast, uncertainty is typically low for sediment concentrations because of higher ambient concentrations and limited transformation potential. The increased use of automated samplers in recent years has magnified the potential for post-collection transformation due to the time delay between sample collection and sample retrieval (Kotlash and Chessman, 1998). Therefore, quality assurance protocols generally focus on sample preservation and storage procedures to minimize physical, chemical, and biological transformation and thus reduce uncertainty. Preservation and storage procedures typically use cold, dark storage environments and/or chemical preservatives; however, container characteristics and filtration methodology have also been shown to influence post-collection sample transformations (Henriksen, 1969; Ryden et al., 1972; Latterell et al., 1974; Skjemstad and Reeve, 1978; Fishman et al., 1986; Kotlash and Chessman, 1998; Maher and Woo, 1998; Haygarth and Edwards, 2000; Jarvie et al., 2002).

12.2.4 Laboratory Analysis

Because laboratory analysis is an important contributor to uncertainty in measured constituent concentrations (Jarvie et al., 2002; Meyer, 2002), the uncertainty introduced by various analytical procedures also receives considerable quality assurance focus (Ramsey, 1998). Recent efforts, such as the North American Proficiency Testing program, have quantifying differences in results across laboratory techniques and locations providing data and insight on analytical uncertainty. The main sample analysis steps that contribute to analysis uncertainty are sample handling, chemical preparation, analytical method and equipment, and calibration standards and reference materials (Robards et al., 1994; Gordon et al., 2000; Ludtke et al., 2000; Jarvie et al., 2002; Mercurio et al., 2002; CAEAL, 2003). The uncertainty contributed by laboratory analysis varies considerably based on constituent type. Uncertainty is typically higher for dissolved and total nutrient concentrations compared to sediment concentrations because of more complex analytical procedures and lower ambient concentrations (Kotlash and Chessman, 1998; Jarvie et al., 2002; Meyer, 2002). Most standard methods books (e.g., APHA et al., 2005) provide excellent information on proper water quality analysis methodology.

12.2.5 Data Processing and Management

Uncertainty resulting from appropriate use of accepted procedures is addressed by the other procedural categories, but uncertainty can also be contributed by equipment malfunction or personnel mistakes. Therefore, the data processing and management procedural category is needed to account for uncertainty contributed by missing and/or incorrect data resulting from equipment failure, processing and data entry mistakes, misplaced samples, inadequate sample volume/mass, and other problems. Data processing and management can introduce low or very high uncertainty depending on the number of missing or incorrect values and the magnitude of the incorrect values. The potential for high uncertainty due to missing or incorrect data emphasizes the importance of frequent preventative maintenance, adequate personnel training, and attention to detail to minimize data processing and management uncertainty.

12.2.6 Comparison of Uncertainty Sources

Whereas typical quality assurance protocols focus on sample preservation, storage, and analysis, recent research by Harmel et al. (2006a, 2009) has shown that all of the procedural categories can introduce substantial uncertainty in measured water quality data. Uncertainty data for each procedural category for storm monitoring on five small watershed sites with various field and laboratory techniques are shown in Figure 12.1. These results should assist personnel in adequately addressing potential

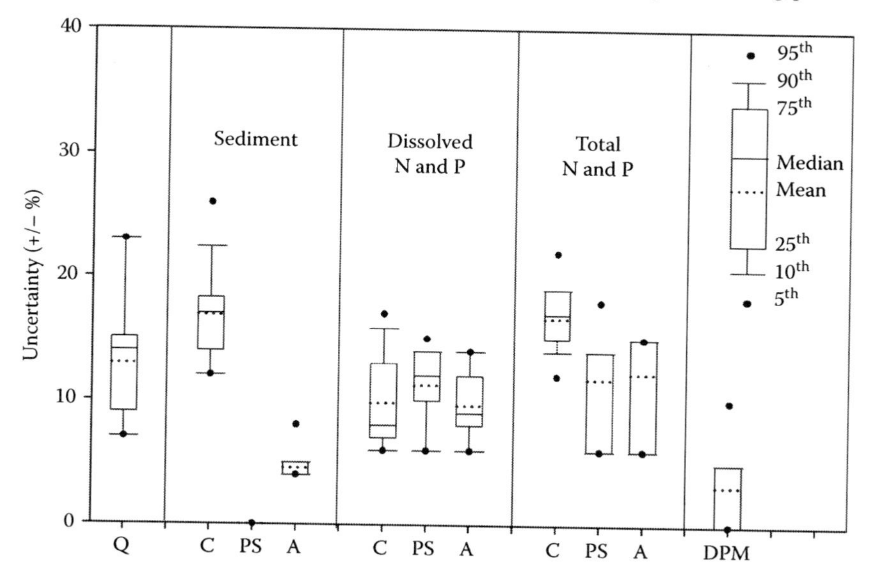

FIGURE 12.1 Uncertainty introduced into storm event discharge and constituent concentration data by the five procedural categories (Q—discharge measurement; C—sample collection; PS—sample preservation and storage; A—laboratory analysis; DPM —data processing and management).

sources of uncertainty for particular constituent(s) of interest and project-specific data requirements (maximum or average concentrations, event mean concentrations [EMCs], annual loads, etc.).

12.3 UNCERTAINTY ESTIMATION (ERROR PROPAGATION)

12.3.1 Choice of Uncertainty Estimation (Error Propagation) Method

The root mean square error (RMSE) propagation method (Topping, 1972) has been previously applied to estimate uncertainty in discharge measurements (Cooper, 2002; Sauer and Meyer, 1992), in sediment volume estimates (Allmendinger et al., 2007), and in pesticide analytical methods (Cuadros-Rodriquez et al., 2002). When the RMSE method was previously applied to water quality data, it was probably selected because of its simplicity. For detailed information on uncertainty estimation, see ISO (1993b) and ISO-based documents (e.g., Taylor and Kuyatt, 1994; EA, 1999). More complex statistical methods could have also been applied, but their application requires procedure-specific distributional information, which is limited for water quality measurement. Thus, unsubstantiated assumptions would be required for application of these methods. Although simplicity is beneficial at the current time when uncertainty is rarely estimated and relevant information is limited, improved understanding might support the future use of complex statistical methods to better represent potential serial correlation, asymmetrical distributions, and value-uncertainty correlation.

12.3.2 Root Mean Square Error Method

With the RMSE method, the uncertainty from each data collection step is propagated to produce a realistic uncertainty estimate, which is best termed "cumulative probable error or uncertainty." This probable error estimate is generally reported with, or instead of, the maximum error, which is the simple sum of the maximum values of component errors (Harmel et al., 2009). The RMSE method assumes that uncertainty is symmetric about the value and thus bi-directional with equal likelihood of over- and under-estimation and that errors for each procedural step are independent (Topping, 1972). Thus, in the absence of contrary data, uncertainties for procedural steps are assumed to be independent and the covariance is omitted. The RMSE, as shown in equation 12.1, was formulated specifically for measured hydrology and water quality data (Harmel et al., 2009):

$$EP_i = \sqrt{\left(E_{Qi}^2 + E_{Ci}^2 + E_{PSi}^2 + E_{Ai}^2 + E_{DPMi}^2\right)} \tag{12.1}$$

where:

EP_i is the cumulative probable uncertainty for each individual measured value (± %)
E_{Qi} = uncertainty in discharge measurement (± %)
E_{Ci} = uncertainty in sample collection (± %)
E_{PSi} = uncertainty in sample preservation/storage (± %)
E_{Ai} = uncertainty in laboratory analysis (± %)
E_{DPMi} = uncertainty in data processing and management (± %)

12.3.3 Application of Root Mean Square Error Method

A recently developed framework and software tool, both of which utilize the RMSE method, were designed to facilitate uncertainty estimation for measured water quality data (Harmel et al., 2006a, 2009). The software tool, Data Uncertainty Estimation Tool for Hydrology and Water Quality (DUET-H/WQ), was developed to be a user-friendly application of the uncertainty estimation framework. This tool is available at no cost online in an open source format at ftp://ftp.brc.tamus.edu/pub/outgoing/bkomar/programs/. Both DUET-H/WQ and its framework-basis classify data collection steps into the five procedural categories discussed in Section 12.2 (discharge measurement, sample collection, sample preservation and storage, laboratory analysis, and data processing and management). The uncertainty contributed by each procedural category (results presented in Section 12.2.6) as well as the uncertainty in the resulting discharge and water quality data (results presented in Section 12.3.4) can then be estimated by propagating the uncertainty from individual data collection steps.

12.3.4 Uncertainty in Measured Data

Several recent studies have utilized the Harmel et al. (2006a) framework and RMSE methodology to estimate the uncertainty in actual monitoring data. Uncertainty in discharge and/or water quality data has been estimated for glacial melt areas in Norway (Cooper, 2002), for tile drained watersheds in Illinois (Gentry et al., 2007), for the Coastal Plain in Florida (Keener et al., 2007), for urban stormwater systems in Australia (McCarthy et al., 2008), for small watershed in Texas, Indiana, and Ohio (Harmel et al., 2009), and for natural gas sites in Texas (Wachal et al., 2008). These studies were conducted under wide-ranging monitoring conditions (e.g., hydrologic setting, land use, and watershed size) and with varying field and laboratory techniques. Uncertainty estimates from these studies are summarized in Table 12.1 and Figure 12.2. While these values represent typical scenarios, much higher uncertainties can occur for extremely low flow and concentration values (Kotlash and Chessman, 1998; McCarthy et al., 2008) and for missing/incorrect data (Harmel et al., 2006a, 2009).

As shown in Table 12.1 and Figure 12.2, the uncertainties for discharge data are typically less than for constituent concentrations or loads. Uncertainty in sediment loads was typically less than in other constituents because of limited post-collection transformation, relatively simple analytical procedures, and high concentration values. The uncertainty in dissolved NO_3-N and PO_4-P loads was typically higher than uncertainty in sediment loads because of post-collection transformation potential, more complex analytical procedures, and lower concentration values, which counteracted the reduced difficulty in sample collection for dissolved constituents as compared to sediment sampling. The uncertainty in total N and P loads was typically higher than for sediment and dissolved N and P loads because of added processing and analytical complexity and increased difficulty in collecting representative particulate samples. McCarthy et al. (2008) published one of the few studies of which we are aware that analyzed the various sources of uncertainty in measured pathogen data (see also Eleria and Vogel, 2005). McCarthy et al. (2008) estimated *E. coli*

TABLE 12.1
Average Values for Measurement Uncertainty as Reported in Recently Published Studies

Data type	Uncertainty	Reference
Discharge	±16%	Cooper (2002)
Discharge	±21%	Wachal et al. (2008)
Discharge	±13%	Harmel et al. (2009)
Discharge	±19%	McCarthy et al. (2008)
Sediment load	±29%	Wachal et al. (2008)
Sediment load	±20%	Harmel et al. (2009)
Sediment concentration	±17%	Harmel et al. (2009)
NO_3-N load	±23%	Harmel et al. (2009)
NO_3-N concentration	±19%	Harmel et al. (2009)
PO_4-P load	±24%	Harmel et al. (2009)
PO_4-P concentration	±20%	Harmel et al. (2009)
Total N load	±26%	Harmel et al. (2009)
Total N concentration	±23%	Harmel et al. (2009)
Total P load	±10%	Gentry et al. (2007)
Total P load	±28%	Keener et al. (2007)
Total P load	±27%	Harmel et al. (2009)
Total P concentration	±27%	Keener et al. (2007)
Total P concentration	±24%	Harmel et al. (2009)
E. coli concentration	±33%	McCarthy et al. (2008)

concentration uncertainties that ranged from ±15–67%, which are higher than for other constituents presumably due to post-collection death/regrowth and increased analytical uncertainty and subjectivity. Regardless of constituent type, it is important to note that the uncertainty associated with measured concentrations is always less than or equal to that of measured loads because discharge measurement and its associated uncertainty is irrelevant in concentration determination.

12.4 SUMMARY

All measured data are uncertain to some degree due to various data collection procedures, which for hydrology and water quality can be categorized as discharge measurement, sample collection, sample preservation and storage, laboratory analysis, and data processing and management. Historically, uncertainty from these sources has not been assessed or included (with some exceptions) when presenting measured data or when comparing measured data and predicted values in hydrology and water quality analyses.

This oversight, however, is changing with recent application of the RMSE propagation method, which provides realistic estimates of uncertainty in measured

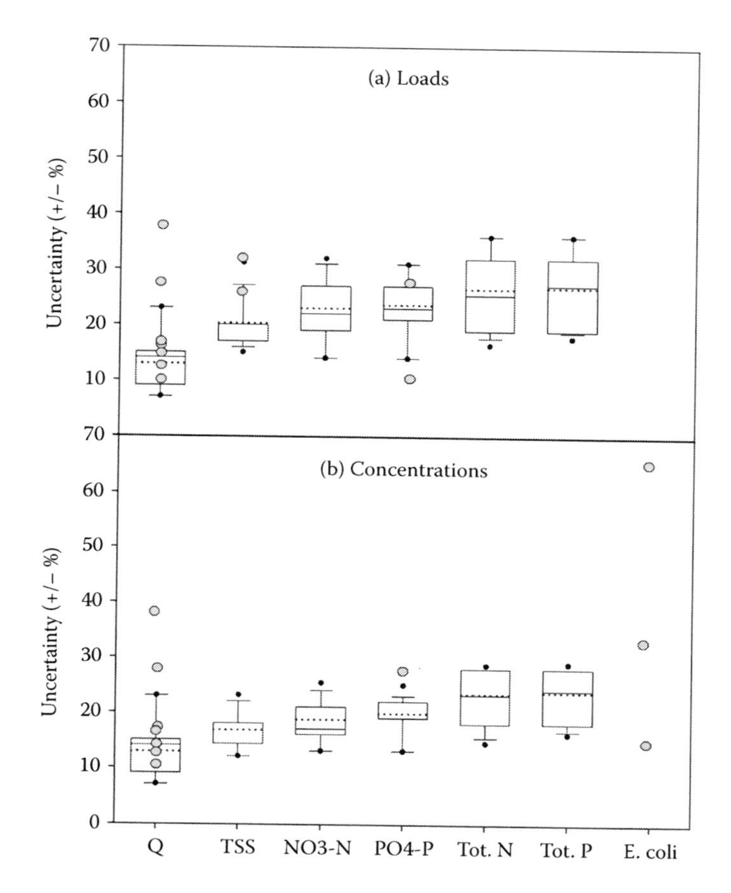

FIGURE 12.2 Uncertainty in discharge and selected water quality constituent loads (a) and concentrations (b). Distributed data (box plots) from Harmel, R.D., D.R. Smith, K.W. King, and R.M. Slade. 2009. *Environ. Modelling Software* 24, 832–842. Individual values (circles) from Cooper, R. J. 2002. Chemical Denudation in the Proglacial Zone of Finsterwalderbreen, Svalbard. Ph.D. thesis, University of Bristol, Bristol, U.K. 184 p; Gentry, L. E., M. B. David, T. V. Royer, C. A. Mitchell, and K. M. Starks. 2007. *J. Environ. Qual.* 36, 408–415; Keener, V. W., K. T. Ingram, B. Jacobson, and J. W. Jones. 2007. *Trans. ASABE* 50(6), 2081–2089; McCarthy, D. T., A. Deletic, V. G. Mitchell, T. D. Fletcher, and C. Diaper. 2008. Uncertainties in stormwater *E. coli* levels. *Water Res.* 42, 1812–1824; and Wachal, D.J., R.D. Harmel, K.E. Banks, and P.F. Hudak. 2008. *Trans. ASABE* 51(6): 1977–1986.

hydrology and water quality data. The expanded realization of the importance of including uncertainty estimates with measured data has produced methods that identify and reduce uncertainty as well as a tool for estimating measurement uncertainty specifically for hydrology and water quality applications (i.e., DUET-H/WQ). These advancements give water quality professionals the needed information and tools to incorporate data uncertainty in hydrology and water quality analyses and thus facilitate realistic, science-based interpretation of water quality monitoring results.

REFERENCES

Allmendinger, N.E., J.E. Pizzuto, G.E. Moglen, and M. Lewicki. 2007. A sediment budget for an urbanizing watershed, 1951–1996, Montgomery County, Maryland, USA. *J. Amer. Water Resources Assoc.* 43(6): 1483–1498.

American Public Health Association (APHA), American Water Works Association, and Water Environment Federation. 2005. Standard Methods for the Examination of Water and Wastewater. Port City Press: Baltimore, MD.

Beven, K. 2006a. On undermining the science? *Hydrol. Process.* 20: 3141–3146.

Beven, K. 2006b. A manifesto for the equifinality thesis. *J. Hydrol.* 320: 18–36.

Bonta, J.V., and B. Cleland. 2003. Incorporating natural variability, uncertainty, and risk into water quality evaluations using duration curves. *J. Amer. Water Resources Assoc.* 39(6): 1481–1496.

Brakensiek, D.L., H.B. Osborn, and W.J. Rawls, coordinators. 1979. Field Manual for Research in Agricultural Hydrology. USDA Agriculture Handbook No. 224. Washington, D.C.: USDA.

Brown, J.D., G.B. Heuvelink, and J.C. Refsgaard. 2005. An intergrated methodology for recording uncertainties about environmental data. *Water Sci. Tech.* 52(6): 153–160.

Buchanan, T.J., and W.P. Somers. 1976. Book 3, Chapter A8: Discharge measurements at gaging stations. Techniques of Water-Resources Investigations of the United States Geological Survey. Washington, D.C.: USGS.

CAEAL. 2003. Policy on the estimation of uncertainty of measurement in environmental testing. Ottawa, Ontario: Canadian Association For Environmental Analytical Laboratories. Available at: www.caeal.ca/P19_CAEAL_Unce_Pol.pdf.

Carter, R.W., and I.E. Anderson. 1963. Accuracy of current meter measurements. *J. Hydraulics Division, Proc. ASCE* 4(1): 105–115.

Carter, R.W., and J. Davidian. 1989. Book 3, Chapter A6: General procedure for gaging streams. Techniques of Water-Resources Investigations of the United States Geological Survey. Washington, D.C.: USGS.

Collins, G., J.N. Kremer, and I. Valiela. 2000. Assessing uncertainty in estimates of nitrogen loading to estuaries for research, planning, and risk assessment. *Environ. Mgmt.* 25(6): 635–645.

Cooper, R.J. 2002. Chemical Denudation in the Proglacial Zone of Finsterwalderbreen, Svalbard. Ph.D. thesis, University of Bristol, Bristol, U.K. 184 p.

Cuadros-Rodriquez, L., M.E. Hernandez Torres, E. Almansa Lopez, F.J. Egea Gonzalez, F.J. Arrebola Liebanas, and J.L. Martinez Vidal. 2002. Assessment of uncertainty in pesticide multiresidue analytical methods: Main sources and estimation. *Anal. Chim. Acta* 454(2): 297–314.

Deutsch, C.V., and A.G. Journel. 1998. *GSLIB-Geostatistical Software Library and User's Guide*. New York: Oxford University Press.

Dickinson, W.T. 1967. Accuracy of discharge determinations. Hydrology Papers, No. 20. Fort Collins, CO: Colorado State University.

Eleria, A., and R.M. Vogel. 2005. Predicting fecal coliform bacteria levels in the Charles River, Massachusetts, USA. *J. Amer. Water Resources Assoc.* 41(5):1195–1209.

European Co-operation for Accreditation (EA). 1999. Expression of the Uncertainty of Measurement in Calibration, EA 4/02.

Fishman, M.J., L.J. Schroder, and M.W. Shockley. 1986. Evaluation of methods for preservation of water samples for nutrient analysis. *Intl. J. Environ. Studies* 26(3): 231–238.

Fitzgerald, G.P., and S.L. Faust. 1967. Effect of water sample preservation methods on the release of phosphorus from algae. *Limnol. Oceanogr.* 12(2): 332–334.

Gentry, L.E., M.B. David, T.V. Royer, C.A. Mitchell, and K.M. Starks. 2007. Phosphorus transport pathways to streams in tile-drained agricultural watersheds. *J. Environ. Qual.* 36: 408–415.

Ging, P. 1999. Water-quality assessment of south-central Texas: Comparison of water quality in surface-water samples collected manually and by automated samplers. USGS Fact Sheet FS-172-99. Washington, D.C.: USGS.

Gordon, J.D., C.A. Newland, and S.T. Gagliardi. 2000. Laboratory performance in the sediment laboratory quality-assurance project, 1996–98. USGS Water Resources Investigations Report 99-4184. Washington, D.C.: USGS.

Haan, C.T. 2002. *Statistical Methods in Hydrology*. 2nd ed. Iowa State Press, Ames, IA.

Harmel, R.D., and K.W. King. 2005. Uncertainty in measured sediment and nutrient flux in runoff from small agricultural watersheds. *Trans. ASAE* 48(5): 1713–1721.

Harmel, R.D., and P.K. Smith. 2007. Consideration of measurement uncertainty in the evaluation of goodness-of-fit in hydrologic and water quality modeling. *J. Hydrol.* 337: 326–336.

Harmel, R.D., K.W. King, J.E. Wolfe, and H.A. Torbert. 2002. Minimum flow considerations for automated storm sampling on small watersheds. *Texas J. Sci.* 54(2): 177–188.

Harmel, R.D., K.W. King, and R.M. Slade. 2003. Automated storm water sampling on small watersheds. *Applied Eng. Agric.* 19(6): 667–674.

Harmel, R.D., R.J. Cooper, R.M. Slade, R.L. Haney, and J.G. Arnold. 2006a. Cumulative uncertainty in measured streamflow and water quality data for small watersheds. *Trans. ASABE* 49(3): 689–701.

Harmel, R.D., K.W. King, B.E. Haggard, D.G. Wren, and J.M. Sheridan. 2006b. Practical guidance for discharge and water quality data collection on small watersheds. *Trans. ASABE* 49(4): 937–948.

Harmel, R.D., D.R. Smith, K.W. King, and R.M. Slade. 2009. Estimating storm discharge and water quality data uncertainty: A software tool for monitoring and modeling applications. *Environ. Modelling Software* 24: 832–842.

Haygarth, P.M., C.D. Ashby, and S.C. Jarvis. 1995. Short-term changes in the molybdate reactive phosphorus of stored waters. *J. Environ. Qual.* 24(6): 1133–1140.

Haygarth, P.M., and A.C. Edwards. 2000. Sample collection, handling, preparation, and storage. In Methods of Phosphorus Analysis for Soils, Sediments, Residuals, and Waters. Southern Cooperative Series Bulletin No. 396. Available at www.soil.ncsu.edu/sera17/publications/sera17-2/pm_cover.htm.

Henriksen, A. 1969. Preservation of water samples for phosphorus and nitrogen determination. *Vatten* 25: 247–254.

Herschy, R. 1995. *Streamflow Measurement*, Second edition. Elsevier Applied Science Publishers.

Hipolito, J.N., and J.M. Leoureiro. 1988. Analysis of some velocity-area methods for calculating open-channel flow. *Hydrol. Sci. J.* 33: 311–320.

Hunsaker, C.T., M.F. Goodchild, M.A. Friedl, and T.J. Case (editors). 2001. *Spatial Uncertainty in Ecology*. New York: Springer Verlag.

International Organization for Standardization (ISO). 1993a. International Vocabulary of Basic and General Terms in Metrology, Second edition (Geneva, Switzerland).

International Organization for Standardization (ISO). 1993b. Guide to the Expression of Uncertainty in Measurement (Geneva, Switzerland).

Isaaks, E.H., and R.M. Srivastava. 1989. *An Introduction to Applied Geostatistics*. New York: Oxford University Press. 561 p.

Jarvie, H.P., P.J.A. Withers, and C. Neal. 2002. Review of robust measurement of phosphorus in river water: Sampling, storage, fractionation, and sensitivity. *Hydrol. Earth System Sci.* 6(1): 113–132.

Johnson, A.H., D.R. Bouldin, and G.W. Hergert. 1975. Some observations concerning preparation and storage of stream samples for dissolved inorganic phosphate analysis. *Water Resources Res.* 11: 559–562.

Kavetski, D., S.W. Franks, and G. Kuczera. 2002. *Confronting Input Uncertainty in Environmental Modelling in Calibration of Watershed Models*, AGU Water Science and Applications Series, S. Duan, H.V. Gupta, S. Sorooshian, A.N. Rousseau, and R. Turcotte (eds). Vol. 6, pp. 49–68.

Keener, V.W., K.T. Ingram, B. Jacobson, and J.W. Jones. 2007. Effects of El-Niño/Southern Oscillation on simulated phosphorus loading in South Florida. *Trans. ASABE* 50(6): 2081–2089.

Kennedy, E.J. 1984. Book 3, Chapter A10: Discharge ratings at gaging stations. Techniques of Water-Resources Investigations of the United States Geological Survey. Washington, D.C.: USGS.

King, K.W., and R.D. Harmel. 2003. Considerations in selecting a water quality sampling strategy. *Trans. ASAE* 46(1): 63–73.

Kotlash, A.R., and B.C. Chessman. 1998. Effects of water sample preservation and storage on nitrogen and phosphorus determinations: Implications for the use of automated sampling equipment. *Water Res.* 32(12): 3731–3737.

Lambert, D., W.A. Maher, and I. Hogg. 1992. Changes in phosphorus fractions during storage of lake water. *Water Res.* 26(5): 645–648.

Latterell, J.J., D.R. Timmons, R.F. Holt, and E.M. Sherstad. 1974. Sorption of orthophosphate on the surface of water sample containers. *Water Resources Res.* 10: 865–869.

Ludtke, A.S., M.T. Woodworth, and P.S. Marsh. 2000. Quality assurance results for routine water analysis in U.S. geological survey laboratories, water year 1998. USGS Water Resources Investigations Report 00-4176. Washington, D.C.: USGS.

Maher, W., and L. Woo. 1998. Procedures for the storage and digestion of natural waters for the determination of filterable reactive phosphorus, total filterable phosphorus, and total phosphorus. *Anal. Chim. Acta* 375(1–2): 5–47.

Martin, G.R., J.L. Smoot, and K.D. White. 1992. A comparison of surface-grab and cross-sectionally integrated stream-water-quality sampling methods. *Water Environ. Res.* 64(7): 866–876.

McCarthy, D.T., A. Deletic, V.G. Mitchell, T.D. Fletcher, and C. Diaper. 2008. Uncertainties in stormwater *E. coli* levels. *Water Res.* 42: 1812–1824.

McIntyre, N., and M. Marshall. 2008. Field verification of bed-mounted ADV sensors. *Proceedings of the Institution of Civil Engineers Water Mgmt.* 161: 199–206.

Mercurio, G., J. Perot, N. Roth, and M. Southerland. 2002. Maryland biological stream survey 2000: Quality assessment report. Springfield, VA: Versar, Inc., and Baltimore, MD: Maryland Department of Natural Resources.

Meyer, V.R. 2002. Minimizing the effect of sample preparation on measurement uncertainty. *LC-GC Europe.* July 2002, 2–5.

Miller, P.S., R.H. Mohtar, and B.A. Engel. 2007. Water quality monitoring strategies and their effects on mass load calculation. *Trans. ASABE* 50(3): 817–829.

Nature. 2005. Responding to uncertainty. 437(7055), 1.

Novak, C.E. 1985. WRD Data Reports Preparation Guide. U.S. Geological Survey Water Resources Division 1985 Edition Open File Report 85–480, 333 pg.

Pappenberger, F., and K.J. Beven. 2006. Ignorance is bliss: 7 reasons not to use uncertainty analysis. *Water Resources Res.* 42(5), doi:10.1029/2005W05302.

Pelletier, P.M. 1988. Uncertainties in the single determination of river discharge: A literature review. *Canadian J. Civil Eng.* 15(5): 834–850.

Ramsey, M.H. 1998. Sampling as a source of measurement uncertainty: Techniques for quantification and comparison with analytical sources. *J. Anal. Atomic Spectrometry* 13: 97–104.

Rantz, S.E. et al. 1982. Volume 2: Computation of discharge. In *Measurement and Computation of Streamflow*, 284–601. USGS Water-Supply Paper 2175. Washington, D.C.: USGS.

Reckhow, K.H. 1994. Water quality simulation modeling and uncertainty analysis for risk assessment and decision making. *Ecol. Modelling* 72: 1–20.

Reckhow, K.H. 2003. On the need for uncertainty assessment in TMDL modeling and implementation. *J. Water Resources Planning Mgmt.* 129(4): 245–246.

Robards K., I.D. Kelvie, R.L. Benson, P.J. Worsfold, N.J. Blundell, and H. Casey. 1994. Determination of carbon, phosphorus, nitrogen, and silicon species in waters. *Anal. Chim. Acta* 287(3): 147–190.

Robertson, D.M., and E.D. Roerish. 1999. Influence of various water quality sampling strategies on load estimates for small streams. *Water Resources Res.* 35(12): 3747–3759.

Rode, M., and U. Suhr. 2007. Uncertainties in selected river water quality data. *Hydrol. Earth System Sci.* 11: 863–874.

Rouhani, S., R.M. Srivastava, A.J. Desbarats, M.V. Cromer, and A.I. Johnson (editors). 1996. Geostatistics for Environmental and Geotechnical Applications. STP 1283. Ann Arbor, MI: ASTM.

Ryden, J.C., J.K. Sayers, and R.F. Harris. 1972. Sorption of inorganic phosphate by laboratory ware: Implications in environmental phosphorus techniques. *Analyst* 97: 903–908.

Sauer, V.B., and R.W. Meyer. 1992. Determination of error in individual discharge measurements. USGS Open File Report 92-144. Washington, D.C.: USGS.

Schmidt, A.R. 2002. Analysis of stage-discharge relations for open-channel flows and their associated uncertainties. PhD. diss. Urbana-Champaign, Ill.: University of Illinois at Urbana-Champaign, Department of Civil and Environmental Engineering.

Shirmohammadi, A., I. Chaubey, R.D. Harmel, D.D. Bosch, R. Muñoz-Carpena, C. Dharmasri, A. Sexton, M. Arabi, M.L. Wolfe, J. Frankenberger, C. Graff, and T.M. Sohrabi. 2006. Uncertainty in TMDL Models. *Trans. ASABE* 49(4): 1033–1049.

Silberstein, R.P. 2006. Hydrological models are so good, do we still need data? *Environ. Modelling Software* 21(9): 1340–1352.

Skjemstad, J.O., and R. Reeve. 1978. The automatic determination of ppb levels of ammonia, nitrate plus nitrite, and phosphate in water in the presence of added mercury II chloride. *J. Environ. Qual.* 7(1): 137–141.

Slade, R.M. 2004. General Methods, Information, and Sources for Collecting and Analyzing Water-Resources Data. CD-ROM. Copyright 2004 Raymond M. Slade, Jr.

Taylor, B.N., and C.E. Kuyatt. 1994. Guidelines for Evaluating and Expressing the Uncertainty of NIST Measurement Results. National Institute of Standards and Technology (NIST) Technical Note 1297. United States Department of Commerce Technology Administration.

Topping, J. 1972. *Errors of Observation and Their Treatment.* 4th ed. London: Chapman and Hall.

U.S. Geological Survey (USGS). 1999. Book 9, Section A: National field manual for the collection of water-quality data. Techniques of Water-Resources Investigations of the United States Geological Survey. Washington, D.C.: USGS.

Wachal, D.J., R.D. Harmel, K.E. Banks, and P.F. Hudak. 2008. Evaluation of WEPP for runoff and sediment yield prediction on natural gas well sites. *Trans. ASABE* 51(6): 1977–1986.

Young, K.B. 1950. A Comparative Study of Mean-Section And Mid-Section Methods for Computation of Discharge Measurements. USGS Open File Report. Washington, D.C.: USGS.

13 Water Quality Statistical Analysis

Kati W. Migliaccio, Joffre Castro, and Brian E. Haggard

CONTENTS

13.1 INTRODUCTION

Water quality data sets are translated into meaningful information through statistical analysis. Common data measures are the first step in any data set review. In completing any water quality data analysis, an understanding of censored values and detection limits is essential to ensure that conclusions are not biased due to chemical analyses techniques or changing detection limits. More sophisticated statistical procedures can also be used to derive additional information from a data set such as load estimations, trend analyses, and principal component analysis. A review of these concepts and their application to water quality data are presented in this chapter.

13.2 COMMON DATA MEASURES

A water quality dataset is generally composed of analytical results from water samples collected using automatic and/or manual sampling techniques over a designated time frame, and thus categorized as a time series dataset. Sample datasets are evaluated using common data measurements to determine the status of a water body in regards to its level of impairment. Impairment is assessed by comparing dataset measurements to relevant criteria (see Chapter 3 for more information on water quality criteria).

13.2.1 MEAN

The most common measurement evaluated for a dataset is probably the arithmetic mean. The arithmetic mean ($\overline{X}$) (or the average value) is calculated as

$$\overline{X} = \sum_{i=1}^{n} \frac{X_i}{n} \tag{13.1}$$

where n is the number of data values and X_i is the ith data value in the dataset. The arithmetic mean provides the average value considering all values; thus, outliers may substantially influence the tabulated result.

The geometric mean is another type of data measurement that differs from the arithmetic mean and requires that all numbers in the dataset be positive and nonzero. The geometric mean ($\overline{X}_G$) is calculated as

$$\overline{X}_G = \sqrt[n]{X_1 X_2 \ldots X_n} \,. \tag{13.2}$$

Geometric mean values are often used in water quality standards for bacteria. The geometric mean may be selected to describe the dataset if the dataset includes a wide range of values with very low numbers and/or very high numbers to reduce the influence of these outliers.

In addition to the arithmetic mean and the geometric mean, there are other types of mean values, such as the weighted mean, harmonic mean, truncated mean, and inter-quartile mean. However, the most commonly used mean values in describing water quality data are the arithmetic and geometric means.

13.2.2 Median

Median is also a common data measurement used to describe central tendencies of a water quality dataset. The median is the value in the center of the dataset if data are organized in an ascending (or descending) numeric order. The median could also be described as the value for which half of the total observations are less than that value and half of the total observations are greater than that value (Haan, 2002). The median is commonly used in assessing water quality data as it is less sensitive to outliers (as compared to the arithmetic mean).

13.2.3 Mode

The mode of a dataset is the value that occurs most frequently and must occur at least twice. There is a possibility that a dataset may have more than one mode or may not have a mode; however, the latter is less common. An example would be a dataset with two modes, termed a *bimodal distribution*, with two peaks or modes.

13.2.4 Range

The dataset range refers to the numeric distance between the smallest dataset value and the greatest dataset value. The range provides an idea of the variability within the dataset that complements other measurements such as variance and standard deviation (see Section 13.2.8 Variance and Standard Deviation).

13.2.5 Distribution

The distribution of a dataset refers to how the data varies in composition. Distributions for water quality data are often graphically depicted using histograms. This method consist of dividing the data into "bins" or ranges and counting the number of occurrence within each bin (Helsel and Hirsch, 2002). This number becomes the value for that bin. Thus, histograms are vertical bar graphs with the bins on the x-axis and the frequency on the y-axis where the bars are plotted so that they touch each other (McBride, 2005; Figure 13.1). Histograms, although common, should be interpreted with caution as selection of bins can influence visual results. Statistical tests of distribution are available (such as the chi-square test) and provide a more sound method for evaluating a distribution. Some common distributions in water quality data analysis are the normal distribution, the lognormal distribution, and the exponential distribution (Figure 13.2).

The distribution of a dataset can also be described by its skewness. Skewness refers to how data is organized around the mean. A dataset that follows the well known bell-shaped pattern has a symmetric distribution. A dataset that is skewed does not have a symmetric distribution around the mean or median (Helsel and Hirsch, 2002).

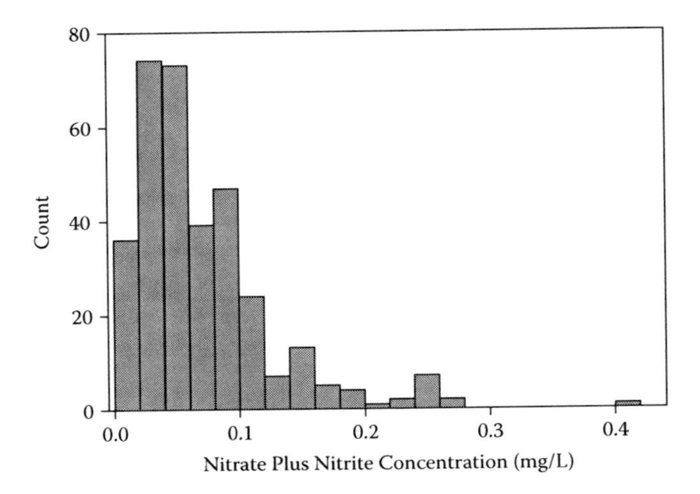

FIGURE 13.1 Example histogram of water quality data.

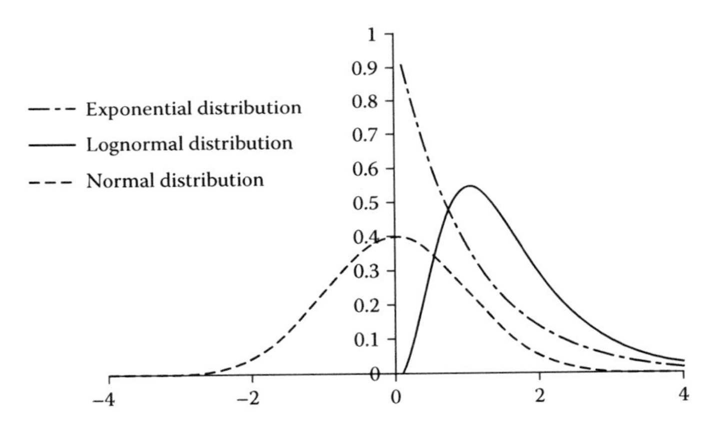

FIGURE 13.2 Example distributions commonly observed in water quality datasets.

There are different skewness measurements; one measure of skewness in a dataset is the coefficient of skewness (g):

$$g = \frac{n}{(n-1)(n-2)} \sum_{i=1}^{n} \frac{\left(x_i - \overline{X}\right)^3}{s^3}. \tag{13.3}$$

Right-skewed data have a positive coefficient of skewness and left-skewed data have a negative coefficient of skewness (Figure 13.3).

13.2.6 Quartiles

Quartiles are measurements of the distribution that divide a dataset into four parts of equal sample sizes. The first quartile refers to the value at which 25% of the data

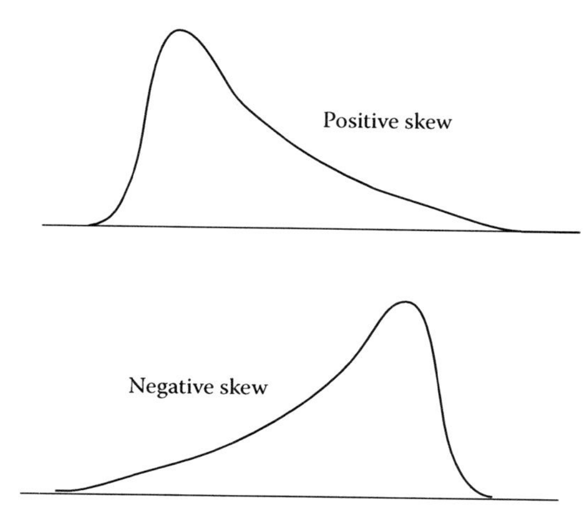

FIGURE 13.3 Example of positive skew and negative skew.

is less than this value and 75% of the data is greater than this value. Similarly, the second (50%) and third (75%) quartiles can be determined. Quartile data are often used in conjunction with probability plots.

13.2.7 Outliers

Another characteristic of a dataset is the presence of outliers. Outliers refer to data points outside the expected range of data. Outliers could be very small numbers or very large numbers. Outliers may be the result of an error (e.g., sampling error, analytical error, or data management error) or may be a natural occurrence outside of the normal data range. Identification and review of outliers should be conducted with all datasets to ensure the values are not the result of a detectable and correctable error.

13.2.8 Variance and Standard Deviation

The variance of a dataset is a measure of the spread or dispersion of the data. The variance value increases as the spread or range of the dataset increases. The variance is directly influenced by the presence of outliers (Helsel and Hirsch, 2002). The sample variance (s^2) is calculated as:

$$s^2 = \sum_{i=1}^{n} \frac{\left(X_i - \overline{X}\right)^2}{(n-1)}. \tag{13.4}$$

Similar to variance, standard deviation is another measurement of dataset spread or dispersion. The standard deviation is more commonly reported than the variance. This is likely due to its use in describing the probability of values occurring if the dataset is normally distributed. For normally distributed datasets, the range of one standard deviation minus the mean ($\overline{X}$) to one standard deviation plus the mean ($\overline{X}$)

includes 68% of all values in the dataset; similarly 95% of all values fall within two standard deviations of the mean ($\overline{X}$). The standard deviation (s) of a sample is the square root of the variance and is calculated as

$$s = \sqrt{\sum_{i=1}^{n} \frac{\left(X_i - \overline{X}\right)^2}{(n-1)}}. \tag{13.5}$$

13.2.9 Correlation

Correlation is a measurement of the relationship between two variables (McBride, 2005) and is generally represented by the Pearson's correlation coefficient (r). This is calculated as

$$r = \frac{\sum_{i=1}^{n}\left[\left(X_i - \overline{X}\right)\left(Y_i - \overline{Y}\right)\right]}{\sqrt{\sum_{i=1}^{n}\left(X_i - \overline{X}\right)^2 \sum_{i=1}^{n}\left(Y_i - \overline{Y}\right)^2}}. \tag{13.6}$$

where X and Y represent the two different variables and n is the number of data values. The range of r is between −1 and 1: the greater the absolute value the stronger the relationship or correlation. A negative r value indicates an inverse (opposite) relationship, so that when one variable increases the other decreases. A positive r indicates a direct relationship; that is, as one variable increases (or decreases) so does the second variable.

13.2.10 Visual Dataset Evaluation

The common data measurements described previously can be visually evaluated using different plotting techniques. Although these techniques are not meant to provide hypothesis testing, they do allow for easy and simple inspection of datasets. Visual evaluation is a powerful way to identify characteristics to select appropriate statistical tests and perform a quality review of the dataset. Visual evaluation also is used to compare datasets, validate statistical findings, and better understand the behavior of the data.

The most common visual dataset evaluation is a time series plot. Time series plots consist of an x-y plot with time on the x-axis and the measured data variable on the y-axis. This quick type of plot can provide some insight into a dataset so that better interpretation of other statistics can be completed. For example, outliers become visible, censored values have a horizontal line pattern, overall ranges and trends are more apparent, and serial correlation can be visually ascertained (Figure 13.4). Time series plots may be used with a smoothing line to add some clarity to patterns in the data set (McBride, 2005). Smoothing techniques are available in most statistical software packages. A common smoothing method used in water quality data

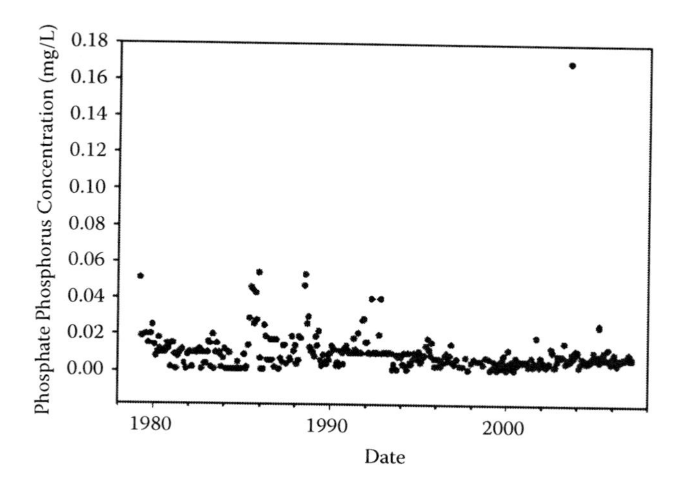

FIGURE 13.4 Time series data set with censored values identifiable by the vertical lines the dots form.

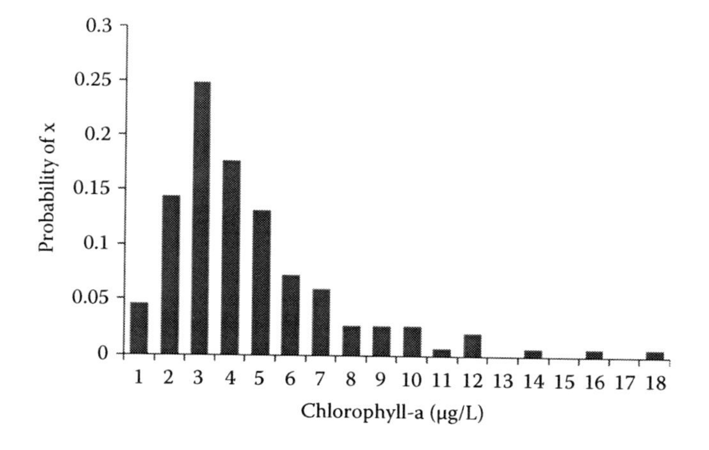

FIGURE 13.5 Example histogram of a water quality dataset.

evaluation is the Loess technique (Cleveland, 1979; Hirsch et al., 1991; Petersen, 1992; Richards and Baker, 2002; White et al., 2004).

Probability plots depict the probability of certain values occurring within a dataset (Figure 13.5).

Probability plots are similar to distribution plots in that they often include vertical bars with the different bins or ranges for each bar on the x-axis and the frequency of occurrence or probability of occurrence on the y-axis. A probability density function (PDF) can be used instead of vertical bars to depict this relationship (Figure 13.6). Probability plots can be shown as (composed of) PDFs or cumulative distribution functions (CDFs; Figure 13.7). Probability plots that depict PDFs are characterized by a y-axis with varying range. Alternatively, probability plots that depict CDFs are

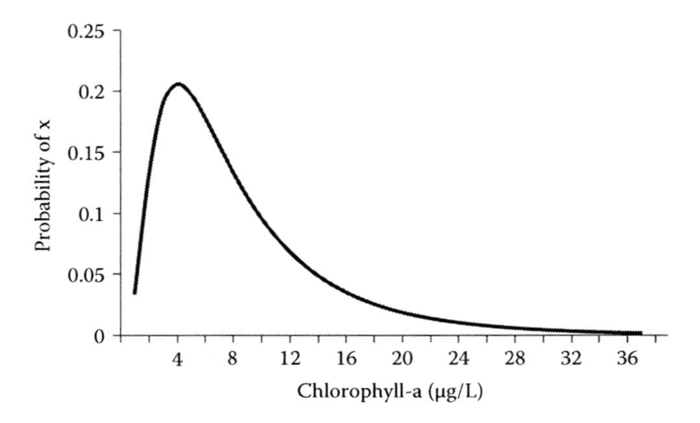

FIGURE 13.6 Probability density function example.

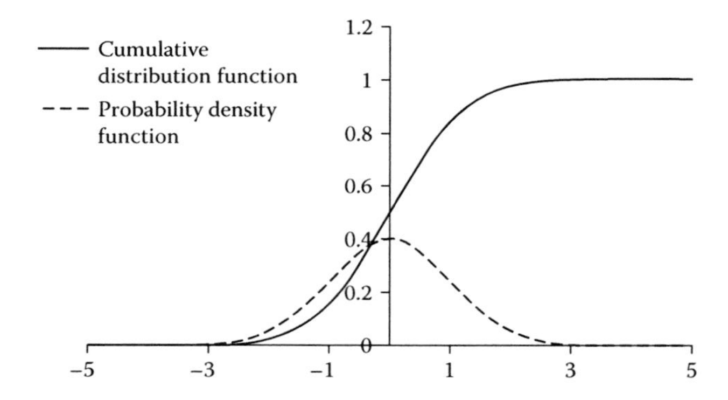

FIGURE 13.7 Example cumulative distribution function (CDF) and probability density function (PDF).

characterized by a y-axis that ranges from 0 to 1. Probability plots are often used to better understand the uncertainty of a measured data point that is inherent in water quality sampling (see Chapter 12).

Boxplots (or box-and-whiskers plots) are used to provide dataset measurement information in a visual format. Boxplots are often used to compare different sites and includes the median, inter-quartile range, quartile skew, and outliers. Boxplots provide information on the skewness of a data set and the range of values. The greatest benefit of a boxplot is the ease it provides in comparing multiple sites (Figure 13.8).

13.3 CENSORED VALUES AND DETECTION LIMITS

Most water quality constituents are best described by their summary statistics, such as mean, standard deviation, and quartiles (see Section 13.2). This simple task could become complicated when the data include censored values. These are values that have been identified by the analytical chemist, who analyzed the sample

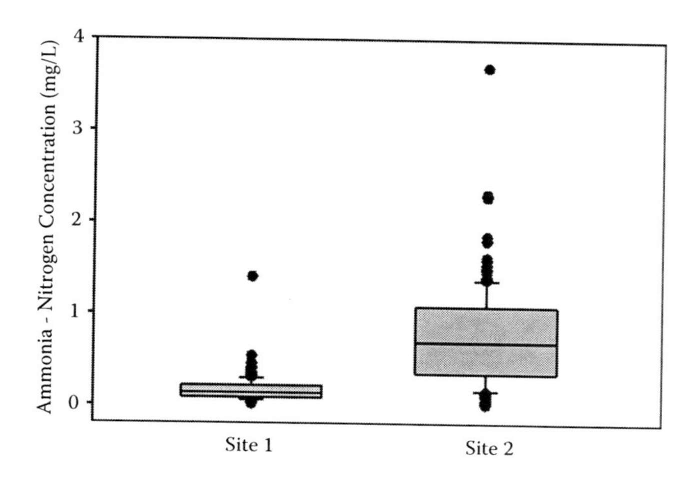

FIGURE 13.8 Box plot example where black dots are outliers, the whiskers are 10 and 90 percentiles, the top and bottom of the box are 25 and 75 percentiles, and the line within the box is the median. Note that box plot symbols do not always represent the same value and thus symbology should always be explained.

at the laboratory, to be below a pre-determined concentration level, such as a MDL (method detection limit). The MDL is the lowest concentration level that can be detected with 99% confidence (EPA, 40 CFR). If a sample concentration is equal to or above the MDL, the concentration value becomes a detect or an uncensored value. If the concentration value is below the MDL, the concentration value becomes a nondetect or a censored value, meaning that the concentration is a value between zero and the MDL but is not precisely known. Because censoring of the water quality data most often occurs at the lower end of the concentration distribution (left side), between zero and the MDL, this type of censoring is referred to as left censoring. By contrast, in survival analyses (medicine) and in failure analyses (engineering) right censoring is typical. It is important to point out that there are other thresholds that are also used to censor water quality data. A review of these other thresholds can be found in Helsel (2005) and USEPA (2007).

The following sections provide a brief review of methods used in the analysis of water quality data with censored values. The purposes of these sections are to introduce the reader to modified methods of determining summary statistics for datasets that include censored values. Traditionally, summary statistics have been calculated for datasets with censored values by replacing the censored values with a value that is equal to one half of the MDL. The traditional method is simple and, thus, is easy to understand and use, which is its main appeal. It may, however, provide incorrect results.

Methods used in the analysis of censored water quality data have received increased attention in recent years. Many water quality practitioners have recognized that simple substitution methods are not adequate enough to deal with censored data (Shumway et al., 2002; Helsel, 2006; Hewett and Ganser, 2007; Antweiler and Taylor, 2008). Since the early 1980s, there has been a gradual improvement on the theoretical development of statistical methods for data with censored values.

Spooner (1991), Wendelberger and Campbell (1994), and Helsel (2005) provide interesting chronological reviews of these methods. Today, modified methods have been developed (Helsel, 2005) and they are now available in most statistical packages. While the theoretical basis of these newer methods is more complex than that of simple substitution methods, their use in environmental sciences has greatly improved due to user-friendly applications and advancements in computer technology.

Nutrient, trace metal, and pesticide data are most likely to have censored values. In a water quality analysis of Everglades National Park's fresh, surface water stations, nearly 1/2 of the constituents studied had censored values. They were mostly trace metals and nutrients—there were no pesticide data. The percentage of censored values ranged from 3% for iron and kjeldahl nitrogen to 86% for zinc. It could be tempting in some situations to ignore (deleted) the censored values when their percentage is small. Unfortunately, there is no rule of thumb indicating when to ignore or not to ignore these censored values. This decision should be based on the purpose and implications of the water quality analysis. For example, regulatory work may require a more rigorous analysis than an environmental assessment. Always keep in mind that data series behave differently. Figure 13.9 illustrates this point by showing errors of estimation of means for two synthetic series. The data for the left-hand side graph is a short series of 20 values randomly generated. The data for the right-hand graph is a long series of 224 ammonia values from stations in Everglades National Park, where the censored values were deleted to create a synthetic ammonia series. The mean of each series was calculated and used to compute errors of estimation for the mean. The two series were left-censored multiple times to create a battery of censored series. The synthetic series was censored ranging from 5% to 30% in increments of 5%, and the ammonia synthetic series was censored at 5% and then between 20% and 80% in increments of 20%. The means of the censored series were estimated by 1-MDL substitution and by MR (these methods are explained later). The error was computed as the percent deviation from the mean of the uncensored series. The results show that the errors increased more rapidly when the 1-MDL

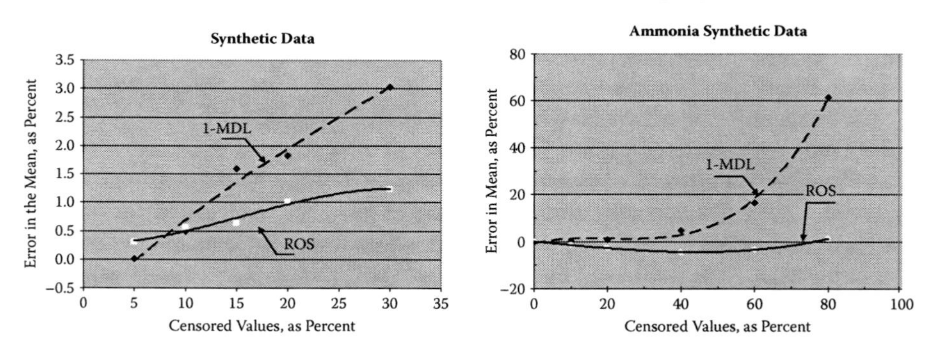

FIGURE 13.9 Errors in estimating the mean for water quality data with censored value. The error is expressed as the percent deviation from the mean of the uncensored series. The data on the left graph is a short series of randomly generated values. The data on the right graph is a synthetic ammonia series from Everglades National Park freshwater stations. Both a 1-MDL substitution and a MR method were used to estimate the means.

substitution method is used than when the MR method is used for greater than 50% censored values (right graph); errors were smaller in the synthetic data than in the synthetic ammonia data for less than 30% censored data; errors were surprisingly small, less than 3%, for the synthetic data (left graph); the zero percent error for the MR method at 75% censoring is a fortuitous result (right graph). Based on these results, it is easy to understand why it is not appropriate to decide whether or not (a) to ignore the censored values just because their percentage is small and (b) to use a substitution method in lieu of a more appropriate method. If anything, the recommendation should be to use methods that are best suited to deal with censored data whenever possible.

Another type of challenge encountered in the analysis of water quality data is the presence of multiple censored levels. Because analytical methods are improving and analytical instruments are becoming more accurate, the MDL levels are getting smaller. It is not uncommon for a water quality constituent to have multiple, decreasing censoring thresholds. In addition, data from different laboratories may have different MDLs. The overall effect is that data series may have multiple MDLs because (a) the MDLs are getting smaller with time and (b) the data comes from different laboratories.

Figure 13.10 shows the evolution of the MDL for sulfate at Everglades National Park. The data are sulfate concentration, in mg/L, of nine fresh, surface water stations. Since the middle 1980s the sulfate MDLs have decreased three times: from 5 mg/L to 2 mg/L, then to 1 mg/L, and finally to 0.1 mg/L. The sensitivity of the sulfate analytical method has increased by a factor of 50 (5/0.1). It is not unreasonable to expect that the sulfate MDL will decrease once more in the near future.

Some of the older methods for censored data analysis, especially the substitution methods, have more difficulty in dealing with multiple MDLs than the newer methods. A common approach for the less sophisticated methods is to accept only one MDL, the maximum MDL. This problem is more evident in trend analysis of multiple censored data.

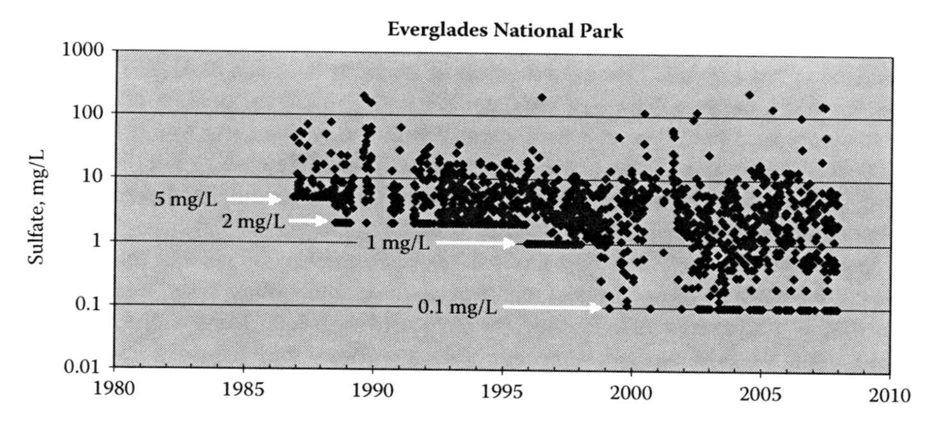

FIGURE 13.10 Chronological changes for the sulfate method detection limit (MDL) in fresh surface-water stations of Everglades National Park.

13.3.1 Censored Methods

Methods used in the analyses of water quality data with censored values fall into two camps: (a) those that use simple substitution techniques and (b) those that do not. The simple substitution methods have been very popular in the past because they are easy to use. The other methods require a better understanding of statistical techniques and are resource intensive. The benefits of using the latter models (b) are that they provide better estimates for statistical properties and now they are easy to use.

13.3.1.1 Simple Substitution

For simple substitution methods, the censored values are replaced by a number that is the product of the MDL times a factor. Whereas the MDL is known, the factor is arbitrarily selected. The factor could range from 0 to 1 and popular choices are 0, 1/2, $1/\sqrt{2}$, and 1. There is no real justification for choosing the factor, and that is the source of the problem with these methods. These substitution values tend to mask and distort the statistical properties of the data. The published literature provides numerous examples of how poorly these techniques perform by comparison to other more reliable methods (Gilliom and Helsel, 1986; Helsel and Cohn, 1988, She, 1997; Helsel, 2006; Hewett and Ganser, 2007). Helsel (2006) makes a clever demonstration of how substitution methods fail to provide a reasonable estimate for six statistic parameters regardless of substitution method (factor) selected. He progressively increased the substitution factor from 0 to 1 in small increments, estimated the six statistics, and compared these results to those of the uncensored series. The results show that the mean and standard deviation were overestimated and underestimated, the correlation coefficient and the regression slope were always underestimated, and the T-test and its p-value were always overestimated. Also, the means increased linearly and the regression slopes decreased linearly, but the other parameters did not vary linearly.

13.3.1.2 Nonsubstitution Methods

These methods can be subdivided into parametric and nonparametric methods. The first group of methods assumes that the data follow a known distribution—usually a normal or log-normal. The second group of methods does not make any assumption regarding the distribution of the data. For both groups, the percent of censored values is an essential piece of information. While it is encouraging to see that there has been significant progress in the development of these methods, it is a bit discouraging (and confusing) to note a lack of consistency in the naming of these methods. Some names and acronyms have changed over the years.

Kaplan-Meier (KM). The KM method has been extensively used in the medical and engineering fields to conduct survival analysis and failure tests, respectively. The KM is a nonparametric method for right-censored data. Most statistical packages include the KM method. Some application software may require a transformation of the water quality data, which is left-censored, to right-censored data (She, 1997). Helsel (2005) recommends subtracting the concentrations from a constant that is greater than the largest observed concentration. The KM method computes a survival probability function—similar to the cumulative distribution. Estimates for statistics (such as mean, standard deviation, standard error, and quartiles) are made on the transformed data. The estimates for the mean and quartiles need to be retransformed to the original units, but the estimates for variance, standard deviation, standard error, and inter-quartile range do not need to be retransformed.

Maximum Likelihood Estimator (MLE and AMLE). The MLE is a parametric method that fits the data to a normal or lognormal distribution (Cohen, 1959) or other known distribution. Estimates of the mean and standard deviation, in log-transformed form, are obtained by maximizing a likelihood function for the observed concentrations and the proportion of censored values. This method is not robust for datasets with a small number of observations (< 30) and datasets that do not follow a lognormal distribution (Helsel, 2005). Most statistical packages include the MLE method and, thus, Cohen's (1959) lookup table is not needed. Cohn (2005) has proposed an AMLE (adjusted MLE) to reduce bias; El-Shaarawi et al. (1989) and Kroll and Stedinger (1996) have used a fill-in method, which predicts the censored values, also to reduce bias from the MLE.

Regression on Order Statistics (ROS and MR). The ROS is a parametric method that regresses the observed values (or logarithms) to their normal scores (Helsel, 2005). The intercept and slope of the line estimate the mean and standard deviation, respectively. A robust ROS (MR) approach is currently used to avoid transformation biases (Helsel and Cohn, 1988). The MR (Multiple-limit Regression) is a semi-parametric method that predicts values (similar to fill-in the MLE) in log-transformed units for the censored observations using a regression line (same as in ROS). These predictions are transformed to their original units and in combination with the uncensored observations are used to estimate the descriptive statistics. These predicted values, as a group, are used to estimate statistics but should not be used as a substitute for each individual censored value.

To demonstrate the methods discussed above, the synthetic ammonia data were used to estimate summary statistics. Table 13.1 shows the results of estimating the mean, standard deviation (STDEV), first quartile (1Q), second quartile (2Q), and third quartile (3Q). As already explained, the synthetic ammonia data were created by combining ammonia data from nine surface water stations in Everglades National Park and deleting all censored values; hence, it is a synthetic series. The summary statistics for the 224 uncensored observations are shown in Table 13.1 as "original." This dataset was then arbitrarily censored at two levels: 0.128 mg/L and 0.198 mg/L. All observations below 0.128 mg/L were censored and only half of the observations

TABLE 13.1
Summary Statistics for the Synthetic Ammonia Data, mg/L

	Original	0-MDL	½-MDL	1-MDL	MR	KM	MLE	AMLE
Mean	0.371	0.310	0.342	0.374	0.350	0.360	0.349	0.349
STDEV	0.332	0.374	0.350	0.329	0.346	0.338	0.399	0.395
1Q	0.160	0.000	0.099	0.198	0.136	0.140	0.124	0.124
2Q	0.232	0.193	0.193	0.198	0.211	0.195	0.230	0.230
3Q	0.448	0.448	0.448	0.448	0.449	0.450	0.426	0.449

below 0.198 mg/L were censored. In all, there were 78 observations (35%) that were censored. Seven methods were used to estimate summary statistics and their performance was evaluated based on an error of estimation. The methods used were substitution of the censored value with 0 (0-MDL), with ½ of the MDL (½-MDL), and with the MDL (1-MDL), and application of MR, KM, MLE, and AMLE. The error was calculated as a percent deviation from the known statistics, which was estimated on the original set. For example, the error for the mean (E) was calculated as

$$E = \frac{\left(\overline{X}(x) - \overline{X}(o)\right)}{\overline{X}(o)} * 100 \tag{13.7}$$

where $\overline{X}(x)$ is the mean estimated by any of the seven methods and $\overline{X}(o)$ is the mean of the original series. The errors of estimation are shown in Figure 13.11.

The results show that the mean was underestimated by most methods, except by the 1-MDL, which overestimated the mean slightly. The standard deviation (STDEV) was overestimated by most methods, except by the 1-MDL, which underestimated the mean. The first quartile's (1Q) estimates have the largest error of all estimated statistics—as expected. It is evident that the substitution methods did not perform well, in particular, when estimating the first (1Q) and second (2Q) quartiles. The third quartile (3Q) was more accurately estimated by the substitution methods than by the other methods. A possible explanation for this fortuitous result may have to do with the censoring level (35%). At this censoring level, most of the information loss occurred at the lower percentiles (below 2Q) more that at the upper percentiles (above 2Q). This exercise was very limited in scope and its results should not be generalized. A different data series would have provided different results. What could be generalized, because it has been

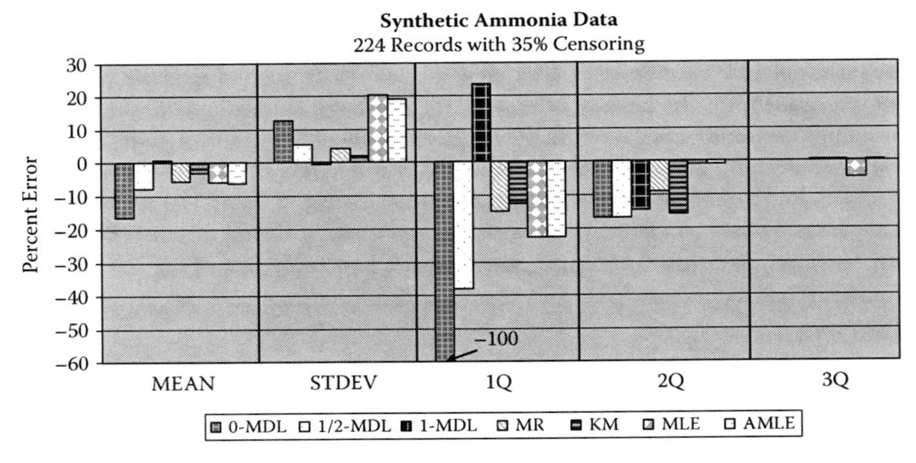

FIGURE 13.11 Error of estimations for mean, standard deviation (STDEV), first quartile (1Q), second quartile (2Q), and third quartile (3Q). Statistics for the synthetic ammonia data were computed with 0-MDL, ½-MDL, 1-MDL, multiple-limit regression (MR), Kaplan–Meier (KM), maximum likelihood estimator (MLE), and adjusted maximum likelihood estimator (AMLE).

extensively documented, is that substitution methods are not accurate. In the example discussed above, substitution method statistics were calculated using Excel, the MR and MLE statistics were made with Minitab, and the AMLE statistics were computed with a routine written for SPlus (Slack et al., 2003).

The use of censored-data methods extends beyond computing statistical summaries. Potential applications of those methods include hypothesis testing, comparison of two groups, comparison of three or more groups, estimation of correlation coefficients, regression analysis, and trend analysis. Additionally, newer techniques are being developed that will expand the applicability of those methods. For example, bootstrapping (Efron, 1981) is used in combination with MLE (Rao et al., 1991) or KM (Helsel, 2005) to obtain better estimates for the mean, median, quartiles, and their confident intervals.

13.4 LOAD ESTIMATIONS

The purpose of this section is to introduce statistical concepts that can be used to estimate constituent loads in streams, rivers, and canals using water quality data that have been collected from monitoring programs where continuous concentration measurement or high-frequency sampling was not feasible. The constituent load in a stream, river, or canal for a given period of time (e.g., annually) is not ever really known, but loads can be estimated with reasonable precision. The use of automated equipment allows the collection of water samples at a high frequency, which can then be used to integrate the product of concentration and discharge for given period of time (i.e., mass integration) to estimate constituent loads:

$$L = \int C(t) \cdot Q(t) dt, \tag{13.8}$$

where L represents the load over a defined period of time (*dt*), C represents concentration at time *t*, and Q represents discharge at time *t*. However, water samples in streams and rivers are often collected at frequencies much less than that required to effectively apply mass integration. The precision of the integration approach decreases substantially with the amount of time in-between the samples, requiring alternative methods to estimate precise constituent loads. Aulenbach and Hooper (2006) categorized the alternative approaches into the following: averaging methods, period-weighted methods, ratio estimators and regression-model (or rating curve) methods; these authors further suggested a composite method to estimate constituent loads, combining aspects of the period-weighted method and regression models. Many studies (e.g., Dolan et al., 1981; Preston et al., 1989; Robertson and Roerish, 1999; Haggard et al., 2003; Robertson, 2003; Zamyadi et al., 2007; Toor et al., 2008) have been completed that evaluated the performance of various load estimation techniques at varying spatial scales (i.e., ephemeral catchment to smaller perennial streams to large river systems). The "best" method to estimate loads often varies among streams and even data records (i.e., years) for the same site (e.g., see Coats et al., 2002). The selection of an appropriate method depends on the number and frequency of water samples collected, concentration relations with exogenous factors (e.g., discharge), distribution of water samples collected across the range of observed stream flows, the temporal scale at which loads will be estimated, and whether an estimate of accuracy and precision (e.g., bias and error) is needed.

13.4.1 Averaging Methods

The concentrations in water samples collected in a given time interval are averaged, and this average concentration (C_{mean}) is multiplied by the total amount of stream flow (i.e., Q_t) that occurred during the time interval (t) to provide the load (L):

$$L = C_{mean} \cdot Q_t \ . \tag{13.9}$$

The average concentration should be flow-weighted, if water samples were not collected at fixed interval or randomly through the interval (Aulenbach and Hooper, 2006). The typical way to flow-weight concentrations would be to multiply the observed concentrations (C_i) by the instantaneous discharge (Q_i) associated with each water sample, and then the sum of these products is divided by the sum of instantaneous discharges, or the total discharge:

$$FWMC = \frac{\sum C_i \cdot Q_i}{\sum Q_i} , \tag{13.10}$$

where FWMC is the flow–weighted mean concentration, and load is calculated as the product of FWMC and the total amount of stream flow during a specified period of time (e.g., annually):

$$L = FMWC \cdot Q_t \ . \tag{13.11}$$

Averaging methods are limited in accuracy, and these methods should only be used to estimate loads when concentrations are not correlated with stream flow or other exogenous variables that show temporal variability. The standard error of the average concentration may estimate the variability of this technique and the precision of the load estimate (Aulenbach and Hooper, 2006).

13.4.2 Period-Weighted Methods

The period-weighted methods basically apply mass integration to datasets that have continuous discharge but do not have continuous concentration data; therefore, assumptions must be made about unknown concentrations in between collected water samples and measured concentrations. The period-weighted methods use a variety of techniques to estimate concentration between the measured data points, including linear interpolation between measured concentrations, measured concentrations applied to the mid-point between consecutive measurements, and measured concentrations applied forward or backward until the next measured concentration. No matter the integration technique, this method requires high-frequency sampling and analysis of many water samples to precisely estimate loads, especially during storm events. This

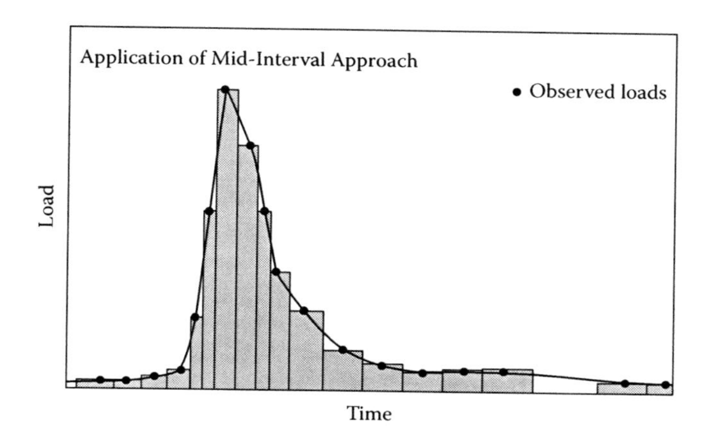

FIGURE 13.12 Example of the mid-interval approach to estimating constituent loads using period-weighted methods on an individual storm event hydrograph (modified from Richards, R.P. 2001. User's guide for INTEGRATOR: A program to calculate pollutant loads by numeric integration including upper and lower bound load estimates; National Center for Water Quality Research, Heidelberg University, Ohio), where measured concentrations are applied to the mid-point between samples at regular intervals. (In this example, notice the missing data point where concentrations were only extended at the regular interval.)

technique uses all the measured concentration data from collected water samples, and it assumes that concentrations do not follow a preconceived model (Aulenbach and Hooper, 2006). An example the mid-interval approach is provided in Figure 13.12, where estimate loads are represented as the rectangles under observed instantaneous loads. These approaches are often adjusted to account for specific hydraulic conditions. For example, concentrations from water samples collected during base flow conditions might only be extended to the beginning of succeeding storm events, and concentration from the first water sample collected during the storm event would be extended backwards to the beginning of the event. These individual approaches might lead to some bias in load estimates, but some software programs (e.g., INTEGRATOR, Richards, 2001) attempt to standardize this approach to load estimation. Shih et al. (1998) have proposed a method to estimate the variance of load estimates using linear interpolation between measured concentrations, or piecewise linear interpolation, based on a semivariogram with coefficients calculated using a cross-validation technique.

13.4.3 Regression Models or Rating Curve Methods

Load estimates from period-weighted methods may be substantially improved upon using empirical models (e.g., regression models) of concentration that account for exogenous variables (e.g., flow or seasonality), when significant relations exists (Smart et al., 1999). The number of water samples and measured concentrations required to estimate loads using regression models, or rating curves, is less than that when using period-weighted methods, and these regression models can be applied to periods of time when samples were not collected (Robertson and Roerish, 1999).

In this method, regression models relating concentration to a continuously measured variable (e.g., discharge) are used to estimate a continuous concentration record at a given time interval (Cohn et al., 1989, 1992). These regression models are used to account for variations in concentration often associated with exogenous variables, such as discharge and seasonality. For example, studies have shown that concentrations of many constituents (such as major ions and nutrients) are significantly correlated to discharge across many different streams (e.g., Cohn et al., 1992; Coats et al., 2002; Vieux and Moreda, 2007). These regression models are often developed using log-transformed data to account for the general log-normal distribution of stream flow and water quality data.

The basic regression models used in load estimation can be represented with these equations:

$$\ln L_d = \beta_0 + \beta_1 \ln Q_d + \varepsilon, \quad (13.12)$$

$$\ln L_d = \beta_0 + \beta_1 \ln Q_d + \beta_2 T + \varepsilon,$$

$$\ln L_d = \beta_0 + \beta_1 \ln Q_d + \beta_2 \sin 2\pi T + \beta_3 \cos 2\pi T + \varepsilon,$$

$$\ln L_d = \beta_0 + \beta_1 \ln Q_d + \beta_2 \sin 2\pi T + \beta_3 \cos 2\pi T + \beta_4 T + \varepsilon.$$

where L_d is the daily load, Q_d is the mean daily discharge, T represents decimal time, the sine and cosine functions are used to adjust for seasonality, and ε represents the residual error; these regression models can be supplemented with additional terms such a lnQ_d^2 and T^2, and some software programs (e.g., LOADEST, Runkel et al., 2004) allow for split regression analysis across flow regimes or time periods. The daily loads are estimated using the selected regression model, and then summed to give monthly, seasonal, or annual load estimates. Regression models often are developed with load data transformed using logarithms, and this can lead to errors (typically underestimation) when load are retransformed to original units. Ferguson (1986) proposed a simple method to account for the transformation bias associated with taking the anti-log of estimated loads:

$$L'_d = e^{\ln L_d + \frac{\sigma^2}{2}}, \quad (13.13)$$

where L'_d is the retransformed daily load with bias correction, and σ^2 is the variance of the residuals of the regression model; this has been referred to as the quasi-maximum likelihood estimate (QMLE; Cohn et al., 1989). Another bias correction is the smearing method (Duan, 1983), which can be represented as

$$K_S = \frac{1}{n} \sum_{i=1}^{n} e^{\varepsilon_i}, \quad (13.14)$$

where K_S is the smearing estimate, n represents the number of observed data, and ε_i are the model residuals (deviations between individual observations and estimated loads). The smearing estimate is essentially the mean of the exponentiated residuals, and it does not require assumptions about the normality of the residuals from the regression model (Koch and Smillie, 1986). Cohn et al. (1989) presented a minimum variance unbiased estimator (MVUE) as an alternative to bias correction. MVUE estimates are difficult to calculate without the use of computer software programs (e.g., LOADEST, Runkel et al., 2004). Cohn et al. (1992) showed that the retransformation technique provided relatively accurate results in a simulation study. The LOADEST program provides load estimates based on three different methods: Maximum Likelihood Estimation (MLE), Adjusted Maximum Likelihood Estimation (AMLE), and Least Absolute Deviation (LAD). MLE is used to fit parameters of the $\ln(L)=f\{Q,T\}$ model and then MVUE is used to adjust for retransformation bias. AMLE is used to fit the model (the parameters themselves are adjusted for bias) and also to adjust for retransformation bias. LAD is used to fit the model and Duan's smearing estimator is used to adjust for retransformation bias. LAD is only computed for completely uncensored data sets because the smearing estimator requires a residual for each observation, and these are not defined for censored values. The load estimation methods available give similar results when the linear model selected is approximately correct, 30 or more observations are used in model calibration, and the water samples were collected to represent concentrations across the range of observed discharge (Cohn, 1995). Overall, the accuracy of load estimates from regression models are sensitive to how well the selected model truly represents changes in concentration, that is, regression models perform well when the concentration, discharge, and/or time regressions are accurate.

13.4.4 Composite Method

The composite method is different from typical log load models, as it regresses concentrations (not log transformed) against discharge. It is a hybrid between typical regression models and period-weighted methods (Aulenbach and Hooper, 2006), where this method adjusts the predicted concentration from the regression model to the measured concentrations when water samples were collected. The composite methods incorporate any structure present in the residual concentrations from the regression models (i.e., predicted minus measured concentrations), using these residuals to adjust the estimated loads between sampling times using piece-wise linear interpolation. This approach may improve load estimates when the residual concentrations express serial autocorrelatation, which suggest that structure exists in the residual concentrations; it has been suggested that this method improves load estimates when serial autocorrelation is 0.2 or greater. When estimating loads, the composite method adapts to short-term deviations in the regression models and is less sensitive to sampling deficiencies than period-weighted methods (Aulenbach and Hooper, 2006). This approach works well for watersheds with a relatively high frequency of data collection (e.g., at least weekly), and it might not be appropriate or applicable for sites sampled at a more typical frequency (e.g., monthly or every other month with supplemental storm chasing).

13.4.5 Ratio Estimators

Ratio estimators (developed by Bealle, 1962) generally use concentration and discharge to estimate an average load specific to the measured data (e.g., daily loads), and then adjust the average load proportionally based on the total discharge not represented in the measured data to estimate total load (e.g., annual load). The ratio estimators follow the form

$$L = Q_{mean}\frac{L_d}{Q_d}\left(\frac{1+\frac{1}{n}\cdot\frac{S_{LQ}}{Q_d L_d}}{1+\frac{1}{n}\cdot\frac{S_Q^2}{Q_d^2}}\right) \tag{13.15}$$

$$S_{LQ} = \frac{1}{n-1}\left(\sum_{i=1}^{n} Q_i L_i - nQ_d L_d\right) \tag{13.16}$$

$$S_Q^2 = \frac{1}{n-1}\left(\sum_{i=1}^{n} Q_i^2 - nQ_d^2\right), \tag{13.17}$$

where L represents the load, Q_{mean} is the mean daily discharge for the year, L_d and Q_d are the average daily load and discharge on days for which concentration data are available, n is the number of days where concentration was measured, S_{LQ} and S_Q^2 represent variance terms, and L_i and Q_i are the individual paired daily discharge and measured concentration (modified from Mukhopadhyay and Smith, 2000). While discharge is the most common exogenous variable used in ratio estimators, other factors that are measured more frequently or continuously may be used in this technique as well. Ratio estimators assume a linear relation passing through the origin between the measured loads and the factor (i.e., discharge) used in adjustment. Ratio estimators are often biased, but some techniques have been developed to reduce any bias associated with this load estimation method. Some studies (e.g., Dolan et al., 1981) have suggested that this approach was best for estimating average annual loads, although ratio estimators are less frequently used than the other methods in the recent literature.

13.5 TREND ANALYSES

The purpose of this section is to introduce the concepts and methods used to evaluate trends in water quality using various statistical techniques, where a trend is defined as a change in water quality over time. There are a myriad of climatic factors that influence water quality over multiple time scales (i.e., short-term, seasonal, and long-term) including precipitation and consequently discharge, temperature, and light availability. Climate variability influences the flow paths of various water quality parameters (e.g., chemical concentrations), groundwater and surface water contributions to stream flow, erosion processes across the landscape and within the fluvial channel, biological activity within the fluvial channel, etc. in complex and diverse ways. The fundamental difficulty in evaluating water quality trends is separating persistent changes in water quality from short-term fluctuations that occur due to climatic variability. Thus, the methods employed to evaluate trends need to account for climatic variations such that we can determine anthropogenic influences on water quality over time. This section touches on simple and more complex statistical approaches to measure changes in water quality.

Statistical methods to determine water quality trends vary from parametric approaches such as simple linear regression between water quality data (e.g., chemical concentrations) normalized for exogenous variables (e.g., stream flow, seasonality, etc.) over time to nonparametric approaches such as the Mann–Kendall Test that are considered distribution free methods. Each of these approaches has advantages and disadvantages. Parametric methods often more completely describe the trend and relation with explanatory variables (advantage), but these methods assume normally distributed and independent errors (disadvantage). Nonparametric methods require no assumptions of normally distributed errors (advantage), but these methods can be influenced by adjustment techniques for explanatory variables, and by censored values (i.e., data that is reported as less than the method detection limit), which are treated as equals, thereby limiting the ability to detect trends. Additionally, parametric methods might be better understood by regional stakeholders (e.g., policy makers, watershed groups, etc.) that have an interest in water quality changes over time.

The characteristics of water quality data (e.g., chemical concentrations) are generally complex showing large variations in the magnitude of constituents that are generally not normally distributed. These characteristics can result from multiple factors including variations in constituents with stream flow and time; variations in sampling frequency, missing data, and programmatic changes; censored data that vary over time with analytical methods, and from lab to lab (which was discussed earlier); and changes in methods such as sample collection, processing, and analyses. The key is to understand the water quality data using exploratory data analyses to help formulate initial conclusions about these water quality characteristics and trends. Furthermore, the trends detected in the water quality data represent simply the period of record used in the analyses, and this aspect should be clearly communicated when reporting trend results. The following subsections will focus on flow adjustment and accounting for seasonal variations while analyzing for trends in water quality constituent concentrations over time. There exist many publications in the scientific literature that may serve as additional resources when evaluating changes in water quality data over time using parametric and/or nonparametric techniques (e.g., see Hirsch et al., 1982, 1991; Alley, 1988; Berryman et al., 1988; El-Shaarawi, 1993; Esterby, 1996; Johnson et al., 2009).

13.5.1 Flow Adjustment of Water Quality Data

Factors other than time (e.g., stream discharge) often have a considerable influence on the water quality data (e.g., chemical concentration), and the variation in chemical concentrations with stream discharge needs to be removed to analyze the impacts of

TABLE 13.2
Examples of Empirical Models Used for Flow Adjustment of Concentrations

Model	General Description or Generic Example
Linear	$C = \beta_0 + \beta_1 \cdot Q + \varepsilon$
Multi–linear	$C = \beta_0 + \beta_1 \cdot Q + \beta_2 \cdot Q^2 + \varepsilon$
Log	$C = \beta_0 + \beta_1 \cdot \ln Q + \varepsilon$
Log–log	$\ln C = \beta_0 + \beta_1 \cdot \ln Q + \varepsilon$
Reciprocal	$C = \beta_0 + \beta_1 \cdot Q^{-1} + \varepsilon$
Hyperbolic	Several available options
LOWESS	Locally weighted scatterplot smoothing using data with or without transformation

Where C represents concentration, Q represents stream discharge, β represents various regression coefficients, and ε represents statistical deviations or noise in the observed data.

changes in the watershed over time. The removal process involves empirical modeling of how chemical concentrations change with stream discharge, and there are several techniques to accomplish this process (Table 13.2). These techniques vary from simple linear regression to log–log regression to LOcally WEighted Scatterplot Smoothing (LOWESS) of raw data or even log-transformed data. The purpose is to use the residual values from an adequate, empirical model as the flow-adjusted concentrations, so that the effects of discharge on chemical concentrations are removed.

When selecting empirical models relying on regression of data with and without transformation, several assumptions are made (Helsel and Hirsch, 2002):

1. The selected model for this particular data is correct, or adequate.
2. The variance of the residuals of this model is homeoscedastic (equal over the range of observed values).
3. The residuals are normally distributed and independent.

It has become common practice to log-transform water quality data, especially chemical concentrations and stream discharge, because this simple transformation often fits the inherent assumptions when using regression analyses (Richards and Baker, 2002). The basic premise is that the model selected to adjust for stream discharge needs to provide a good fit to the data, where there is no discernable pattern in the residual values or flow-adjusted concentrations over the range of observed concentration. Figure 13.13 show three flow adjustment models and the residual values or flow-adjusted concentrations.

Perhaps the most common method used to account for variations in chemical concentration with discharge is the use of LOWESS (Cleveland, 1979; Cleveland and Devlin, 1988) on log-transformed concentration (C) and discharge (Q). LOWESS combines the simplicity of local regressions to provide a nonlinear fit to the data based on the proportion of data influencing the local regression (i.e., sampling proportion; Cleveland, 1981), and its curve generally fits the concentration–discharge relation better than a fixed relation (e.g., log–log regression). Visual analysis to evaluate the fit of

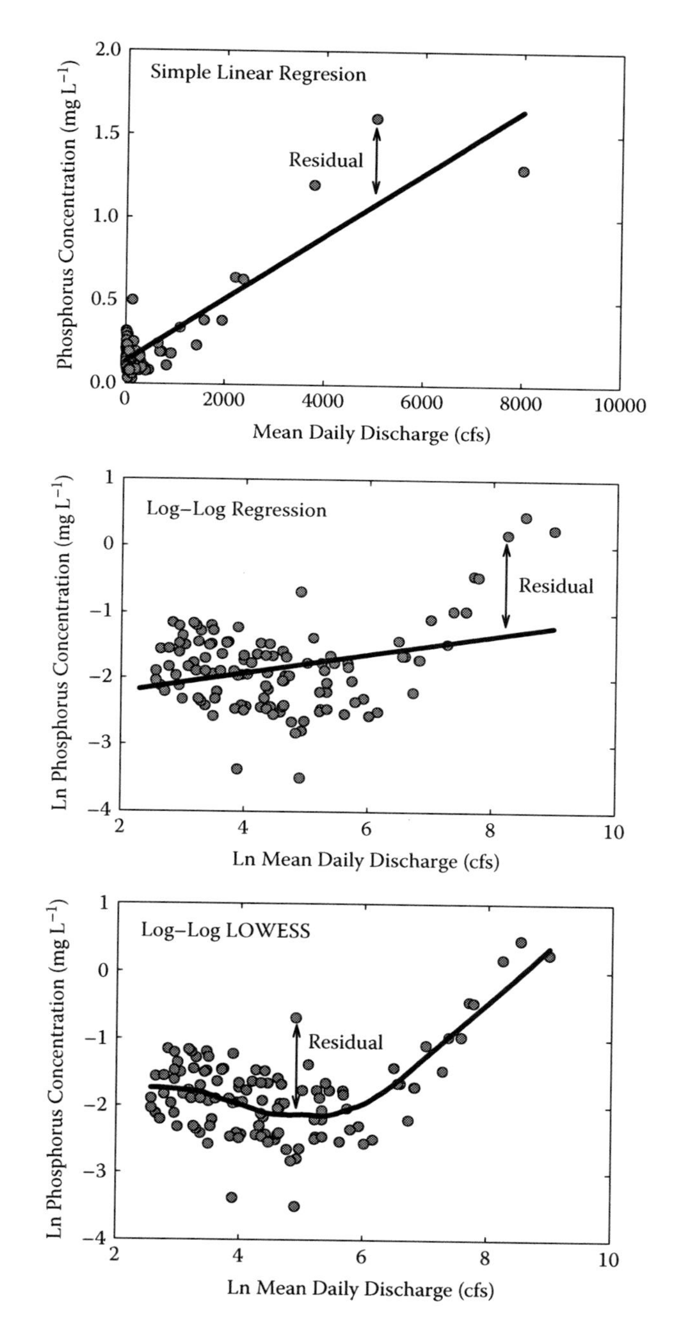

FIGURE 13.13 Three approaches (e.g., simple linear regression, log–log regression, and log–log LOWESS) that may be used to compute flow adjusted total phosphorus concentrations; data were from Spavinaw Creek near Cherokee, Oklahoma, from October 2001 through September 2008.

various estimated relations (e.g., LOWESS curves with different sampling proportions) is an effective method (Marron and Tsybakov, 1995). However, Bekele and McFarland (2004) used statistical optimization techniques and suggested that a sampling proportion of approximately 0.5 (which is the default in most statistical software packages) is adequate for reducing variability in chemical concentrations with stream discharge using LOWESS. Figure 13.14 shows that the LOWESS function using natural logarithm transformed data was the visual best option to account for variations in total phosphorus concentrations with discharge; this example used a sampling proportion of 0.5 for LOWESS. Several studies evaluating long-term trends in water quality data have used the LOWESS approach to adjust concentrations with changes in stream discharge (e.g., see Richards and Baker, 2002; White et al., 2004; Richards et al., 2008).

After selecting the appropriate flow-adjustment procedure, the residual values or flow-adjusted concentrations are then evaluated to see if significant changes occur over time. The parametric approach would involve using simple linear regression between the flow-adjusted concentrations and time, where the general assumptions of linear regression must be met. An example of this approach can be seen in Figure 13.14 where total phosphorus concentrations at Spavinaw Creek near Cherokee, Oklahoma, were flow adjusted using log–log LOWESS, and the flow-adjusted concentrations show a significant decrease over the time period evaluated. The flow-adjusted concentrations from any technique may also be used with non–parametric methods such as the Mann–Kendall Test.

13.5.2 Adjustment for Seasonal Variability in Water Quality Data

Seasonal changes in water quality data (e.g., chemical concentrations) are often a major source of short-term or annual variability, which can result from changes in

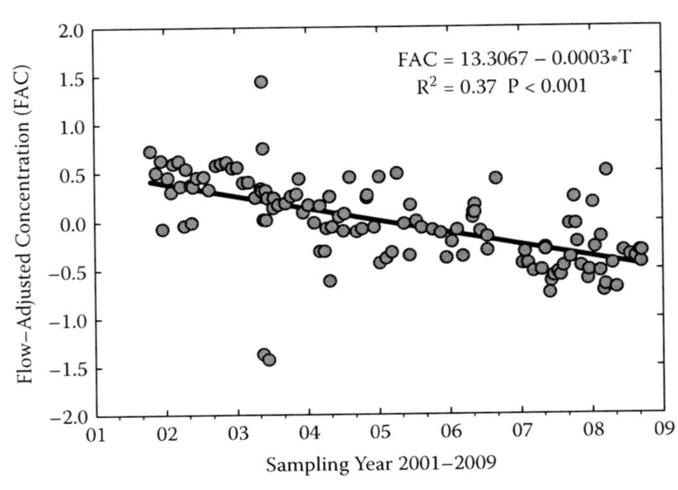

FIGURE 13.14 Flow–adjusted total phosphorus concentrations (FAC) over time (T) at Spavinaw Creek near Cherokee, Oklahoma, from October 2001 through September 2009; simple linear regression is used to evaluate changes in concentration over the time period.

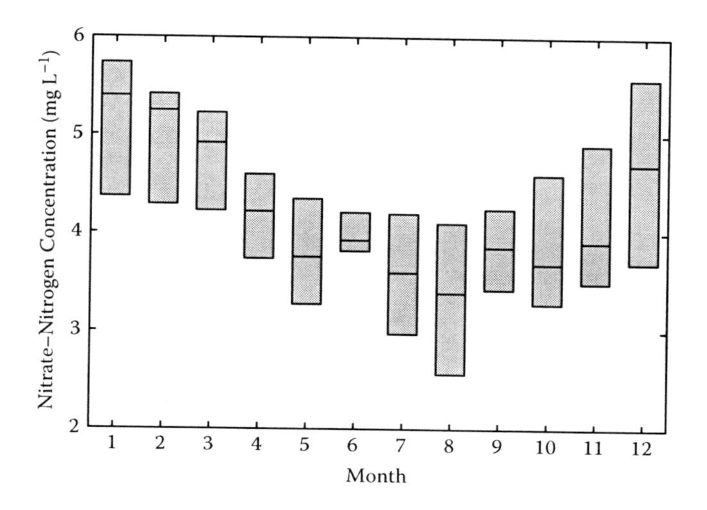

FIGURE 13.15 Nitrate–nitrogen concentrations during base flow conditions show a distinct seasonal pattern where concentrations are generally less from late spring through fall; data was from October 2002 through September 2008 at Spavinaw Creek near Cherokee, Oklahoma.

stream discharge and water source (e.g., snowmelt, groundwater, and/or surface runoff), biological activity, and anthropogenic activities (e.g., manure and/or fertilizer applications, recreation, and other human activities within the catchment). The concentrations of many parameters show some seasonal pattern, such as nitrate–nitrogen concentrations in streams as presented in Figure 13.15.

The basic premise is that accounting for such seasonal variations in water quality data increases the ability to detect changes over time or long-term trends. The two most common techniques to account for seasonal variation include stratification of the water quality data and the use of mathematical functions to approximate seasonal changes. Stratification involves separating the data into periods of time throughout the annual cycle, where these separate periods are expected to have similar characteristics (e.g., dormant and growing seasons). Many software programs may account for seasonality by stratification using dummy variables, representing the different periods of time that match the seasons in the regression. This approach has some advantages such as exploratory analyses, but one disadvantage is that the final model might not be adequate because stratification represents seasonality as step changes in water quality data. There are nonparametric approaches to trend analysis that evaluate changes in chemical concentrations across predefined seasons or data stratifications, e.g., modified Seasonal Kendall Test (Hirsch et al., 1982). The Seasonal Kendall Test evaluates water quality trends in pre-determined seasons (months, quarters of the year, etc.), where the focus is on increases or decreases in concentrations between consecutive years for each season. This technique does require that concentrations from multiple water samples collected during a month or season be collapsed down into one value, such as the median concentration. However, if there is a systematic change in sampling frequency such as shifting from monthly to quarterly (e.g., seasonal) water sampling then the seasons need to be defined based on the sampling minimum (e.g., see Helsel and Hirsch, 2002). For this example, Helsel and Hirsch (2002) would suggest that the monthly data closest to the mid-point of the defined seasons be used, not the median concentration during the season within a year.

Seasonality may also be accounted for in trend analyses using continuous functions like the Fourier Series, where decimal time is used in conjunction with sine and cosine variables:

$$Ln(C) = \beta_0 + \beta_1 T + \beta_2 \sin(2\pi T) + \beta_3 \cos(2\pi T) + \varepsilon \,, \tag{13.18}$$

where C represents concentration, T represents decimal time, β represents various regression constants, and ε represents residual error. Additional sine and cosine terms (e.g., $\beta_3\sin(4\pi T) + \beta_4\cos(4\pi T)$) can be added to help represent additional minima and maxima in the observed data over the cycle. This approach is relatively straightforward and easy to use, often fits most water quality data (e.g., chemical concentrations), and represents a well-accepted technique. However, this approach does require that the normality assumption in linear regression be met, and it does not fit the seasonal patterns observed in all water quality data (e.g., pesticides). An alternative to trigonometric functions in representing seasonal variability in select water quality data is the Seasonal Wave as described by Vecchia et al. (2008); this technique is mathematically complex but relatively easy to use within applicable software programs (e.g., S–Plus with the USGS library). Overall, the selection of the appropriate method to account for seasonal variation in water quality data is critical to increasing the power in detecting long-term trends.

13.5.3 Monotonic and Step Changes in Trend Analyses

Many of the techniques in trend analyses, whether parametric or nonparametric approaches, should only be applied to situations where water quality is changing monotonically over time. This type of change can be gradual or more sudden, but it is systematic over the period of record. However, many changes in water quality from various human activities occur at some specified time during the study period, for example, a municipal wastewater treatment plant decreases its effluent phosphorus concentrations discharge to the receiving stream. Step trends should be evaluated under two specific cases: (1) the data records represent two distinct periods of time with a relatively long gap in between, and (2) an abrupt change in water quality management within the watershed occurs within the period of record (Hirsch et al., 1991). Hirsch et al. (1991) further suggest that the decision to use step trends should be made prior to any examination of the data, or based on some prior knowledge of changes within the watershed that influenced water quality. The basic approaches to evaluating step trends are based on the assumption that the data from the different periods of record represent different populations, and these differences in the location (mean or median) of the populations can be evaluated using parametric procedures as simple as the two sample t test or nonparametric procedures such as the Wilcoxon–Mann–Whitney test.

The methods described for monotonic trends (e.g., simple linear regression or the Kendall's τ approach) may still suggest that there is a significant change in water quality over time. But, the slope of the overall trend across the period of record may not represent the magnitude of the increase or decrease following a specific or step change in water quality. Several studies have used LOWESS to evaluate changes in water quality over time, and the LOWESS fit may qualitatively suggest increases and/or decreases over different time periods (e.g., Robson and Neal, 1996; Reynolds et al., 1997; Renwick et al., 2008). The LOWESS fit over time series data may better portray the nonlinearity of the trends throughout the period of record.

13.6 PRINCIPAL COMPONENT ANALYSIS

Principal component analysis (PCA) is a procedure that takes a large, multi-variable dataset and evaluates this dataset with the goal of reducing the dimensionality while retaining as much of the variation as possible. PCA results in identification of latent variables or principal components that can be used to describe the dataset but cannot be directly measured. Latent variables in water quality analysis might be allochthonous effect, nature and development of biomass effects (Parinet et al., 2004), biological activity effect, dilution effect (Petersen et al., 2001), discharge effect, seasonal effect, waste water effect (Haag and Westrich, 2002), or another phenomenon that is difficult or impossible to measure directly. PCA has been used since the early 1900s to evaluate large datasets in a variety of disciplines (Jolliffe, 2002). It has recently become more common place in evaluating large water quality datasets. This is likely due to the increasing numbers of water quality sampling sites and constituents sampled and the inability of other tools to adequately characterize such large datasets.

PCA is conducted through a multistep process. The first step is to identify the dataset and the variables for analysis. The different variables included in the analysis are likely to represent different types of measurements with different units. This bias can be minimized by standardizing each variable as follows:

$$S_i = \frac{X_i - \overline{X}}{s}, \tag{13.19}$$

where S_i is the standardized value of X_i and s is the standard deviation. The new standardized dataset is input into a software program that will calculate the covariance matrix. Alternatively, a correlation matrix can be used instead of standardizing the dataset and calculating a covariance matrix. The use of standardization or the correlation matrix is based on the assumption that all variables are considered equally important (Chartfield and Collin, 1980).

The matrix is then decomposed (using computer software as the calculations are tedious) into a scores matrix and a loadings matrix by determining the eigenvectors and eigenvalues (Meglen, 1991). Eigenvectors and eigenvalues represent the direction and length, respectively, of each principal component (Burstyn, 2004). Each principal component is a new axis or dimension in space that is derived from linear combinations of the original variables. These combinations are common in water quality datasets as many variables are characterized by natural associations with other measured variables. Thus, PCA is an attractive technique for evaluating water quality datasets.

The eigenvalue divided by the number of variables in the analysis provides the percentage of variability explained by a principal component (Burstyn, 2004). Thus, eigenvalues are oftentimes presented as a percentage. The number of principal components generated from the matrix is equal to the number of variables in the original dataset. Principal components and their eigenvalues and eigenvectors can be presented visually for a better understanding. The first principal component is the new dimension (or latent variable) that includes the most variability. Each remaining principal component represents a dimension with variability less than the previous principal component (Burstyn, 2004).

An example dataset was evaluated for PCA and eigenvalues are provided in Table 13.3. This dataset consisted of 12 variables that were standardized before evaluating. The 12 variables were nitrate plus nitrite, ammonium, total nitrogen, total inorganic nitrogen, total organic nitrogen, total phosphorus, soluble reactive phosphorus, chlorophyll-a, total organic carbon, temperature, dissolved oxygen, and turbidity. The next step in PCA is to identify the "important" principal components. There are different recommendations for this procedure. Burstyn (2004) suggested two approaches: (1) select principal components that have eigenvalues greater than 1.0 or (2) evaluate a plot of principal component numbers versus eigenvalue (i.e., a scree plot) and identify the location where the plot flattens out. Considering these two techniques for the eigenvalues presented in Table 13.3, the first 6 principal components were selected as important.

PCA also returns loading values for all variables for each principal component; loadings for the first 6 principal components are presented in Table 13.4. Loadings

TABLE 13.3
Eigenvalues for the 12 Principal Components from a Principal Component Analysis

Principal Component	Eigenvalue
PC1	1.658
PC2	1.585
PC3	1.372
PC4	1.146
PC5	1.034
PC6	0.938
PC7	0.849
PC8	0.788
PC9	0.667
PC10	0.610
PC11	0.007
PC12	0.002

TABLE 13.4
Loadings for the First 6 Principal Components from a Principal Component Analysis

Variable	PC1	PC2	PC3	PC4	PC5	PC6
Nitrate plus nitrite	0.182	0.454	−0.202			0.117
Ammonium	0.397	0.345	−0.182	−0.184		−0.111
Total nitrogen	0.487	−0.137		0.419	−0.167	
Total inorganic nitrogen	0.373	0.443	−0.217	−0.146		
Total organic nitrogen	0.420	−0.237	0.122	0.463	−0.173	
Total phosphorus		−0.299	−0.459	−0.107		−0.158
Soluble reactive phosphorus		−0.191	−0.289	−0.300	−0.477	−0.132
Chlorophyll-a		−0.311	−0.484		0.134	
Total organic carbon	0.300	−0.305	−0.169	−0.147	0.142	
Temperature	0.349	−0.269	0.257	−0.419	0.419	0.670
Dissolved oxygen	−0.190	0.102	−0.461	0.467	−0.113	0.661
Turbidity			−0.154	0.182	0.796	−0.171

represent the correlation of each variable with the respective principal component. Thus, loadings provide a measure of how each variable contributes to each principal component. The loadings are directly related to the eigenvalue, as the sum of the squared loadings equals the variance explained by each the particular principal component (Dunteman, 1989).

The most common procedure following PCA is factor analysis (FA). FA, similar to PCA, is also computationally challenging and therefore is computed with the assistance of computer software. FA is conducted using the number of principal components selected as "important" from the PCA. The selected principal components are then subjected to an axis rotation that allows for easier interpretation of the data. The most commonly used rotation is the varimax rotation. The objective for rotating the axis is to maximize the sum of the variance of the loading vectors. The result should be that variables will be heavily loaded on as few principal components as possible (Mazlum et al., 1999). FA was performed on the example dataset and results are presented in Table 13.5. Comparison of principal component loadings and factor loadings indicate that factor loadings do provide an easier quick assessment of which variables contribute the greatest loadings by factor. This can be seen when comparing PC1 (from Table 13.4) and factor 1 (from Table 13.5). The loadings for PC1 range from −0.190 to 0.487 with eight different variables contributing. The loadings for factor 1 range from 0.178 to 0.990 with four different variables contributing. Thus, factor loadings for factor 1 indicate that the first factor represents a latent variable that is a product of predominately ammonium and total inorganic nitrogen with two additional components of lesser importance—nitrate plus nitrite and total nitrogen. Results of the factor analysis also have total phosphorus, soluble phosphorus, and chlorophyll-a's greatest loadings as contributing to factor 3, which could be identified as the limiting nutrient latent variable effect.

TABLE 13.5
Factor Loadings for the First 6 Principal Components from a Principal Component Analysis

Variable	Factor 1	Factor 2	Factor 3	Factor 4	Factor 5	Factor 6
Nitrate plus nitrite	0.421		–0.121	0.886	0.114	
Ammonium	0.990	0.102				
Total nitrogen	0.178	0.971				0.137
Total inorganic nitrogen	0.917			0.386		
Total organic nitrogen		0.987				0.139
Total phosphorus			0.787			
Soluble reactive phosphorus			0.344	–0.155		
Chlorophyll-a			0.746			0.195
Total organic carbon		0.202	0.322		–0.105	0.830
Temperature		0.213			–0.275	0.289
Dissolved oxygen			0.154		0.892	
Turbidity			0.130			

13.7 SUMMARY

Methods for statistically evaluating water quality datasets are actively being used. Depending on the methodology applied, statistical results can be misleading. This chapter introduces many concepts that should be included in statistical analysis of a water quality dataset. Water quality datasets often differ from other scientific data as they contain censored data, may be characterized by seasonal variability, and are likely not normally distributed and therefore require special attention for proper evaluation. Most importantly, water quality data statistical results should be assessed to ensure their appropriateness, particularly if results will be used for management and planning of water resources.

REFERENCES

Alley, W.M. 1988. Using exogenous variables in testing for monotonic trends in hydrologic time series. *Water Resources Research* 24:1955–1961.

Antweiler, R.C., and H.E. Taylor. 2008. Estimating mean concentrations under transformation for environmental data with detection limits. *Technometrics* 31(3):347–356.

Aulenbach, B.T., and R.P. Hooper. 2006. The composite method: An improved method for stream-water solute load estimation. *Hydrological Processes* 20: 3029–3047.

Bealle, E.M.L. 1962. Some uses of computers in operational research. *Industrielle Organisation* 31:51–52.

Bekele, A., and A. McFarland. 2004. Regression-based flow adjustment procedures for trend analyses of water quality data. *Transactions of the American Society of Agricultural Engineers* 47:1093–1104.

Berryman, D., B. Bobee, D. Cluis, and J. Haemmerli. 1988. Nonparametric tests for trend detection in water quality time series. *Water Resources Bulletin* 24:545–556.

Burstyn, I. 2004. Principal component analysis is a powerful instrument in occupational hygiene inquiries. *Annals of Occupational Hygiene* 48(8):655–661.

Chartfield, C., and A.J. Collin. 1980. *Introduction to Multivariate Analysis*. Chapman Hall in Association with Methuen, Inc.: New York.

Cleveland, W. S. 1979. Robust locally weighted regression and smoothing scatter plots. *Journal of the American Statistical Assoc.* 74(368): 829–836.

Cleveland, W.S. 1981. LOWESS: A program for smoothing scatterplots by robust locally weighted regression. *The American Statistician* 35:54.

Cleveland, W.S. and S.J. Devlin. 1988. Locally weighted regression: An approach to regression analysis by local fitting. *Journal of the American Statistical Association* 83: 596–610.

Coats, R., F. Liu, and C.R. Goldman. 2002. A Monte Carlo test of load calculation methods, Lake Tahoe Basin, California–Nevada. *Journal of the American Water Resources Association* 38:719–730.

Cohen, A. C. 1959. Simplified estimators for the normal distribution when samples are singly censored or truncated. *Technometrics* 1:217–237.

Cohn, T.A. 1995. Recent advances in statistical methods for the estimation of sediment and nutrient transport in rivers. *Reviews of Geophysics* (Supplement) 33:1117–1123.

Cohn, T.A. 2005. Estimating contaminated loads in rivers: An application of adjusted maximum likelihood to type I censored data. *Water Resources Research* 41, W07003, doi:10.1029/2004WR003833.

Cohn, T.A., L.L. Delong, E.J. Gilroy, R.M. Hirsch, and D.K. Wells. 1989. Estimating constituent loads. *Water Resources Research* 25:937–942.

Cohn, T.A., D.L. Caulder, E.J. Gilroy, L.D. Zynjuk, and R.M. Summers. 1992. The validity of a simple statistical model for estimating fluvial constituent loads: An empirical study involving nutrient loads entering Chesapeake Bay. *Water Resources Research* 28:2353–2363.

Dolan, D. M., A. K. Yui, and R. D. Geist. 1981. Evaluation of river load estimation methods for total phosphorus. *Journal of Great Lakes Research* 7:207–214.

Duan, N. 1983. Smearing estimate—a nonparametric retransformation method. *Journal of the American Statistical Association* 78:605–610.

Dunteman, G.H. 1989. *Principal Components Analysis*. U.K.: SAGE University Press.

Efron, B. 1981. Nonparametric estimates of standard error: The jackknife, the bootstrap and other methods. *Biometrika* 68:589–599.

El-Shaarawi, A.H. 1993. Environmental monitoring, assessment and prediction of change. *Environmetrics* 4:381–398.

El-Shaarawi, A.H., P.B. Kauss, M.K. Kirby, and M. Walsh. 1989. Inferences about the variability of means from censored data. *Environmental Monitoring and Assessment* 12: 295–304.

Esterby, S.R. 1996. Review of methods for the detection and estimation of trends with emphasis on water quality applications. *Hydrological Processes* 10:127–149.

Ferguson, R.I. 1986. River loads underestimated by rating curves. *Water Resources Research* 21:74–76.

Gilliom, R.J., and D.R. Helsel. 1986. Estimation of distributional parameters for censored trace level water quality data, 1., estimation techniques. *Water Resources Research* 22:135–146.

Haag, I., and B. Westrich. 2002. Processes governing river water quality identified by principal component analysis. *Hydrological Processes* 16:3113–3130.

Haan, C. 2002. *Statistical Methods in Hydrology*. 2nd ed. Ames, Iowa: Iowa State Press.

Haggard, B.E., T.S. Soerens, W.R. Green, and R.P. Richards. 2003. Using regression methods to estimate stream phosphorus loads at the Illinois River, Arkansas. *Applied Engineering in Agriculture* 19:187–194.

Helsel, D.R. 2005. *Nondetects and Data Analysis, Statistics for Censored Data.* Wiley-Interscience, John Wiley & Sons.

Helsel, D.R. 2006. Fabricating data: How substitution values for nondetects can ruin results, and what can be done about it. *Chemosphere* 65:2434–2439.

Helsel D.R., and T.A. Cohn. 1988. Estimation of descriptive statistics for multiply censored water quality data, *Water Resour. Res.* 24(12):1997–2004.

Helsel, D.R., and R.M. Hirsch. 2002. Statistical Methods for Water Resources, Chapter A3, Techniques of Water-Resources Investigations of the United States Geological Survey, Book 4, Hydrologic Analysis and Interpretation, http://water.usgs.gov/pubs/twri/twri4a3/.

Helsel, D.R., and R.M. Hirsch. 2002. Techniques of Water-Resources Investigations of the United States Geological Survey. Book 4, Hydrologic Analysis and Interpretation. Chapter A3 Statistical Methods in Water Resources. United States Geological Survey.

Hewett, P., and G.H. Ganser. 2007. A comparison of several methods for analyzing censored data. *Annals of Occupational Hygiene* 51(7):611–632.

Hirsch, R.M., R.B. Alexander, and R.A. Smith. 1991. Selection of methods for the detection and estimation of trends in water quality. *Water Resources Research* 5(5): 803–813.

Hirsch, R.M., J.R. Slack, and R.A. Smith. 1982. Techniques of trend analyses for monthly water quality data. *Water Resources Research* 18:107–121.

Johnson, H.O., S.C. Gupta, A.V. Vecchia, and F. Zvomuya. 2009. Assessment of water quality trends in the Minnesota River using non-parametric and parametric methods. *Journal of Environmental Quality* 38:1018–1030.

Jolliffe, I.T. 2002. *Principal Component Analysis.* Springer, New York.

Koch, R.W., and G.M. Smillie. 1986. Bias in hydrologic prediction using log-transformed regression models. *Water Resources Bulletin* 22:717–723.

Kroll, C.N., and J.R. Stedinger. 1996. Estimation of distributional moments and quantiles for censored data. *Water Resources Research* 32:1005–1012.

Marron, J.S., and A.B. Tsybakov. 1995. Visual error criteria for qualitative smoothing. *Journal of the American Statistical Association* 90:499–507.

Mazlum, N., A. Ozer, and S. Mazlum. 1999. Interpretation of water quality data by principal components analysis. *Turkish Journal of Engineering and Environmental Sciences* 23:19–26.

McBride, G.B. 2005. *Using Statistical Methods for Water Quality Management: Issues, Problems and Solutions.* New Jersey: John Wiley & Sons.

Meglen, R.R. 1991. Examining large databases: A chemometric approach using principal component analysis. *Journal of Chemometrics* 5: 163–179.

Mukhopadhyay, B., and E.H. Smith. 2000. Comparison of statistical methods for estimation of nutrient load to surface reservoirs for sparse data set: Application with a modified model for phosphorus availability. *Water Research* 12:3258–3268.

Parinet, B., A. Lhote, and B. Legube. 2004. Principal component analysis: An appropriate tool for water quality evaluation and management—application to a tropical lake system. *Ecological Modelling* 178:295–311.

Petersen, W., L. Bertino, U. Callies, and E. Zorita. 2001. Process identification by principal component analysis of river water-quality data. *Ecological Modelling* 138:193–213.

Petersen, J. C. 1992. Trends in stream water quality data in Arkansas during several time periods between 1975 and 1989. Water Resources Investigations Report 92–4044. Little Rock, Ark.: U.S. Geological Survey.

Preston, S. D., V. J. Bierman Jr., and S. E. Silliman. 1989. An evaluation of methods for the estimation of tributary mass loads. *Water Resources Research* 25:1379–1389.

Rao, S.T., J.Y. Ku, and K.S. Rao. 1991. Analysis of toxic air contaminant data containing concentrations below the limit of detection. *Journal of the Air and Waste Management Association* 41:442–448.

Renwick, W.H., M.J. Vanni, Q. Zhang, and J. Patton. 2008. Water Quality Trends and Changing Agricultural Practices in a Midwest U.S. Watershed, 1994–2006. *Journal of Environmental Quality* 37:1862–1874.

Reynolds, B., M.R. Enshaw, T.H. Sparks, S. Crane, S. Hughes, S.A. Brittain, and V.H. Kennedy. 1997. Trends and seasonality in stream water chemistry in two Moorland catchment of the Upper River Wye, Plynlimon. *Hydrology and Earth System Science* 1:571–581.

Richards, R.P. 2001. User's guide for INTEGRATOR: A program to calculate pollutant loads by numeric integration including upper and lower bound load estimates. National Center for Water Quality Research, Heidelberg University, Ohio.

Richards, R.P., and D.B. Baker. 2002. Trends in Water Quality in LEASEQ Rivers and Streams (Northwestern Ohio), 1975–1995. *Journal of Environmental Quality* 31:90–96.

Richards, R.P., D.B. Baker, J.P. Crumrine, J.W. Kramer, D.E. Ewing, and B.J. Merryfield. 2008. Thirty-year trends in suspended sediment in seven Lake Erie tributaries. *Journal of Environmental Quality* 37:1894–1908.

Robertson, D.M. 2003. Influence of differing temporal sampling strategies on estimating total phosphorus and suspended sediment concentration and transport in small streams. *Journal of the American Water Resources Association* 39:1281–1308.

Robertson, D.M., and E.D. Roerish. 1999. Influence of various water quality sampling strategies on load estimates for small streams. *Water Resources Research* 35:3747–3759.

Robson, A.J., and C. Neal. 1996. Water quality trends at an upland site in Wales, UK, 1983–1993. *Hydrological Processes* 10:183–203.

Runkel, R.L., C.G. Crawford, and T.A. Cohn. 2004. Load Estimator (LOADEST): A FORTRAN program for estimating constituent loads in streams and rivers. U.S. Geological Survey Techniques and Methods, Book 4, Chapter A5, 69 p.

She, N. 1997. Analyzing censored water quality data using a non-parametric approach. *Journal of the American Water Resources Association* 33: 615–624.

Shih, G., X. Wang, H.J. Grimshaw, and J. Van Arman. 1998. Variance of load estimates derived by piecewise linear interpolation. *Journal of Environmental Engineering* 124:1114–1120.

Shumway R.H., R.S. Azari, and M. Kayhanian. 2002 Statistical approaches to estimating mean water quality concentrations with detection limits. *Environmental Science and Technology* 36: 3345–53.

Slack, J.R., D.L. Lorenz, and others. 2003. USGS library for S-PLUS for Windows, release 2.1. U.S. Geological Survey Open-File Report 03–357. http://water.usgs.gov/software/library.html.

Smart, T.S., D.J. Hirst, and D.A. Elston. 1999. Methods for estimating loads transported by rivers. *Hydrology and Earth System Sciences* 3:295–303.

Spooner, J. 1991. Censored data values: Description and eVect of censoring on statistical trend analyses. NWQEP NOTES. The NSCU Water Quality Group Newsletter No. 4. http://www.bae.ncsu.edu/programs/extension/wqg/issues/48.html.

Toor, G.S., R.D. Harmel, B.E. Haggard, and G. Schmidt. 2008. Evaluation of regression methodology with low-frequency sampling to estimate constituent loads for ephemeral watersheds in Texas. *Journal of Environmental Quality* 37:1847–1854.

USEPA, 2007. Report of the Federal Advisory Committee on Detection and Quantification Approaches and Uses in Clean Water Act Programs; http://www.epa.gov/waterscience/methods/det/faca/final-report-200712.pdf.

Vecchia, A.V., J.D. Martin, and R.J. Gilliom, 2008. Modeling Variability and Trends in Pesticide Concentrations in Streams. *Journal of the American Water Resources Association* 44:1308–1324.

Vieux, B.E., and F.G. Moreda. 2007. Nutrient loading assessment in the Illinois River using a synthetic approach. *Journal of the American Water Resources Association* 39:757–769.

Wendelberger, J., and K. Campbell. 1994. Non-detect data in environmental investigations. Los Alamos National Laboratory. LA-UR-94-1856.

White, K.L., B.E. Haggard, and I. Chaubey. 2004. Water Quality at the Buffalo National River, Arkansas, 1991–2001. *Transactions of the American Society of Agricultural Engineers* 47:407–417.

Zamyadi, A., J. Gallichand, and M. Duchemin. 2007. Comparison of methods for estimating sediment and nitrogen loads from a small agricultural watershed. *Canadian Biosystems Engineering* 49:127–136.

14 Examples of Water Quality Monitoring

Qingren Wang and Yuncong Li

CONTENTS

14.1 INTRODUCTION

The objectives of this chapter are to provide examples of water quality monitoring (instrumentations, procedures, and techniques) in relation to general environmental concerns. The water quality monitoring experiences of the University of Florida's Tropical Research and Education Center (TREC) serve as the basis for many of these examples. These examples are intended to provide a practical model to those who are new to monitoring water quality, as well as to provide additional information to improve the techniques and efficiencies in monitoring water quality in ongoing projects. Although water quality monitoring can encompass a wide range of water body types (e.g., streams, lakes, oceans, aquifers), this chapter focuses on monitoring surface water quality of canals and wetlands. Readers may apply the approaches presented here to similar projects, provided that fundamental requirements are met. Since projects often differ widely and since environmental systems differ from one setting to another, it is impossible to find an example that can serve as a perfect model for each specific new project. Therefore, practical procedures included in this chapter can be used as a guideline for developing individual water quality monitoring projects. This chapter covers the following topics: development of standard operating procedures (SOPs); site selection and station establishment; environmental parameters to be monitored; essential instruments; assembly and installation of equipment; program development and application; state-of-the-art technology, such as the telecommunication system; sampling and site maintenance; transportation; chemical analysis; and data processing. Detailed sampling guidelines and discussions can be found in Chapter 5 for surface water, Chapter 6 for groundwater, Chapter 7 for pore water, and Chapter 11 for emerging contaminants.

14.2 EXAMPLE FOR MONITORING WATER QUALITY IN CANALS

Components for a water quality monitoring program of canals are presented in this section.

14.2.1 Developing a Standard Operating Procedure (SOP)

The standard operating procedure (SOP) consists of detailed written instructions to achieve uniformity of performance in sample collection, handling, processing, analysis, and documentation. Because one project varies from another, a site-specific SOP should be developed before starting a project. A series of SOPs for water quality monitoring have been developed by the USEPA (2003, 2007) and FDEP (2008), such as FS 2000 and FS 2100, which are available online (http://epa.gov/quality/sops.html). These SOPs may serve as examples or boilerplate documents as SOPs differ to some extent for each project, depending on the purpose of the sampling, parameters to be measured, type of statistical analysis to be performed, and other project-specific details. Additional guidance on SOPs is provided in Chapter 4. The SOPs related to examples explored in this chapter are provided in Appendix 14.1. Specific equipment and commercial products mentioned should be considered as examples only and not recommendations.

14.2.2 Site Selection and Station Establishment

After the SOP for canal water quality monitoring has been developed, appropriate monitoring sites should be selected to meet the project objectives. Once a site has been selected, a structure is erected to house sampling equipment and provide sampling personnel access. For canal sampling in south Florida, an ideal structure is a wooden dock.

Such docks were constructed for a monitoring project conducted by the University of Florida in south Florida. For the project, the canal sampling dock was a 1.5 × 1.5 m platform with 1-m high metal hand railing and 50 cm wide walkway. The platform was used to support monitoring instruments including a metal cabinet for an automatic sampler and accommodate activities such as sampling, instrument installation, and maintenance. A dock walkway provided access to the platform from the canal bank. The platform was designed to be at least 50 cm above the highest expected water level (Figure 14.1).

Dock platforms were modified to secure Data Sonde field monitoring equipment. Platforms were retrofitted for this using 10-cm diameter PVC pipes extending 1 m

FIGURE 14.1 Building a surface water sampling dock and platform at a canal site.

FIGURE 14.2 A PVC pipe with holes at the bottom to let water run through, for Hydrolab installation on a platform at a canal sampling site.

above the platform. Prior to installation, holes (~2 cm diameter) positioned below the water surface were drilled into the PVC to allow water to pass through (Figure 14.2). The PVC was driven into the canal sediment and securely fastened to the platform. Data Sonde sensors were placed in the PVC and secured with a locking cap. For an automated flow proportional sampling, the velocity and stage sensor was attached to a 5-cm diameter metal bar that was driven into the canal sediment and protruded vertically through the water, adjacent to the edge of the platform, and facing the opposite canal bank. A manual stage reading device (staff gauge) was also installed for verification of equipment accuracy. A metal pole extending about 3 m above the platform was installed to support solar power panels.

14.2.3 Flow Proportional Sampling for Water Quality Monitoring

As detailed in Chapter 5, automated flow proportional sampling requires a flow meter, automatic sampler, and data logger that communicate with each other. However, since the transverse section of a canal usually has an irregular shape, the flow velocity differs substantially in different sections of the canal. Thus, the flow rate must be calculated as the total volume of water passing through the measuring section of

FIGURE 14.3 The SonTek RiverCAT for canal profile measurement.

the canal per unit time. This total volume of water passing the transverse sampling section is the product of water velocity and cross sectional area of the transverse section. The cross sectional area can be calculated based on canal profile measurements. The water level can be automatically measured by an electronic depth sensor as a separate instrument or combined with a flow meter.

There are different methods for measuring a canal profile. One common method is the use of an acoustic Doppler current profiler (ADCP) system, which emits preprogrammed ultrasonic beams for measuring speed, direction, and depth of the water current. ADCP-based equipment, such as RiverCat or RiverSurveyor, can be attached to a boat and dragged slowly from one side of the canal to the other to measure the canal profile (Figure 14.3). The output data from such instruments can be used to delineate the cross-sectional area for estimating the total volume of flow (by measuring stage and velocity) in order to conduct the flow proportional sampling.

14.2.4 Instrument Assembly

Instrument requirements depend on project objectives and water quality parameters to be measured. The example presented here focused on total phosphorus (TP) with flow-weighted sampling in a canal located in south Florida. Electrical power was supplied by solar panels connected to 12 V rechargeable batteries. Other water quality parameters measured in the field (see Chapter 8) included pH, EC, dissolved oxygen (DO), specific conductivity, and temperature.

The following instruments and supplies were used:

1. Flowmeter: SonTek Argonaut-SL has three ultrasonic beams: 2 horizontally-projected beams for measuring water velocity and 1 vertically projected beam for measuring the water level.
2. Data logger: CR-10 X (Campbell Scientific Inc., 815 W 1800 N, Logan, UT 84321). The data loggers require a site-specific program to record data, calculate, and operate the automatic sampler. A flow proportional sampling

program for Campbell Scientific CR-10 X was developed for this project and is provided in Appendix 14.2 as an example.
3. Hydrolab: MS 5 (Hach Environmental Inc., P.O. Box 389, Loveland, CO 80539).
4. Auto-sampler: ISCO 6700 (AMJ Equipment Corp., 5101 Great Oak Drive, Lakeland, FL 33815).
5. Solar panel: Sunsei 12 V solar battery charger (ICP Global Technologies Inc., www.sunsei.com).
6. Digital modem: Raven GPRS G3210 (AirLink Communications Inc., www.airlink.com).

All the above instruments were connected to the data logger with an SDI-12 (series data interface), which was powered by the solar panel connected to two rechargeable 12 V batteries that provided power during the night or cloudy periods of the day. The data logger has additional ports to supply different voltages to instruments as required. The power requirements of each instrument must be carefully checked during assembly. A metal cabinet with a side opening door was installed to hold an auto-sampler and a weather-proof plastic box was used to protect the data logger and modem. Additional information on instrument requirements and their assembly for an automatic monitoring station setup is provided in Wang et al. (2004).

14.3 EXAMPLE FOR MONITORING WATER QUALITY IN WETLAND

There are many similarities between canal and wetland water quality monitoring but some specific sampling techniques and precautions should be adopted to successfully implement a wetland water quality monitoring project. Features common to both canal and wetland water quality monitoring that have already been described will be omitted in this section.

14.3.1 DEVELOPMENT OF A SOP

General requirements and protocols for most parts of the SOP developed for canal water quality monitoring are applicable to wetlands. However, a wetland monitoring SOP should include precautions for transport safety and dealing with wildlife, as well as the specialized application of certain sampling and monitoring equipment.

14.3.2 TRANSPORTATION AND SITE SETUP

In many cases the main issue to be addressed for wetland water quality monitoring is transportation to the monitoring site. An airboat may be used if the water covers the entire path for the sampling period. Alternatively, a helicopter may be used. Helicopter operation is limited by thunderstorms, low visibility, and high cost. An advantage to using helicopter transport is the minimum disturbance to the wetland sampling site as compared to airboat transport. Wetland monitoring sites require a highly visible sign or other landmark to mark the location for repeat sampling visits. A GPS instrument should also be used in conjunction with the visible sign.

14.4 EXAMPLES OF WATER SAMPLING

Water sampling frequency in a canal or wetland depends primarily on the project objectives, water sources, and biweekly sampling frequencies for canal and wetland, respectively. The following examples of canal and wetland water sampling are presented to outline the steps that should be followed.

14.4.1 Canal Water Sampling

Based on the SOP developed, two types of canal water samples can be collected: grab samples and composite samples from an auto-sampler with the flow proportional sampling scheme. Sampling should be scheduled early in the week with Tuesday being the best choice as preparations can be made on Monday, and if there is an unexpected event, such as a severe weather, postponement to Wednesday or Thursday is still possible. End-of-week sampling is not recommended unless the laboratory operates on weekends. Generally, canal water sampling includes the following procedures:

14.4.1.1 Sampling Preparations

The instruments and sampling supplies required to assist the laboratory staff and field crew are the following:

1. A water container (5 L) filled with distilled or deionized (DI) water
2. A tool box with keys to access the site and a string with a hook for bailer to collect grab samples
3. Bailers within a capped PVC pipe protected with a plastic bag from contamination
4. Sampling bottles (250–500 mL); sampling bailers and bottles need to be washed, soaked in 10% HCl overnight, rinsed well with DI water, and air-dried in a pollutant-free environment before use
5. A cooler with ice for storing samples
6. Reagent grade or pure concentrated acid (H_2SO_4, > 99.99%) to preacidify samples
7. Waders with a container
8. A notepad with self-adhesive labels preprinted and record data sheet for sampling
9. A waterproof notebook or travel log
10. A pen
11. Wasp and hornet spray, and mosquito repellant if necessary
12. Raincoats
13. A first-aid kit

14.4.1.2 Transportation and Departure

For transportation, a four-wheel drive vehicle is necessary for use on unimproved trails and unpaved roads because in most cases, the monitoring site is located in a remote area. The driver must make certain that the vehicle is in a good condition, has been recently serviced, has properly inflated tires (including a spare), sufficient oil, transmission fluid, coolant and gasoline or diesel, a jack, wrenches, and tools for emergency repair. The preparation list must be checked prior to departure. In some cases, authorization to access the site is required, and the authorization should be obtained prior to departure.

14.4.1.3 Grab Sample Collection

Upon arrival at the site, ascertain that conditions are safe to work. Anything unusual must be identified and assessed before collecting the first sample. An equipment and field blank need to be collected at each site where actual samples are collected.

1. Equipment blank sample collection: For grab samples with a bailer as sampling equipment, fill the bailer with DI water through the open top and collect it from the bottom end as an equipment blank sample. The bailer used for the equipment blank must be clean and dry because rinse is not allowed. The same bailer can be used later for collecting actual samples, provided that it should be rinsed with sample water.
2. Field blank sample collection: At one of the sampling sites, a field blank sample is required to verify that the DI water was not polluted during transportation from the laboratory to the sampling site. To collect a field blank sample, simply fill one of the sample bottles without rinsing it, tightly cap, and keep it in the same cooler with other water samples.
3. Grab sample collection: Attach the bailer to a string, remove the water release tip from the bottom, and slowly lower the bailer into the water until it is about 50 cm below the surface. Retrieve the bailer and release the water from it to rinse the sample bottle and bailer. Repeat the above steps to finally fill the sample bottle up to the neck (Figure 14.4), cap tightly, and place sample in a cooler and cover with ice.
4. Replicate sample collection: At one of the sites, a replicate sample should be taken to verify that no error has been introduced through the sampling

FIGURE 14.4 Bailer used for surface water grab samples at canal sampling site.

procedure. To do this, two sample bottles should be rinsed and filled with the water from the same bailer with one labeled as a replicate.

5. Split sample collection: To verify the uniformity of a water sample, a split sample should be collected by repeating the same procedure as used for replicate sample collection. Fully fill one bottle and half fill the second bottle, then pour water from the full bottle into the half-filled one, shake and pour water back into the first bottle, and mix the contents. Repeat this procedure several times to assure that the two split samples are identical.
6. Sample labeling: Once a sample is collected, label it immediately with the preprinted self-adhesive label or label the bottles before collecting the samples.
7. Avoid cross-contamination: A dedicated clean individual bailer is used for each site. Therefore, once a bailer has been used for an actual sample collection, it should be stored separately from clean ones.

14.4.1.4 Composite Sample Collection

A composite sample is collected by the auto-sampler based on predetermined time intervals or flow events if flow proportional sampling has been applied. An auto-sampler, such as ISCO 6712, has four jars in different positions (Figure 14.5) with a motor that drives the distributor from one jar to the other in rotation. The sequential sampling events and total volume of sample for each jar can be programmed based on information of the water flow rate for each specific site. The sampler has a function to purge the tubing before pumping water from the canal to the sample jars for each sampling event. Therefore, any sample collected is fresh from the water body since the residual water left in the tubing or suction line has been dispelled. To collect composite samples, one must make certain that the auto-sampler is in a

FIGURE 14.5 Containers in the auto-sampler of ISCO 6712.

functional status, that the power supply is normal, and that the program is correct. An LCD panel displays the sample number, total number of samples, exact time of sample collection, power consumed, and warnings, such as "Replace pump tubing." To collect samples from an auto-sampler with a flow proportional sampling technique, the following steps are necessary:

1. Open the auto-sampler cover and read the information on the LCD panel to see if any samples have been collected. Label all the sample bottles needed for the specific site.
2. Unsecure the top part of the auto-sampler, lift it up, and remove the metal frame holding sample jars in their correct positions. Take out the water-filled sample jars. Fill the sample bottle by taking the similar ratio of the total volume of water from each of the jars. In other words, the more water present in the sample jars, the more water needs to be transferred into the sample bottle. Remember, usually only one composite sample is collected from all these four jars and each of them has a capacity of 3.78 L.
3. Shake the jars and discard the contents before rinsing each jar very well with DI water, use only a small amount of water for each time but rinse them at least three times.
4. Equipment blank sample collection: Choose one of the rinsed jars, fill with DDI water, and transfer into a sample bottle to serve as an equipment blank for the auto-sampler.
5. Pre-acidification to preserve the composite samples: Use a pipette to add a small volume of concentrated H_2SO_4 proportional to the volume of water to be collected to each jar if samples are to be collected. For example, if the jar is to be filled with water, 2–3 mL of acid per jar is usually optimal. Over-acidification ($pH < 1.5$) can cause analytical errors in the determination of total P, as explained in the SOP.
6. Place all jars in position and attach the top, restart the program, place the sampler cover on, and lock in the metal box.

14.4.2 Wetland Water Sampling

1. Transportation and precautions: In addition to the items required for canal water sampling, a wader or a pair of rubber boots depending on water level in the wetland sampling site is required. For transportation to sampling sites, a helicopter is usually superior to an airboat. Three types of helicopters are commercially available. The Long Ranger, being the largest and most expensive, can accommodate four to five passengers and has ample space, including a cabinet for sampling equipment and supplies. The Robinson 42, being the smallest and least expensive, accommodates three passengers but lacks of sufficient space for sampling gears. In many cases, the Jet Ranger is the best choice, since it accommodates three passengers and has enough space for sampling equipment. Also the Jet Ranger has a cabinet for a cooler box, a DI water container, and a wader box. Two people are required for sampling and the third passenger seat in the Jet Ranger can be

used for a volunteer helper. Safety is always the top priority! Every passenger must scrupulously follow all of the pilot's instructions. Always buckle up safety belts and secure all interior cabinets and items before taking off. Remember, it is absolutely imperative for all field team members to avoid the vicinity of the tail rotor, and to keep all objects such as meter sticks and antennae away from the rotors. In most cases, the helicopter engine continues to run after it has landed at the sampling site. If more than 30 min are needed for sampling, the pilot should be asked to shut down the engine. The GPS position for each site must be available at all times for communication to the pilot. Although every helicopter is equipped with a GPS device, the pilot expects the sampling team to provide the coordinates of each site and the order in which they should be visited. In accordance with safety protocols, a licensed pilot usually keeps radio contact with headquarters to report current position, status, and plans. Again, a permanent and clear landmark positioned at each site is helpful for quickly locating each sampling site.

2. Wading: If a wader is necessary, put it on before take-off, since space is limited on board. To save time and for convenience, sample bottles should be labeled before embarking and kept in organized groups by sampling site to accommodate multiple sampling sites.
3. Field blank sample collection: A single field blank sample as a control should accompany the sampling team to assess any transportation effects on the samples. To do this, an unrinsed sample bottle is filled with DI water, capped, and stored in the cooler with the actual samples.
4. Equipment blank sample collection: An equipment blank sample must be taken in an unrinsed sample bottle (part of the equipment). This equipment blank provides an assessment of bottle cleanliness and cross-contamination.
5. Sampling site selection and number of samples collected: When transported by a helicopter to the sampling site, sampling personnel should wade 30–50 m away from it to avoid sampling water churned by the rotor. In some alligator-infested areas of south Florida, caution must be exercised, especially if the water is deep. After wading from the helicopter, select an undisturbed area with at least 10 cm deep that is representative of the water body and devoid of modification or disturbance by birds or animals. Samples should be taken at the corners of an isosceles triangle (about 5 m between spots). From one of these spots, one replicate and one split sample should be collected simultaneously.
6. Actual sample collection: Take 3 labeled sample bottles plus a fourth if a replicate is required, remove the lids and submerge the bottles in the water at a 60° angle, and allow the bottles to fill up to the neck. Cap them tightly and surround them with ice in the cooler box. When a split sample is required from the sampling spot, pour water from one of the three filled bottles with the correct label into the split sample bottle, shake it, and pour the water back into the original bottle to ensure thorough mixing. Repeat this mixing of the contents of the two bottles at least three times to create two identical

samples. After the sampling is complete, carefully arrange and secure all items in the helicopter. Before take-off, double check that all external doors of the helicopter and interior cabinets are locked.

14.5 EXAMPLE OF TELECOMMUNICATION SYSTEM

For both surface water and groundwater monitoring, state-of-the-art techniques and instrumentation have been developed rapidly in recent years; for example, digital modems have replaced their analog forerunners and the telecommunication can be conducted through the Internet. Automatic data transfer and remote control have brought great convenience. Under some circumstances, video and other information can be transferred through a wireless system from remote field sites to the laboratory, which facilitates dynamic monitoring of water quality.

14.5.1 Devices

To assemble a telecommunication system, one simply needs to connect a digital modem (e.g., AirLink, EarthLink, Raven) to an antenna and 12 V power supply using a data series cable. Obtain and install a communication card with corresponding IP address from a service company (e.g., AT&T, Verizon). Configure the system on the computer with a unique IP address for each site, select data collection intervals, and follow the service company's instructions.

14.5.2 Installation and Program Setup

The digital modem should be installed in a box shared with the data logger and connected to an antenna attached to a pole that extends ≥ 2 m upward by a cable through the weather-proof box. Twelve volts of power can be obtained through the data logger and a data serial cable (Campbell Scientific Inc.) is used to connect the modem to the data logger.

An example for Logger Net 2.1c (a common data logger software) is as follows. After installing the appropriate software on the computer using the option SETUP from the main menu, add the IP Port, select communication enable, and input the corresponding IP address for each site where an individual modem has been installed. Then, define each site number under the IP Port, and select respective items such as hardware, schedule to collect the data, and arrangement of the final storage of the data collected from the site by following the option menu on the screen. Once the setup has been completed, each site will be connected automatically to receive data based on the schedule selected.

14.5.3 Interface and Precautions

When the program is in operation, the status of each site can be accessed with the status monitor option, which displays the current status, connection error, last attempted collection time, and the next data collection time. To view the data collected, simply press VIEW and select OPEN from the file menu and then select the site number, whereupon the system will automatically display a data file. Because the data file can only be viewed and printed from this software, processing of the data requires conversion to an Excel or other appropriate file format.

14.6 SAMPLE HANDLING AND ANALYSES

The proper handling of samples is critically important. Once a sample has been collected, one must ensure that the sample bottle is capped tightly and surrounded with ice during transport at a temperature of 4 ± 0.3°C as monitored by a certified thermometer in the cooler box. A decision to reject samples may be made by a laboratory staff if the temperature has deviated from the above range when the samples arrive at the laboratory. When accepting samples, the laboratory staff sign the data sheet, which records sample condition, temperature, date, and arrival time.

14.6.1 Sample Storage and Chain of Custody Requirements

Once samples have been accepted by the laboratory, a chain of custody must be developed by first assigning a unique serial laboratory number to each individual sample. Immediately following the establishment of the chain of custody, a 20 mL aliquot from each sample must be filtered through a certified analyte-free filter paper for water dissolved elements (e.g., Whatman 42) for ortho phosphorus (PO_4-P). If samples cannot be analyzed immediately, they are stored in a refrigerator at 4°C. However, the maximum storage time for P is 48 h even under optimal conditions. Samples for total element content, such as TP, must be acidified to pH <2.0, and stored at ≤6°C for a period not to exceed 28 days.

14.6.2 Sample Preparation and Chemical Analyses

For water dissolved elements, such as PO_4-P, filtered samples can be analyzed directly by a colorimetric method with specified equipment. To assay total elements, samples must be digested with certified acid under certain temperature and pressure conditions according to standard methods such as EPA 365.1 for TP. After the samples have been pretreated, analysis is carried out by certified methods for both water soluble and total elements according to the laboratory quality manual and instrument instructions, using instruments such as the Auto-analyzer (AA-3, Bran+Luebbe, Germany) or the Lachat Auto-analyzer (Hach Company, CO). Detailed procedures for chemical analyses are discussed in Chapter 10.

14.7 EXAMPLE OF DOCUMENTATION

High quality documentation is of utmost importance, allowing evaluation of the quality of the project by anyone regardless of their familiarity with the project. In addition, the documentation is the final product to be kept on record and submitted to the project sponsors. Moreover it may be used to resolve legal issues. Other guidelines for consideration are detailed as follows.

14.7.1 Original Sampling Data Sheet

The data sheet records a clear and unique project name, date, names of personnel involved, sample collection time, location(s), unique field sample ID numbers, water depth, clear definitions of abbreviations, time of arrival of samples at the laboratory, cooler number, sample condition, cooler temperature, and the signature of the person who officially accepted the samples. If the samples are rejected by the laboratory personnel because of improper storage or transportation, an additional specific document needs to explain the basis for rejection and affirms that this decision has immediately been communicated to the field sampling staff.

14.7.2 Records of the Travel Log

In addition to the sample collection data sheet, a waterproof log book should record all the relevant activities or events including transportation, personnel, time of departure and return, sites visited, sampling status, and weather conditions for each sampling trip.

14.7.3 Site Visit Report

Based on the travel log, an official site visit report, which should include the file series number, date of the visit, personnel, name or a serial number of sites visited, purposes of the visit, sampling events, and a summary of the trip, must be developed upon return to the office.

14.7.4 Project Reports

The project report is usually written to meet the periodic requirements of the project sponsors and should serve as a clear, concise presentation of the monitoring project's findings. The report generally includes an official cover page, which indicates the project title, report duration, serial number, names of persons responsible, contact information, name of individual to whom the report is submitted and his or her affiliation, and date of the report.

The body of the report provides a general description of the project, site locations with GPS coordinates, water quality parameters monitored, site status, methodology, and optional solutions and results including tables and charts. Some documents, such as laboratory chemical analysis reports, a program applied to each individual site, canal profile measurements, laboratory certificate, and qualifications to conduct a water quality monitoring project, can be attached as appendices.

14.8 SUMMARY

This chapter provides an overview and examples of water quality monitoring techniques and protocols. For more information, see Appendix 14.1 for a specific standard operating procedure (SOP) for a surface water quality monitoring project. See Appendix 14.2 for a data-logger program with CR-10X for recording flow rate, temperature, dissolved oxygen, pH, and controlling auto-sampler for flow proportional sampling.

REFERENCES

FDEP (Florida Department of Environmental Protection). 2008. FS 2000. General Aqueous Sampling. ftp://ftp.dep.state.fl.us/pub/labs/assessment/sopdoc/2008sops/fs2000.pdf. (accessed January 12, 2010)

FDEP. 2008. FS 2100. Surface Water Sampling. ftp://ftp.dep.state.fl.us/pub/labs/assessment/sopdoc/2008sops/fs2100.pdf. (accessed January 12, 2010)

USEPA (U.S. Environmental Protection Agency). 2003. Requirements for Quality Assurance Project Plans, EPA QA, EPA/240/B-01/003.

USEPA. 2007. Guidance for preparing standard operating procedures (SOPs), EPA QA/G-6. http://www.epa.gov/quality/qs-docs/g6-final.pdf.

Wang, Q., Y. Li, T. Obreza, and R. Munoz-Carpena. 2004. Monitoring stations for surface water quality. EDIS. Soil and Water Department, Florida Cooperative Extension Service, Institute of Food and Agricultural Sciences, University of Florida. Fact Sheet SL218, pp. 1–6. http://edis.ifas.ufl.edu.

APPENDIX 14.1: STANDARD OPERATING PROCEDURES (SOPs) FOR SURFACE WATER SAMPLING OF PROJECT X

Prepared by: Qingren Wang, Project Manager
University of Florida, IFAS
Tropical Research and Education Center
Homestead, Florida

____ ____________
Yuncong Li, Director, Soil and Water Research Laboratory — Date

____ ____________
Qingren Wang, QA officer, Soil and Water Research Laboratory — Date

Section No.: SOP-SWS-01-09-01
Revised No.: SOP-SWS-01-09-01
Effective Date: 12/01/2009
Authorized signature:
Total pages: 26

Introduction and Scope

1.1. This section presents standard operating procedures to be used to consistently collect representative surface water samples from the Everglades, Florida. Each collection event must be performed so that samples are neither contaminated nor altered from improper handling.

1.2. All personnel have to follow the procedure in sampling unless otherwise specified or authorized.

1.3. The following topics include acceptable equipment selection and equipment construction materials, and standard grab, flow-proportional surface water sampling techniques.
1.4. For flow proportional sampling, ISCO samplers with transducers are installed and used to perform sampling according to programming.
1.5. It is worth it to mention that this specific SOP was developed based on general requirements proposed by USEPA (2003, 2007), FS 200 and FS 2100 proposed by FDEP (2008).

General Precautions

1.1. According to the Florida DEP, a minimum of two people should be assigned to a field team.
1.2. When using helicopter, take samples away from the helicopter disturbed (wind) area.
1.3. If wading is necessary, collect samples upstream from the water body.
1.4. Avoid disturbing sediments in the immediate area of sample collection.
1.5. Collect surface water samples from downstream toward upstream.

Equipment and Supplies

1.1. Polyethylene bailer (Ben-Meadows) shall be used for grab sampling.
1.2. The sample container size is 250 or 500 mL, preserve the samples at pH <2 with holding time <28 days at 4°C.
1.3. For sample equipment cleaning requirements, refer to Sections 4.2 and 4.3 below.
1.4. For documentation requirements, refer to Section 6.

Cleaning or Decontamination

1.1. Performance criteria
  1.1.1. The cleaning/decontamination procedures must ensure that all equipment that contacts a sample during sample collection is free of P and constituents that would interfere with the analytes of P.
  1.1.2. The detergents and other cleaning supplies should not contain P or cause interferences unless effectively removed during a subsequent step in the cleaning procedure.
  1.1.3. Equipment blanks that monitor potential contamination from cleaning products should always report non-detected values.
1.2. Cleaning reagent
  1.2.1. Recommendations for the types and grades of various cleaning supplies selected to ensure that the cleaned equipment is free from any detectable contamination are outlined below.
  1.2.2. Detergents: Use Liqui-Nox (or a non-phosphate equivalent) recommended by EPA.
  1.2.3. Acids: Use reagent grade hydrochloric acid to soak the containers.

- 1.2.4. Analyte (P)-free water sources, double distilled (DDI) water.
- 1.2.5. Analyte-free water is water in which P and all interferences are below method detection limits.
- 1.2.6. Maintain documentation (such as results from equipment blanks) to demonstrate the reliability and purity of analyte-free water source(s).
- 1.2.7. The source of the water must meet the requirements of the analytical method and must be free from P. Deionized and distilled water (DDI) will be used to rinse the sample containers.
- 1.2.8. Use P-free water for blank preparation and the final decontamination water rinse.
- 1.2.9. In order to minimize long-term storage and potential leaching problems, obtain or purchase P-free water just prior to the sampling event. If obtained from a source (such as a laboratory), fill the transport containers and use the contents for a single sampling event. Empty the transport container(s) at the end of the sampling event.
- 1.2.10. Discard any P-free water that is transferred to a dispensing container (such as a wash bottle) at the end of each sampling day.

1.3. Acids

- 1.3.1. Reagent grade hydrochloric acid: 10% hydrochloric acid (1 volume concentrated hydrochloric plus 10 volumes deionized water).
- 1.3.2. Freshly prepared acid solutions may be recycled during the sampling event or cleaning process. Dispose appropriately at the end of the sampling event, cleaning process, or if acid is discolored or appears otherwise contaminated (e.g., floating particulates).
- 1.3.3. Transport only the quantity necessary to complete the sampling event.

Sampling Procedures

The sampling procedures below are used for instruction in surface water sampling as a standard operating procedure (SOP). Nobody is allowed to make any change without formal authorization.

1.1. General considerations

- 1.1.1. Cross contamination: Special effort should be made to prevent cross contamination or environmental contamination when collecting samples. To do this, a separate sampler will be used for each sampling site.
- 1.1.2. Protective gloves
  - 1.1.2.1. General considerations: Gloves protect the sample collector from potential exposure to sample constituents and/or sample preservatives. Gloves also minimize accidental contamination of samples by the collector.
  - 1.1.2.2. Glove use: Use clean, new, unpowdered, and disposable gloves for all sample collection for the Everglades water

quality monitoring project. Do not allow gloves to come into contact with the sample or with the interior or lip of the sample container. Properly dispose of all used gloves.

1.1.2.3. Change gloves after preliminary activities such as carrying or replacing some instruments, after collecting all the samples at a single sampling point, if torn, or used to handle extremely dirty or highly contaminated surfaces.

1.2. Sampling depth

1.2.1. Sampling depths that must be recorded may vary based on physical condition of the site.

1.2.2. In most cases, grab samples of surface water are collected at a depth of 0.5 m from the surface of the water. If the total depth is less than 1 m, the sample is to be collected at half depth.

1.2.3. Depth is to be measured from the surface of the water down, where the surface of the water is zero.

1.3. Sample collection order

Unless field conditions justify other sampling regimens, collect samples in the following order: First collect samples from the autosampler and then collect grab samples without disturbing the strainer for the autosampler.

1.4. Sample labeling

To ensure the integrity of samples collected, each sample bottle must be labeled with a unique identifier. For routine district samples, this is usually done by assigning a project code, field number, and a code for each bottle. For prelogged in samples, the computer will generate labels with this information. For manual labeling, use project assigned field number assigned prelogin number and write these along with other unique identifying information on the label. Labels must be preprinted or written using indelible ink on waterproof material. Note any changes on the label using single cross-through line and initials. Include the time of sampling on the labels. Labels must be affixed firmly on the sample container.

Preprinted sample labels

Sample type

Site#

Date

Sample #

1.5. Sample rejection criteria during field sample collection

The field sample collection personnel are responsible for visually inspecting the sample and rejecting it as necessary. Questions concerning the rejection of any sample should be initially directed to the project manager or field supervisor. The laboratory personnel receiving the sample are responsible for inspecting the condition of the sample upon receipt and rejecting it as necessary. Established criteria are detailed below.

1.5.1. General considerations

Proper and thorough documentation explaining the reason for rejecting any sample must be carried out. Notes concerning factors that directly affect the quality of the sample must be placed in the note book and on the header sheet in the comment section for the affected sample. Upon return to the office, the sampling personnel must notify the project manager or field supervisor of the problem.

1.5.2. Grab sample rejection criteria

In most cases, if an initial grab sample is unacceptable because it is not representative of the water body being sampled, the original sample should be discarded and another attempt should be made to collect a representative sample. If it is not possible to obtain a representative sample, contact the unit supervisor (or project manager if supervisor is unavailable) and request further guidance as to how to proceed. If it is not possible to receive guidance as to whether or not to collect:

1.5.2.1. Process the most representative sample possible.

1.5.2.2. Document the problem with the sample collection on the header and in the field notes.

1.5.2.3. Inform the field supervisor or project manager upon return from the field.

1.5.3. Autosampler sample rejection criteria: Reject samples that meet any of the following or similar criteria:

1.5.3.1. Samples that are obviously contaminated by outside or foreign matter (e.g., dead animals or insects).

1.5.3.2. Samples where the pH is below 1.0 from either insufficient sample volume and/or excessive acid addition.

1.5.3.3. Collection line contamination from excessive plant accumulation around the intake, or the intake being submerged in, or directly in contact with, bottom sediments (this may not be readily observable in the field.)

1.5.3.4. Cracked or broken sample collection container.

1.5.3.5. Samples obtained from an autosampler that failed during operation such as incomplete program, overflowed containers, etc.

1.6. Sample collection procedures

1.6.1. Equipment and container rinsing:

When collecting water samples, rinse the sample collection equipment with a portion of the sample water three times before collecting and processing the actual sample. Sample collection equipment shall be rinsed twice with analyte-free water at the sample site immediately following collection and processing of the sample. Sample containers are to be rinsed once with sample water.

1.6.2. Precautions for grab water sampling

Surface water grab sampling procedures (except autosamplers) are presented below. The following special precautions are observed when applicable:

1.6.2.1. If a helicopter is used, the sample is taken from the area free of wind disturbance.

1.6.2.2. When wading, the sample is collected upstream from the collector.

1.6.2.3. Care is taken not to disturb the sediment in the immediate sampling area.

1.6.2.4. Pre-preserved containers are not used as collection containers.

1.6.2.5. Intermediate containers are inverted, immersed mouth down to the appropriate depth, and turned upright pointed in the direction of flow, if applicable.

1.6.2.6. Samples are preserved according to USEPA requirements (see Chapter 6, Table 6.1).

1.6.3. Grab sampling steps

1.6.3.1. Polyethylene bailer or bottle (depends on the depth of water) is used for sampling.

1.6.3.2. Lower the bailer or the open bottle straight into the water, rinse and repeat twice.

1.6.3.3. Lower the sampler (bailer or bottle) to the desired depth and wait until filled with water.

1.6.3.4. Gently remove the sampler from the water.

1.6.3.5. Rinse and fill a 250 or 500 mL polyethylene bottle, cap, label, and store in the cooler with ice.

1.6.3.6. Complete documentation of relevant observations.

1.6.3.7. Transport the samples to the laboratory as soon as possible.

1.6.3.8. Complete chain of custody.

1.6.3.9. Filter the samples if necessary; one set of subsamples is used for ortho-P analysis by AA3 within 48 h, the other set of samples shall be transferred to a 20 mL vial and then add 1 drop of DD (double distilled) H_2SO_4 to acidify the sample (pH <2).j). Store the samples at <4°C for total P analysis (<28 days).

Note: The bottle is capped and shaken after which a small amount of the sample is poured onto a narrow range pH (0–3 pH units) test strip to ensure pH <2. If pH is not <2, additional acid is added drop-wise, the bottle is capped and shaken, and the pH is tested again. This procedure is followed until pH <2. *Care must be taken not to over acidify the samples. Over-acidification (pH < 1.5) may cause low bias in some analyses including total P.* For filtered samples, the acid is added after filtration following the procedure outlined for unfiltered samples. Preservatives are taken into the field in polyethylene dropper bottles that are in good physical condition.

1.6.4. Sampling from autosamplers

Autosamplers are installed at specific location based on the project design. Autosamplers located at the desired sites on canals can be triggered automatically by up- and downstream flow meters connected to triggering computers by transducers with flow-proportional techniques. Equipment performance, including volume checks and recalibrations, are conducted weekly or at time of sample pick-up to ensure accuracy and consistency of samples. If an autosampler fails to perform, the field team is trained to troubleshoot or solve problems as necessary or replace the autosampler and return the defective unit to the laboratory for repair. Sampler intake tubing is dedicated to a collection site and is replaced at a minimum of every quarter or when the autosampler is first deployed. The tubing may be replaced sooner if algal growth is observed in and around the inflow tubing. The silicone pump tubing is changed as needed between the normal quarterly maintenance events as pumping degrades the tubing integrity. Sampler intakes at water control structures are usually located 0.5 m below the historic low mean water level.

1.6.4.1. Autosampler sample preservation

Preacidified autosampler containers will be used to perform the autosampling and a short period of preservation (a week).

1.6.4.2. Discrete autosamplers

A discrete automatic sampler is programmed depending on the project requirements to purge the tubing following collection. Approximately 1 mL of 50% sulfuric acid is added to the discrete autosampler bottles before sample collection to maintain pH between 1.5 and 2 after sample collection. Routinely, samples are poured into sample bottles at the collection site, preservation is checked, and samples are immediately placed on ice for transport to the analytical laboratory. Additional drops of acid are added at time of pick-up, if necessary, to bring the pH to 1.5 to 2. For projects that require manual compositing from discrete bottles, measured aliquots of samples are transferred to a properly cleaned plastic bucket, mixed thoroughly, and a composite sample is transferred into a pre-labeled container. Autosampler discrete bottles must be thoroughly mixed prior to pouring into the intermediate container to ensure homogeneity of the sample.

1.6.4.3. Procedures for sampling from autosamplers

Autosampler equipment blank sample (DDI water passed through the autosamples) will be taken before the sampler is installed or after new tubing is replaced on the site. Take out sample jars, thoroughly mix before rinsing and pouring about 1/4 of the contents from each of four jars into each pre-cleaned polyethylene containers (500 or 250 mL), cap and label the containers before preserving them in a cooler with ice.

1.6.4.3.1. Empty the jars but do not disturb the sampling site when discarding the remainder of sample and then replace them in the sampler after adding 1 mL H_2SO_4.

1.6.4.3.2. Take a field blank (FB) for each sampling date in the field by filling a 500 or 250 mL sample bottle with DDI water.

1.6.4.3.3. At the same site, taking a grab sample is performed: A clean polyethylene bailer is used for equipment blank sampling at each sampling date. To do this, fill the bailer with DDI water from the top, then release the water from the bottom into a sampling bottle, rinse the sampling bottle twice with the sampling water in the bailer and then fill up the bottle with sampling water.

1.6.4.3.4. Insert the bailer into the surface water at the canal site to a depth of ~0.5 m, gently pull it up, rinse the sample bottle twice and fill up with water from the bailer. Repeat the steps to obtain a replicate sample.

Wetland Sampling by Helicopter

This section is specific to sample collection when transport by helicopter is necessary due to inaccessibility by other means. Samples are collected in 250 or 500 mL polyethylene sample bottles for each site and each QC sample for each sampling date. When collecting samples, avoid alligator holes, airboat trails, and other non-representative areas.

1.1. Helicopter safety

Exercise extreme caution when approaching a helicopter while the rotor is turning. Approach the helicopter from the front and avoid the tail area. Keep all objects such as meter sticks and vehicle antennae away from the rotors.

1.2. Surface water collection using grab sampler
   1.2.1. After landing, walk out away from helicopter disturbance area, preferably upstream if flow is visible and/or downwind.
   1.2.2. Be careful to avoid weeds and the creation of turbidity.
   1.2.3. Measure and record the depth of water using a long, rigid, graduated pole.
   1.2.4. If water depth > 20 cm, immerse intermediate sample container(s) in an undisturbed area at middle depth upstream from the sampling personnel, rinse three times, and fill to the brim.
   1.2.5. No sampling is required when water depth is <10 cm. Cap and label the container(s) and place on ice.
   1.2.6. Gather equipment, walk back to the helicopter, and place samples in a cooler.
1.3. Sample processing and preservation
   Within 4 hr of sample collection, aliquots are transferred from the large intermediate containers into more appropriate containers for laboratory analyses and preserved according to the procedure required for sample preservation.

Documentation and Record Keeping

1.1. Major types of documentation
   1.1.1. Field notes
      Field notes are recorded using a permanent marker by field staff in a bound water proof notebook, known commonly as a "black book." These books are stored in a filing cabinet for access by field staff and the project manager. The project manager is responsible for reviewing the field notes for accuracy immediately after a sample event.
   1.1.2. Sample header sheets including data from in situ measurements
      Header sheets contain all field information about the samples collected (see Section 6.2 for further details). Original header sheet forms are retained by the laboratory.
      If samples are collected and sent to another laboratory by common carrier for analysis, the field technician provides a copy of the header sheet to the project manager and the quality assurance (QA) officer for data review and entry purposes. The field project manager reviews all header sheets for accuracy. The project manager or field supervisor reviews the data against the header sheet for accuracy.
   1.1.3. Equipment/instrument calibration, maintenance, and troubleshooting logs
      Laboratory and field staff must record all calibration information on properly designated calibration logs. Field equipment calibration logs are maintained by the field personnel and are kept in their office area. Field instrument calibration

logs are maintained by the field personnel and are kept at the designated calibration area. Maintenance and troubleshooting information are entered in instrument logbooks maintained by the field group.

1.1.4. Standard operating procedures

Laboratory and field managers have written SOPs for certain tasks including some specific, complex field sampling projects, laboratory analysis, instrument calibration and maintenance, data review, etc. Original SOPs are maintained by individual managers. At least one copy of every SOP revision must be maintained for future reference. Any justification, description, and validation data package for changing SOP must be submitted to the QA officer for approval. The QA officer maintains a history of all method changes.

1.1.5. Quality assurance plans and quality manual

Several copies of the current Field Sampling Quality Manual are distributed to project, laboratory, field, and QA staff. At least one copy of annual revisions is stored either in the facility or with relevant records in the off-site storage facility. This manual is effective for a period of one year from the date of approval and is subject to annual updates. A copy of the annual revision is retained in-house for future reference.

1.1.6. Validation studies

Documentation on any method change, new method, or other validation studies, including the description of the study and validation data, are maintained by the QA officer and are kept for the duration of the project.

1.1.7. Audit reports and corrective actions

Original completed audit checklists, audit reports, and responses from audited entities (contract laboratories and field collection groups) are maintained by the QA officer.

1.1.8. Administrative records

General administrative records, including qualifications and performance appraisals, training records, and records of demonstration of capability for each collector, are kept in individual personnel files. A log of names, initials, and signatures for all individuals who are responsible for signing or initialing any laboratory records must be kept and maintained by the QA officer.

1.2. Sampling documentation procedure

A verifiable trail of documentation for each sample must be maintained from the time of sample collection to the submittal of samples to a laboratory for analysis. A header sheet must accompany samples submitted to the laboratory, which is a form of custody tracking. The header sheet tracks the samples from collection to submittal to the laboratory and identifies the persons responsible for sample collection. It must be signed, dated, and reviewed for accuracy before transmittal.

The collector must use sampling header sheets to document pertinent field data information such as, but not limited to, station codes, date and time of sample collection, and field ID numbers. The collector should sign the form when relinquishing samples to the laboratory. All corrections are crossed out with a single line and initialed and dated by the person making the correction. Information on the label includes project code, date and time the sample was taken, and the sample number. Sample numbers are unique sequential numbers generated by the project manager.

1.3. Transport

Following sample collection and proper preservation (if required), the bottles are capped, labeled, and placed on ice in a sturdy, rigid cooler. The field technician responsible for filling out the header sheet during the sampling event should have another technician review the sheet for accuracy and sign off on the appropriate line. The cooler is transported to TREC-Soil and Water Research Laboratory for analysis. The header sheet is kept in a file for record checking. When shipping to another laboratory, the samples must be placed in a plastic bag to prevent leakage of the ice water. The cooler is completely sealed with shipping tape to ensure that it will not open during transport. The original header sheet is placed in a clear plastic ziplock bag and included in the cooler for laboratory use.

1.4. Submittal

Field sample collectors personally transport samples for processing and chemical analysis to the laboratory where they are placed in a designated refrigerator. The person bringing the samples to the laboratory signs the chain of custody form with the time clock. The chain of custody form is given to the person responsible for logging in the sample at the laboratory. For samples that are shipped using a carrier service, login personnel at the laboratory should inspect the cooler and samples to ensure that the bottles are on ice at the proper temperature, containers are not broken or leaking, the chain of custody form has accompanied the samples, and the appropriate number of samples have been received.

1.5. Standard operating procedure (SOP) for field documentation

This section describes the minimum guidelines and requirements for field documentation. This SOP is written for the purposes of standardizing field reportable data and dialogue so that the intermediate and end-users can more readily access, comprehend, and utilize it. All TREC-Soil and Water Research Laboratory staff responsible for collecting samples and field measurements shall follow this SOP. Accuracy, consistency, and legibility are key factors that will be enhanced by the utilization of these SOPs. Printing instead of using cursive writing enhances legibility. Field documentation must be sufficient and clear to allow history tracking for any sample collected or any measurement performed.

1.5.1. Chemistry field data log (header sheet) entries

A chemistry field data log or header sheet must accompany all samples collected and submitted to the laboratory. This sheet must be legible, accurate, and complete to render the sampling trip and the samples valid. The following instructions apply to the use of laboratory header sheets. Information that is repeated in every sampling event for a specific project may be entered in log-in templates or added during prelogin. In cases when an external laboratory requires the use of its chain-of-custody forms, the same codes and information must be entered.

1.5.2. Field logbook entry instructions

1.5.2.1. Relevant field observations are noted in a bound waterproof notebook, hereafter referred to as a field logbook or field book, which is specific to the specific field project.

1.5.2.2. Entries shall be made into the field logbook with a waterproof ink pen.

1.5.2.3. To avoid any confusion, entries for the number 0 will have a diagonal slash to differentiate them from the letter O, particularly in alphanumeric fields.

1.5.2.4. Each field logbook must be clearly labeled on its cover and spine with the project name.

1.5.2.5. The first few pages of the field book should contain information such as full project name, project start date, logbook start date, sites/stations covered by the project, SOP revision date, contact person (usually the field supervisor), and abbreviations commonly used within the field book, etc. Maps, directions, and a condensed version of the project SOP are examples of additional information that could appropriately be added to this section.

1.5.2.6. Each field logbook entry for a given project day will adhere to the following guidelines

1.5.2.6.1. Each trip of the project will cover, at a minimum, one page of the logbook. In other words, at no time will more than one trip be included on the same page of a field logbook.

1.5.2.6.2. At the top of the first page for a given project day, the following information will be noted:

- Project name (e.g., STA6)
- Purposes or trip type (e.g., SW grabs)
- Full date, including year
- Collectors' initials. Spell out the entire name for first-time entries of collectors.

- Corresponding responsibilities for collectors (e.g., ABE—grabs, processing; CDF—Hydrolab, books). Note if an individual is new to the project or in training.
- Weather at the first sample site or beginning of the project day. Items to include here are temperature, wind, sky conditions, and any prevailing weather phenomena such as "Tropical storm warning today" or "apparent heavy rain recently." Any changes in weather conditions throughout the course of that project day shall be logged accordingly at the site and time that those changes become apparent.
- Acid—the acid(s) used for sample preservation on that project.
- Laboratories—if any laboratory other than TREC-SWRL is to be used for sample analysis, it must be annotated here (e.g., "organics to DEP, inorganics to TREC-SWRL"). Record number of coolers shipped, the carrier's information, and bill # for tracking purposes.

Notes—any general notes that could apply to that project day, or that could affect the chemistry of the sample, but may not be site specific, should be noted. For example, "Construction ongoing at south end of the project, with heavy trucks traversing along the interior levee roads." If applicable, metering equipment ID and calibration information should be noted here as well.

1.5.2.6.3. For subsequent pages of a project day, the following minimum items must be included at the top of those additional pages: the project name, date, collectors' initials, and "continued from page ___" block.

1.5.2.6.4. Entries shall be made into the logbook chronologically and sequentially by sample number, unless noted with explanation otherwise. For example, "Note: sampling out of numerical sequence due to unforeseen adverse weather conditions."

1.5.2.6.5. Comments/observations. Loggers need to write comments/observations on site if there is something unusual. Maintain consistency of format and keep the record clean and neat, keeping in mind the intermediate and end users who will access this information. (NOTE: The following guidelines apply to the current field "black" books. In the future, a more specific type of book may be implemented. These guidelines will be updated at that time.)

- The sample number is the same ID number that is noted on the sample tag and on the data log sheet.
- The site/station name is the same as noted on the data log sheet. A note should be placed in parentheses here if, for example, it is an autosampler sample at the same site where a grab sample was also taken.
- Time is logged in 24hr format (i.e., 1430) and corresponds to the same time recorded on the sample tags and data log sheet. Writing "hrs" after the time is not required in this format.
- Comments and observations shall be comprehensive yet concise and as objective as possible. They shall include information about sample description and surrounding conditions including such things as flow and stage conditions, sample color, amount of suspended particulates, odor, ambient conditions such as "Station is choked with water hyacinth" or "Ash fallout from crop burning to NE," abnormal animal activity such as "Lots of birds flew off as we arrived," type and amount of acid added to each bottle, equipment ID numbers (if applicable to that site), and visitors or persons other than sampling personnel at that site. Generally, the information noted here should accurately describe the sample, sampling activities, and surrounding sampling conditions so that a future reader would clearly understand them as if he/she were actually there and could determine if there was anything unusual.

1.5.2.6.6. At the bottom of a logbook page, the logger will sign and date the page in the appropriate spaces.

1.5.2.6.7. Should additional pages be required, the "Continued on page ___" and "Continued from page ___" blocks shall be filled accordingly, with each page signed and dated by the logger.

1.5.2.6.8. At a later time, another collector from the same field unit shall read the logbook entries, checking for both accuracy and comprehension.

1.5.2.6.9. If any corrections or deletions are necessary, a single line shall be drawn through the undesired material and the correcting person will place his/her initials, date, and, if appropriate, comments adjacent to it.

1.5.2.6.10. Upon fully reading, agreeing to, and understanding the entries for that project day, the collector will sign and date the appropriate "Read and understood by" spaces of each page.

1.5.2.6.11. Any portion of an unused logbook page shall be struck through, signed or initialed, and noted "No further entries this page" or words to that effect.

1.5.2.6.12. Arrows may be drawn down for repetitive items, such as QC sampling at the same site, but at all times shall be clear and accurate.

1.5.2.6.13. Standard abbreviations such as "EB" and "FCEB" generally follow a station name when that QC sampling is performed for that sample number. However, comments should still be recorded here. For instance, "SITExyz (EB)," time, and equipment, etc., and equipment includes Niskin, bucket, syringe, and filter. The QC samples and codes must correspond to the entries on the TREC-SWRL sheet.

1.5.2.6.14. When a field logbook is full, both the beginning and ending dates are noted on the cover.

1.5.3. Field instrument calibration log

Field multiparameter probe calibrations and calibration checks are documented on the calibration logbook. The information is entered by the laboratory login staff. The calibration system qualifies entries as P (pass), F (fail), or N/A (not applicable) based on quality requirements stated in the Quality Manual. The quality assurance unit reviews the report and follows up to ensure that data entry was correct and project manager is informed to avoid future occurrences. Comments referring to problems with and maintenance of the field parameter instrumentation must be entered in the comments section of the calibration log sheet.

1.6. Field documentation review

Each sampling staff is responsible for ensuring the accuracy and completeness of the header sheet and field notes. The field project supervisor or project manager is responsible for reviewing these documents within 1 week of sampling. The project manager ensures that the header sheet and field notes agree with each other; that collection, measurements, and documentation were executed properly; and checks for any entries or observations that could question the validity of the sample or measurement. These observations and other comments must be entered by the field PM (project manager) into the sampling logbook. The field PM also scans the header sheets and files them according to protocol in an assigned server location.

Any changes, deletions, or additions to the header sheet must be made by one of the members of the field sampling team. After header sheets have been submitted to the laboratory, any changes must be supported by an alternative document and/or approved by the supervisor, field PM, or QA officer.

Calibration data must be available for QA and project managers' review. The current protocol for routine collection is to record this information on the header sheet so that it can be entered into the computer. Alternatively, calibration information may be sent separately, but with identifying information that must link it to a specific trip and equipment.

Further review of field documentation may be conducted by QA staff during a comprehensive review of the data sets.

1.7. Document control and records retention

The laboratory's sampling groups must maintain the required and necessary documentation for all data generated and other relevant information pertinent to the operation of the organization for the entire life of the project and for 3 years thereafter. Records are kept so that historical reconstruction of all relevant field and laboratory activities is possible. Vital records, including field notes, header sheets, and laboratory analytical reports, are also filed electronically by scanning. Sample header sheet information and other relevant notes are scanned for electronic filing and ready reference for data users and reviewers.

Standard operating procedures, quality manuals, monitoring plans, quality assurance project plans, and other technical documents must be controlled to prevent alteration or use of outdated copies. Supervisors must ensure that their staff is supplied with the latest controlled copies and that old copies are removed from work areas. At least one copy of each revision of these documents must be kept in an organized file.

Sample Custody

1.1. Sample custody
   1.1.1. Document all activities related to a sampling event, including sample collection and transport, equipment calibration, and cleaning.
   1.1.2. The required information related to each activity is specified in each of the samplings.

1.2. Legal chain of custody (COC) procedures are used to demonstrate that the samples and/or sample containers were handled and transferred in such a manner to eliminate possible tampering. When a client or situation requires legal COC, use the following procedures to document and track all time periods and the physical possession and/or storage of sample containers and samples from point of origin through the final analytical result and sample disposal.

   When legal COC is used, samples must be in the actual possession of a person who is authorized to handle the samples (e.g., sample collector, laboratory technician); in the view of the same person after being in his/her physical possession; and secured by the same person to prevent tampering or stored in a designated secure area.

   1.2.1. Use a Chain of Custody form to document sample transfers. Other records and forms may be used to document internal activities.
   1.2.2. Limit the number of people who physically handle the sample.
   1.2.3. Legal COC begins when the precleaned sample containers are dispatched to the field.
   1.2.4. The person who relinquishes the prepared sample kits or containers and the individual who receives the sample kits or containers must sign the COC form unless the same party provides the containers and collects the samples.
   1.2.5. All parties handling the sample are responsible for sample custody (i.e., relinquishing and receiving) and documentation except when the samples or sampling kits are relinquished to a common carrier.
   1.2.6. Delivering samples to the laboratory
      1.2.6.1. All individuals who handle the samples, sample containers, or shipping containers with samples must sign (and relinquish) the COC form. The legal custody responsibilities of the field operations end when the samples are relinquished to the laboratory.
      1.2.6.2. Chain of custody seals
         1.2.6.2.1. Use tamper-indicating tape or seals on all shipping containers that are used to transfer or transport shipping containers and samples. Place the seal so the transport container cannot be opened without breaking the seal.
         1.2.6.2.2. The individual who affixes any tamper-indicating seal to any shipping or sampling

container must record on the seal, the time, calendar date, and signatures of responsible personnel affixing the seal and note in permanent records what containers were sealed.

1.2.6.2.3. The individual who receives any sealed container (whether shipping cooler or individual sample container) must note in permanent records whether or not the seals are intact.

1.2.6.2.4. The individual who breaks a seal must note whether the seal was intact and the time and date that the seal was broken.

1.2.6.2.5. Once a seal is broken, the seal may be discarded. If a client requires seals on all containers at all times, affix a new seal after a seal has been broken.

Field Quality Control

Data quality assessment is based on the precision and accuracy checks in the field and laboratory.

1.1. Field quality control checks

The field QC procedures confirm the precision of the sampling techniques, the cleanliness of the equipment, and address possible effects of the sample handling and transport. All QC samples are preserved, handled as and submitted to the laboratory along with routine samples for a given trip. Field QC requirements are applied on a trip and/or project basis. Additional QCs, other than minimum required, may be collected to satisfy specific project requirements. The field QC check samples consist of the following:

1.1.1. Field blank (FB)

1.1.1.1. Collected by pouring P-free water directly into the sample container on site, preserved and kept open until sample collection is completed for the routine sample at that site.

1.1.1.2. The time when the bottle is filled is recorded for this blank.

1.1.1.3. The FB is required only if no other blank is collected for the sampling event.

1.1.1.4. FB may also be taken if there are concerns of environmental contamination during sample collection or processing.

1.1.1.5. The sample container should be rinsed once prior to filling.

1.1.2. Replicate sample (RS)

1.1.2.1. Collect replicates by repeating (simultaneously or in rapid succession) the entire sample acquisition technique that was used to obtain the routine sample.

1.1.2.2. Collect, preserve, transport, and document replicates in the same manner as the samples.

1.1.2.3. Two replicates are collected per trip at the same site for the longest parameter list.

1.1.2.4. Two replicates are collected each time new staff is trained on new sample collection technique (1 site only).

1.1.2.5. A single replicate set (e.g., one routine sample and two replicates) may count for more than one project if the replicate set and the projects in question are sampled during the same sampling event by the same sampling crew with the same sampling equipment.

1.1.2.6. Replicates should be submitted for the same parameters as the associated samples. The parameter list from the site with the most analytes for the associated project(s) should be used if at all possible.

1.1.2.7. RS data are used to evaluate sampling precision. RS data can also be used to evaluate field variability.

1.1.3. Split sample (SS)

Split samples (SS) collected by processing the routine sample plus one split sample from the same sample collection effort. The result should be two chemically identical samples.

1.1.4. Equipment blank (EB)

1.1.4.1. The equipment blank is collected to evaluate the effectiveness of laboratory decontamination. The equipment is not rinsed prior to collection of the EB.

1.1.4.2. One EB is collected per trip.

1.1.4.3. A single EB may count for more than one site if the EB and the sites in question are sampled during the same sampling event by the same sampling crew with the same sampling equipment.

1.1.4.4. EBs are collected for the longest parameter list if possible.

1.1.4.5. EBs are collected before sample collection begins.

1.1.4.6. EBs are prepared by pouring P-free water into the sample collection container and through each piece of sampling equipment and collecting the rinsate.

1.1.4.7. Filter the water for dissolved parameters. The filter must be rinsed with a minimum of 30 mL of water prior to collection of the EB.

1.1.4.8. For trips requiring more than 5 L of water, the volume required to fill the sample bottles may be used. In cases when a peristaltic pump is used, the water should be pumped through the entire sampling train in accordance with the project SOP, then collected as an EB.

1.1.4.9. The EB is preserved and handled as a routine sample.

1.1.5. Field cleaned equipment blank (FCEB)

1.1.5.1. FCEBs are prepared by pouring approximately 1 L P-free water into the sample collection container and through each piece of sampling equipment. For trips requiring more than 5 L of water, the volume required to fill the sample bottles may be used.

1.1.5.2. Collect the FCEB after the equipment has been decontaminated in the field.

1.1.5.3. FCEBs are collected at any site during the sampling event before the first sample collection.

1.1.5.4. Filter the water for dissolved parameters. The filter must be rinsed with a minimum of 30 mL of water prior to collection of the FCEB.

1.1.5.5. The FCEB is preserved and handled as a routine sample.

Health and Safety

Implement all local, state, and federal requirements relating the health and safety.

Field Waste Disposal

All field generated wastes and purge waters are disposed of properly in a manner that will not contaminate the sampling site. The TREC-Soil and Water Research Laboratory does not sample hazardous waste sites so the only field generated wastes are used gloves or tissues, which should be disposed of properly in order to avoid polluting the sites.

APPENDIX 14.2: DATALOGGER PROGRAM WITH CR-10X

An actual datalogger program with CR-10X for recording flow rate, temperature, dissolved oxygen, pH, and controlling autosampler for proportional sampling is following:

```
--------------------------------------------------------------
; {CR-10X}
 1: Z=F x 10^n (P30)
    1: 1.6 F
    2: 00 n, Exponent of 10
    3: 21 Z Loc [ Version ]
 2: Batt Voltage (P10)
    1: 1 Loc [ CR10batt ]
;We measure Sontek SL
 3: If time is (P92)
    1: 1 Minutes (Seconds --) into a
    2: 5 Interval (same units as above)
    3: 30 Then Do
    4: SDI-12 Recorder (P105)
       1: 0 SDI-12 Address
       2: 0 Start Measurement (aM!)
```

```
        3: 1 Port
        4: 9 Loc [ Temp_Cel ]
        5: 1.0 Mult
        6: 0.0 Offset
 5: End (P95)
;6: Z=X*F (P37)
; 1: 17 X Loc [ Flow_m3_s ]
; 2: -1 F
; 3: 17 Z Loc [ Flow_m3_s ]
 6: Z=X*F (P37)
    1: 17 X Loc [ Flow_m3_s ] ;Flow m3/s
    2: 300 F
    3: 18 Z Loc [ Flow_5Min ] ;Flow m3/5minutes
 7: Z=X+Y (P33)
    1: 19 X Loc [ FlowSumTr ]
    2: 18 Y Loc [ Flow_5Min ]
    3: 19 Z Loc [ FlowSumTr ]
 8: Z=X+Y (P33)
    1: 20 X Loc [ Flow24Sum ]
    2: 18 Y Loc [ Flow_5Min ]
    3: 20 Z Loc [ Flow24Sum ]
 9: Running Average (P52)
    1: 1 Reps
    2: 17 First Source Loc [ Flow_m3_s ]
    3: 22 First Destination Loc [ H_L_Trig ]
    4: 3 Number of Values in Avg Window
10: Z=F x 10^n (P30)
    1: 29.30 F ;Low Flow variable
    2: 00 n, Exponent of 10
    3: 24 Z Loc [ L_Trigger ]
11: Z=F x 10^n (P30)
    1: 65.79 F ; High Flow variable
    2: 00 n, Exponent of 10
    3: 23 Z Loc [ H_Trigger ]
;Test Criteria
12: If (X<=>F) (P89)
    1: 22 X Loc [ H_L_Trig ]
    2: 4 <
    3: 5 F ;If moving average flow is < 5 m3/s then do
       instruction
    4: 30 Then Do
13: Z=X (P31)
    1: 24 X Loc [ L_Trigger ]
    2: 25 Z Loc [ Trigger ]
14: End (P95)
15: If (X<=>F) (P89)
    1: 22 X Loc [ H_L_Trig ]
    2: 3 >=
    3: 1 F ;If moving average flow is >= 1 m3/s then do
       instruction
    4: 30 Then Do
```

```
16: Z=X (P31)
    1: 23 X Loc [ H_Trigger ]
    2: 25 Z Loc [ Trigger ]
17: End (P95)
18: If (X<=>Y) (P88)
    1: 19 X Loc [ FlowSumTr ]
    2: 3 >=
    3: 25 Y Loc [ Trigger ]
    4: 30 Then Do
19: Set Port(s) (P20)
    1: 9999 C8..C5 = nc/nc/nc/nc
    2: 9959 C4..C1 = nc/nc/100ms/nc
20: Do (P86)
    1: 72 Pulse Port 2
21: Z=Z+1 (P32)
    1: 26 Z Loc [ Triger_ct ]
;We save the data before we rest counter
22: Do (P86)
    1: 10 Set Output Flag High (Flag 0)
23: Set Active Storage Area (P80)^9475
    1: 1 Final Storage Area 1
    2: 411 Array ID
24: Real Time (P77)^21034
    1: 1220 Year,Day,Hour/Minute (midnight = 2400)
25: Sample (P70)^20939
    1: 1 Reps
    2: 21 Loc [ Version ]
26: Sample (P70)^13139
    1: 1 Reps
    2: 25 Loc [ Trigger ]
27: Resolution (P78)
    1: 1 High Resolution
28: Sample (P70)^14144
    1: 1 Reps
    2: 19 Loc [ FlowSumTr ]
29: Z=X*F (P37)
    1: 20 X Loc [ Flow24Sum ]
    2: .1 F
    3: 20 Z Loc [ Flow24Sum ]
30: Sample (P70)^25277
    1: 1 Reps
    2: 20 Loc [ Flow24Sum ]
31: Z=X*F (P37)
    1: 20 X Loc [ Flow24Sum ]
    2: 10 F
    3: 20 Z Loc [ Flow24Sum ]
32: Resolution (P78)
    1: 0 Low Resolution
33: Sample (P70)^26324
    1: 1 Reps
    2: 26 Loc [ Triger_ct ]
```

```
34: Z=F x 10^n (P30)
    1: 0.0 F
    2: 00 n, Exponent of 10
    3: 19 Z Loc [ FlowSumTr ] ;
;We reset Discharge total to 0 and begin sequence again.
35: Do (P86)
    1: 20 Set Output Flag Low (Flag 0)
36: End (P95)
37: If time is (P92)
    1: 0 Minutes (Seconds --) into a
    2: 59 Interval (same units as above)
    3: 30 Then Do
;YSIydrolab
;SDI address = 1
;Port = C3
;pin out
38: SDI-12 Recorder (P105)
    1: 1 SDI-12 Address
    2: 0 Start Measurement (aM!)
    3: 3 Port
    4: 2 Loc [ Temp_C ]
    5: 1.0 Mult
    6: 0.0 Offset
39: End (P95)
;**************************************************************
;********************FINAL STORAGE OUTPUT**********************
;**************************************************************
;Final Storage, every 60 Minutes:
40: If time is (P92)
    1: 0 Minutes (Seconds --) into a
    2: 60 Interval (same units as above)
    3: 10 Set Output Flag High (Flag 0)
41: Set Active Storage Area (P80)^8907
    1: 1 Final Storage Area 1
    2: 60 Array ID
42: Real Time (P77)^3975
    1: 1220 Year,Day,Hour/Minute (midnight = 2400)
43: Sample (P70)^32284
    1: 1 Reps
    2: 1 Loc [ CR10batt ]
44: Sample (P70)^4922
    1: 1 Reps
    2: 21 Loc [ Version ]
45: Average (P71)^24879
    1: 3 Reps
    2: 9 Loc [ Temp_Cel ]
46: Resolution (P78)
    1: 1 High Resolution
47: Z=X*F (P37)
    1: 20 X Loc [ Flow24Sum ]
    2: .1 F
    3: 20 Z Loc [ Flow24Sum ]
```

```
48: Sample (P70)^3126
    1: 1 Reps
    2: 20 Loc [ Flow24Sum ]
49: Z=X*F (P37)
    1: 20 X Loc [ Flow24Sum ]
    2: 10 F
    3: 20 Z Loc [ Flow24Sum ]
50: Resolution (P78)
    1: 0 Low Resolution
51: Sample (P70)^9432
    1: 1 Reps
    2: 26 Loc [ Triger_ct ]
52: Sample (P70)^32738
    1: 7 Reps
    2: 2 Loc [ Temp_C ]
53: Do (P86)
    1: 20 Set Output Flag Low (Flag 0)
54: End (P95)
-------------------------------------------------------------
```

15 Training Video for Water Quality Sampling and Analysis

Pamela J. Fletcher and Sapna Mulki

CONTENTS

15.1 INTRODUCTION

This chapter is a compilation of selected narratives from the Water Quality Concepts, Sampling, and Chemical Analysis workshop of April 14 to 18, 2008 (http://conference.ifas.ufl.edu/ufwq/index.htm), sponsored by the University of Florida's Tropical Research and Education Center in Homestead, Florida. The course content included water regulations, monitoring and sampling techniques, Best Management Practice (BMP) program initiatives, and laboratory methods related to water quality of coastal water, surface water, and groundwater. The training combined class lectures, field tours, and hands-on field and laboratory activities. Presentations were given by environmental professionals from various governmental, nongovernmental, and educational institutions in Florida. The participants' evaluations reflect the usefulness of the workshop content and hands-on activities that resulted in their ability to improve water quality analysis in their laboratories. A video of the workshop is included in this book.

15.2 PRESENTATIONS

15.2.1 Water Quality Policy in the United States and Florida

Presenter: Dr. Kati Migliaccio, Assistant Professor, Agricultural and Biological Engineering Department at the Tropical Research and Education Center, University of Florida, Homestead, Florida.

DVD Time: 20 minutes

Discussion overview: The presentation outlines the evolution of environmental laws pertaining to water quality in the United States, with a focus on Florida. It includes historical events that led to policy change for preserving and rehabilitating water resources.

15.2.2 Water Quality Monitoring Technology

Presenter: Dr. Yuncong Li, Professor, Soil and Water Science Department at the Tropical Research and Education Center, University of Florida, Homestead, Florida.

DVD Time: 30 minutes

Discussion overview: Water quality monitoring technologies for surface water, groundwater, and soil/pore water are described using examples from Florida. Emphasis is placed on three criteria: (1) adopting reliable water sampling techniques, (2) conducting accurate laboratory analysis, and (3) generating sound data analysis/interpretation.

15.2.3 Water Quality in the Everglades National Park: Trends and Water Sources

Presenter: Joffre Castro, Civil Engineer, Everglades National Park, Homestead, Florida.
DVD Time: 20 minutes
Discussion overview: This session provides a brief historical overview of the changes in hydropatterns and water quality in Everglades National Park.

15.2.4 South Florida Water Management District (SFWMD) Regional Environmental Monitoring (Hydro-Meteorological and Water Chemistry Data Collection)

Presenter: David Struve, Division Director for Water Quality Analysis Division, South Florida Water Management District, West Palm Beach, Florida.
DVD Time: 50 minutes
Discussion overview: The presentation describes how the South Florida Water Management District (SFWMD) conducts its environmental monitoring within the Greater Everglades Ecosystem using (1) hydro-meteorology, (2) water quality, and (3) biology. Hydro-meteorology assesses the physical flow of water inputs and outputs primarily from water control structures such as pumps, weirs, culverts, and spillways. Water quality focuses on chemical elements, most commonly pesticides and nutrients. Biological monitoring investigates flora and fauna interactions within the fresh and marine waters associated with the Everglades.

15.2.5 Miami-Dade Department of Environmental Resources Management (DERM) Water Quality Monitoring Program

Presenter: Forrest Shaw, Restoration and Enhancement Project, Miami-Dade Department of Environmental Resources Management, Miami, Florida.
DVD Time: 30 minutes
Discussion overview: This session describes the projects within the Department of Environmental Resources Management (DERM) focused on restoration, enhancement, and monitoring in the uplands and coastal areas of Miami-Dade County, Florida. Management of 110 water quality monitoring stations and reporting of those sites is reviewed. Concluding remarks illustrate how DERM provides substantial amounts of information to assist resource managers in decision-making and regulatory action where appropriate within county jurisdiction.

15.2.6 National Oceanic and Atmospheric Administration (NOAA) Coastal Water Quality Issues

Presenter: Dr. John Proni, Director (retired), Ocean Chemistry Division at The National Oceanic and Atmospheric Administration's Atlantic Oceanographic and Meteorological Laboratory on Virginia Key in Miami, Florida.

DVD Time: 30 minutes

Discussion overview: The presentation contains information on the Ocean Chemistry Division's science-based research associated with coastal and ocean water quality in southeast Florida. The overview consists of site selection, sampling techniques, and data analysis examining nutrients and microbial concentrations near inlets and ocean outfall pipes along southeast Florida in Miami-Dade, Broward, and Palm Beach Counties. The results are used to assess the cumulative effects of inlets and outfalls in near shore waters as part of the National Oceanic and Atmospheric Administration's (NOAA) long-term research program.

15.2.7 Molecular Microbial Water Quality Assessment for Coastal Ecosystems

Presenter: Dave Wanless, Research Associate, Ocean Chemistry Division at NOAA's Atlantic Oceanographic and Meteorological Laboratory and the University of Miami's Cooperative Institute for Marine and Atmospheric Science on Virginia Key in Miami, Florida.

DVD Time: 1 hour and 15 minutes

Discussion overview: The Environmental Microbiology Program seeks to create better tools to assess coastal water quality. The laboratory is developing new methods to provide information on sewage pollution, human pathogens, and harmful algae in a timely, accurate, and cost effective process. The program is testing molecular assays and sensors to detect microbial contaminants in coastal waters by measuring DNA or RNA signatures that can be tracked to their sources.

15.2.8 U.S. Geological Survey (USGS) Water Quality Monitoring

Presenter: Lee Massey, Hydrological Technician, USGS, Florida Integrated Science Center, Fort Lauderdale, Florida.

DVD Time: 30 minutes

Discussion overview: The presentation describes USGS methodologies for measuring salinity and nutrient concentrations in South Florida waters. Four USGS South Florida monitoring projects are described. They are (1) groundwater salt front monitoring, (2) salinity and nutrient monitoring in Florida Bay, (3) salinity monitoring in the St. Lucie and Loxahatchee Rivers, and (4) sediment and nutrient monitoring in the C51 Canal in West Palm County, Florida.

15.2.9 Understanding Standard Operating Procedures for Water Sampling and Handling

Presenter: Dr. Qingren Wang, Research Scientist, Soil and Water Science Department at the Tropical Research and Education Center, University of Florida, Homestead, Florida.

DVD Time: 30 minutes

Discussion overview: This presentation describes Standard Operating Procedures (SOPs) needed to consistently sample water and ensure Quality Assurance and Quality Control (QA/QC). SOP requirements reviewed are (1) contamination prevention, (2) sample collection order, (3) protective gloves, (4) container and equipment rinsing, (5) sample preservation, and (6) documentation to authenticate the chain-of-custody for samples.

15.2.10 Water Quality Data Analysis

Presenter: Dr. Kati Migliaccio, Assistant Professor, Agricultural and Biological Engineering Department at the Tropical Research and Education Center, University of Florida, Homestead, Florida.

DVD Time: 30 minutes

Discussion overview: This session illustrates a variety of ways that water quality data can be analyzed and used in reporting. The techniques described are summary statistics, trend analysis, load analysis, principal component analysis, and Water Quality Indices.

15.3 HANDS-ON DEMONSTRATIONS

15.3.1 Collecting Soil/Pore Water Samples

DVD Time for 15.3.1–15.3.3.3: 26 minutes

Presenters: Dr. Kati Migliaccio, Assistant Professor, Agricultural and Biological Engineering Department at the Tropical Research and Education Center, University of Florida, Homestead, Florida; Michael Gutierrez, Laboratory Technician at the Tropical Research and Education Center, University of Florida, Homestead, Florida.

Overview: This hands-on presentation describes how to collect water samples using bucket lysimeters (Figure 15.1).

15.3.2 Collecting Well Water Samples

Presenter: Dr. Qingren Wang, Research Scientist, Soil and Water Science Department at the Tropical Research and Education Center, University of Florida, Homestead, Florida.

Overview: This demonstration illustrates how water quality samples are collected from groundwater wells using a peristaltic pump and a bailer (Figure 15.2).

FIGURE 15.1 Workshop participants collected soil pore water samples.

FIGURE 15.2 Workshop participants collected water samples using a bailer.

15.3.3 Collecting Surface Water Samples

15.3.3.1 Canal Water Sampling

Instructor: Dr. Qingren Wang, Research Scientist, Soil and Water Science Department at the Tropical Research and Education Center, University of Florida, Homestead, Florida.

Discussion overview: This presentation describes how to collect and preserve grab samples from surface water.

15.3.3.2 Wetland Water Sampling

Instructor: Dr. Qingren Wang, Research Scientist, Soil and Water Science Department at the Tropical Research and Education Center, University of Florida, Homestead, Florida.

Discussion overview: The demonstration shows methods for collecting water samples from wetland areas.

15.3.3.3 Wetland Water Sampling from Airboat

Instructor: Dr. Qingren Wang, Research Scientist, Soil and Water Science Department at the Tropical Research and Education Center, University of Florida, Homestead, Florida.

Discussion overview: Methods for collecting water samples from an airboat are shown in this video (Figure 15.3).

15.3.3.4 Tour of Water Quality Research Vessel

DVD Time: 25 minutes

Instructor: Miguel McKinney, Operations Coordinator for the Research Vessel *Walton Smith*, University of Miami, Miami, Florida.

Discussion overview: This presentation is a tour of the research vessel Walton Smith owned by the University of Miami. The vessel is operated by a crew of six and can accommodate up to 12 resident scientists. The *Walton Smith* is used to support oceanographic and atmospheric research consisting of, but not limited to, water quality monitoring, marine geology, chemistry, and biology, and it can deploy and retrieve remotely operated vehicles. It is a multipurpose vessel that can be reconfigured to meet the needs of each expedition for both deckside operations and inside laboratory functions. Water quality monitoring equipment presented includes over-the-side gauges and collection bottles that can be lowered to varying depths for real-time monitoring and

FIGURE 15.3 Workshop participants collected wetland water samples from an airboat.

sampling. In addition, through-hull sampling, Doppler current profiling, and side scan sonar can be configured to meet the needs of researchers. More information on the *Walton Smith* research vessel can be found on the Internet at http://www.rsmas.miami.edu/support/mardep/cat/.

15.3.4 Water Sample Processing for qPCR

Instructor: Dave Wanless, Research Associate, Ocean Chemistry Division at NOAA/AOML and the University of Miami's Cooperative Institute for Marine and Atmospheric Science located on Virginia Key in Miami, Florida.

DVD Time: 40 minutes

Discussion overview: This session illustrates the procedures for conducting Real-time Quantitative Polymerase Chain Reaction (qPCR). qPCR is an experimental technique that uses DNA and RNA from water samples to identify the presence of viruses and pathogens and to trace the origin of bacteria from the intestinal tracts of mammals. This analysis can be performed within hours (newer equipment takes 30 minutes) and can accurately identify species-specific bacteria. qPCR is being tested against traditional methods for timeliness and reliability of molecular analyses and source tracking to aid in the discrimination of microbial contaminants in coastal waters. If successful, it should identify the sources of microbial pathogens in coastal waters (Figure 15.4).

15.3.5 Laboratory Chemical Analysis

15.3.5.1 Quick Testing Kits, pH, and Electrical Conductivity Meters

Instructor: Dr. Yun Qian, Postdoctoral Scientist, Soil and Water Science Department at the Tropical Research and Education Center, University of Florida, Homestead, Florida.

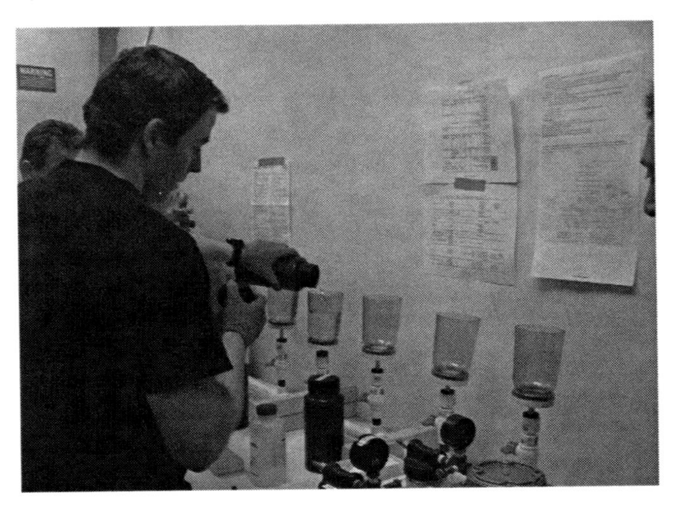

FIGURE 15.4 NOAA scientist demonstrated costal water analysis.

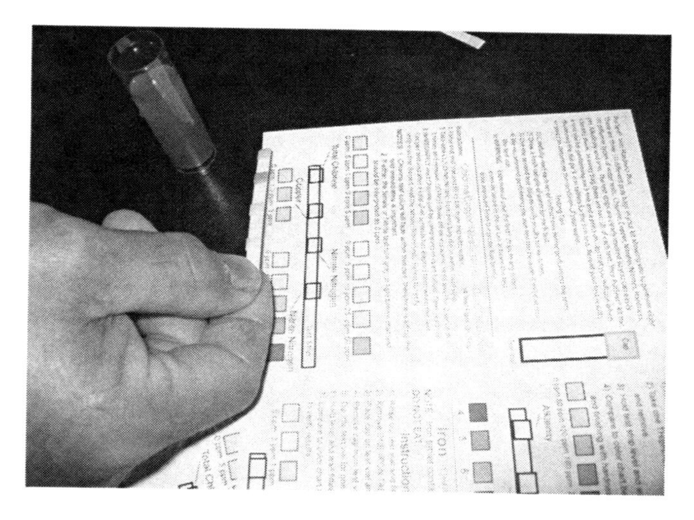

FIGURE 15.5 Workshop participants measured water quality with a quick testing kit.

DVD Time: 20 minutes

Discussion overview: This hands-on demonstration shows how to use quick testing kits to analyze water samples. This method can be used to simultaneously test for chlorine, copper, nitrate, and nitrite and can be combined to test for pH, alkalinity and hardness, iron, ammonia, and phosphate. The session also describes how to test pH and electrical conductivity (EC) using other methods, such as a card reader (Figure 15.5).

15.3.5.2 Ion Chromatograph

Instructor: Dr. Ed Hanlon, Professor, Soil and Water Science Department at the University of Florida's Southwest Florida Research and Education Center, Immokalee, Florida.

DVD Time: 15 minutes

Discussion overview: This presentation illustrates how an ion chromatograph separates ions and polar molecules in water samples. Samples are filtered and injected into a detector column of high pressure ranging from 2400 psi to 2500 psi. Ions are separated between low charge density/high hydrated radii and high charge density/low hydrated radii. Ions create a current when passing through an electronic conductivity detector, thus forming peaks on the instrument reading, first for fluoride, chloride, and then for phosphate.

15.3.5.3 Auto-Analyzer

Instructor: Dr. Kelly Morgan, Assistant Professor, Soil and Water Science Department at the University of Florida's Southwest Florida Research and Education Center, Immokalee, Florida.

DVD Time: 3 minutes

Discussion overview: This presentation shows how to use an auto-analyzer to measure phosphorous concentrations in water samples. There are five basic units in the auto-analyzer: (1) auto-sampler, (2) pump unit, (3) three analytical channels that measure phosphorus, nitrate, and ammonium from the same sample simultaneously, (4) colorimeter, and (5) the computer showing standards and results.

15.3.5.4 Atomic Absorption Spectrometer

Instructors: Dr. Gurpal Toor, Assistant Professor, Soil and Water Science department at the Gulf Coast Research and Education Center, University of Florida, Wimauma, Florida; Dr. Guodong Liu, Postdoctoral Scientist at the University of Florida, Tropical Research and Education Center, Homestead, Florida.

DVD Time: 20 minutes

Discussion overview: The hands-on demonstration illustrates how an atomic absorption spectrometer is used to detect metal concentrations of potassium and iron in water samples. The spectrometer can measure metal concentrations dissolved in a liquid or solid by analyzing the rate at which atoms absorb light. The machine is made up of four units: (1) autosampler, (2) flame, (3) furnace, and (4) computer for reviewing results and analysis (Figure 15.6).

15.3.5.5 Discrete Analyzer

Instructor: Rick Armstrong, Lee County Environmental Lab in Fort Myers, Florida.

DVD Time: 10 minutes

Discussion overview: The presentation describes how a discrete analyzer uses basic chemical reactions to identify the presence of orthophosphate and nitrate. The fully automated process for orthophosphate begins when a sampler probe

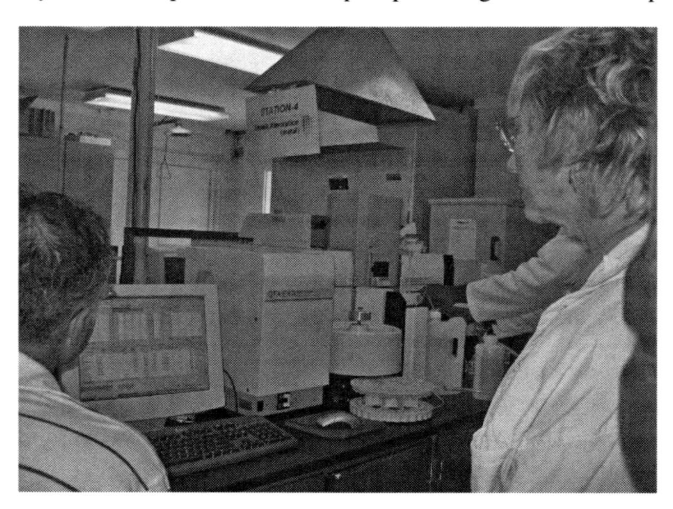

FIGURE 15.6 Workshop participants analyzed water samples with an atomic absorption spectrometer.

picks up a reagent, adds ascorbic acid and molybdate, and is then heated in a chamber. When the three agents are combined and transferred to a cuvette beamed with light (880 nanometers) a blue hue is created because of the presence of phosphomolybdate blue. The intensity of the blue color indicates the orthophosphate concentration with higher levels causing darker shades of blue and lower concentrations a lighter blue color. The output is retrieved by a software program that measures the proportion of orthophosphate and represents it as a calibrated curve for analysis and reporting.

15.4 PROGRAM EVALUATION AND PARTICIPANT TESTIMONIES

Pre- and posttests were administered to assess the knowledge change of the attendees. On average, the increase in knowledge was 73%, where *Knowledge Gained = [pre-test mean – post-test mean)/(pre-test mean)] * 100*. In lecture series, a typical number for knowledge gained is usually in the 9% range. For good lecturing with some hands-on components, the number can be as high as 50%. It appears that lectures, hands-on activities, and ample discussion time helped considerably with participant achievements. One of the participants wrote: "Congratulations on an outstanding job of organizing the Water Quality Training. This was one of the best I have attended in recent years . . . demonstrated through the classroom lectures, field sampling, tour of pump stations and Everglades National Park, airboat collecting trip, and finally the great water analysis lab sessions. Even some of us old hands learned new things." More comments on the program can be viewed from the DVD.

Index

D

E

N

O

P